Henry Jones and Octavia Lanchester, painted by their daughter Mary (probably about 1905).

THE LANCHESTER LEGACY

A Trilogy of Lanchester Works

Volume Three – A Celebration of Genius

Edited by

John Fletcher

COVENTRY UNIVERSITY

VOLUME THREE – A Celebration of Genius

PUBLISHED TO CELEBRATE THE CENTENARY OF THE FIRST LANCHESTER CAR OF 1895.

Coventry University Enterprises Ltd.
City of Coventry
United Kingdom

ISBN 0–905 949–47–1

ANY NEW INFORMATION OR AMENDMENTS RELATING TO THIS WORK WILL BE GRATEFULLY RECEIVED BY THE EDITOR, VIA THE PUBLISHERS.

Printed and bound in Great Britain by
Butler & Tanner Ltd, Frome and London

CONTENTS

Isometrograph
Pendulum governor
Gas engine starter
Gas calorimetry
Perfect radiator
Pendulum accelerometer
Worm gear
Colour photography
Petrol tank level gauge
Device for firing machine guns
   through aeroplane propeller
Radial cursor

CONTRIBUTORS

John ACKROYD	Senior Lecturer, Aerospace Engineering Division, Manchester School of Engineering, University of Manchester.
John BINGEMAN	Frank Lanchester's grandson
Ken BOWEN	Visiting Professor of Operational Research, Royal Holloway, University of London
June DAWSON	Engineering Librarian, Lanchester Library, Coventry University
Norman DYSON	Retired Lecturer, Department of Physics, University of Birmingham
Fred FITZKE	Head, Laboratory of Physiological Optics, Institute of Ophthalmology, University of London
John FLETCHER	Retired University Librarian, Lanchester Library, Coventry University
Joe HASTINGS	Structural engineer, Lanchester and Lodge, architects
Barry HOBSON	Consultant, aerospace education and training
Mrs Elaine LANCHESTER	George Lanchester's daughter-in-law
Ken McNAUGHT	Lecturer in Operational Research, Royal Military College of Science, Cranfield University
Norman RICKETTS	Aerospace training consultant
Yoichi TAKEDA	Lanchester Systems, Fukuoka, Japan
David THOMAS	Freelance computer programmer
Roy THOMAS	Retired electrical engineer
Barry WEST	Deputy Librarian, Lanchester Library, Coventry University

ACKNOWLEDGMENTS

Every editor is very aware of the debt he owes to his contributors, and I am no exception. Their enthusiasm was infectious, and their devotion to producing good quality scripts was exemplary, despite the pressures of their many other responsibilities.

I wish to record my special thanks to Mr Yoishi Takeda, who has provided financial support for the Lanchester Collection at Coventry University over many years. His generosity has enabled the Lanchester Library to add some very significant items to the Collection, and to ensure its safe keeping for the future.

My thanks go also to the staff of Lanchester Library, and especially to June Dawson and Barry West for the support they have given me in working with the Collection. To Chris Clark go my thanks for starting *The Lanchester legacy*, and keeping it going.

Roy Thomas deserves my special gratitude: he came to the Lanchester Collection seeking local history information, but became fascinated by the technical material it contained, and developed a lasting, and productive interest in it.

Finally, my thanks to my wife Barbara for her tolerance of a task which became more time consuming of our retirement than either of us had anticipated.

REFERENCES

References to published items are given in the text in the form:

> ... (Smith, 1915) ... or ... Smith (1917) ...

and at the end of each chapter in the form:

> Smith, J., (1917) *Aeronautics*, London, Collins
> Smith, J., (1918) A Note on aeronautics. *Aeronautical Journal*, 10, 357–359

References to unpublished sources in the Lanchester Collection, in the Lanchester Library of Coventry University are given in the form:

LC. Baxter 1/3	for items in the first collection indexed by Eric Baxter, the first Librarian. The first number refers to broad subject areas, the second is a sequential document number.
LC. vol 4–2	for items added to the Collection since the Baxter index was completed. The first number is a volume, the second a document number in that volume.
LC. SB 1–32R	for items in the sketchbooks. The first number is that of the sketchbook, the second is the page (Left or Right)

Unpublished sources in the Birmingham Museum of Science and Industry are noted (B.M.S.I.)

EDITOR'S NOTES

To mark the centenary of the first all-British, four-wheel, petrol-engined motor car, Coventry University agreed to publish the two-volume history of the marque, written by Chris Clark. Dr. Michael Goldstein, the University's Vice-Chancellor, expressed his unease that Dr Fred Lanchester's other achievements would be forgotten. I was therefore asked to edit a third volume to cover these interests, which were wide, and varied.

Frederick William Lanchester was an incredible genius, an article on him by Karen Gold in *The Times Higher Educational Supplement* of 17 February 1984 was entitled "The British Leonardo", and this does not seem to be media hype. Lanchester is well remembered for the motor cars which carried his name (though many models were in fact designed by his brother George) and for his many engineering interests. In a completely different sphere his theories of aerodynamics were some of the first to be set out in detail.

In this volume I have tried to bring together experts in the different subjects to write about what Fred Lanchester said, wrote, and did in their area, and to assess his importance in retrospect. This volume, then, is a composite, a "sweeping-up" of what might have been forgotten about the other interests of this brilliant man. The result must, by definition, be a mixture of subjects, and of styles, and as editor, I make no apology for this. It is inherent in a "festschrift".

To put the man himself into perspective, we begin, in Chapter 1 with a brief look at his siblings, then, in more detail in Chapter 2, I give a short biography of Frederick Lanchester, concentrating on his life and other interests, leaving the description of his motor car developments to Chris Clark. Family memories of Frank Lanchester, the third brother whose life was devoted to the Lanchester cars, are contributed by John Bingeman, his grandson, in Chapter 3. In Chapter 4, Mrs Elaine Lanchester gives a brief biography of her father-in-law, George Lanchester.

The first of the technical chapters is "Lanchester's *Aerodynamics*", Chapter 5, contributed by Dr John Ackroyd, based on the Lanchester Lecture which he gave to the Royal Aeronautical Society meeting at Coventry in 1992. In Chapter 6 he deals with the second volume of Lanchester's important work on aerial flight, *Aerodonetics*, and in the next chapter John Ackroyd gives a review of Lanchester's work on the design and construction of aircraft. Barry Hobson in the next, related, chapter reviews Fred Lanchester's ideas on the strategic use of aircraft.

Fred Lanchester did not limit his study of military strategy to aerial warfare, and the following chapter, Chapter 9, by Prof. Ken Bowen and K.R. McNaught, examines the development of Lanchester's mathematics as applied to warfare since his seminal work, *Aircraft in warfare: the dawn of the fourth arm* (London, Constable, 1916). Lanchester's "n-square law" became one of the foundation stones of the new science of operational research developed during and after the Second World War. Many mathematicians, and some non-mathematicians, have seen analogies between the "law" and the structure of other events. Yoichi Takeda is one of the foremost figures in developing the n-square law into a marketing strategy for business, and he contributes Chapter 10.

Fred Lanchester himself applied the n-square law to other military situations, and at the beginning of the Second World War wrote an article on the defence of Gibraltar, which he offered for publication. It was refused, probably, Fred thought, because it conflicted with official government policy. The Lanchester Collection includes the typescript of the full version of this article, and a summary version. I felt it could now see the light of day, and the summary is reproduced for the first time as Chapter 11.

The idea of using a combination of petrol engine and electric dynamotor to drive a vehicle was not new, but Lanchester was asked by Daimler to work on it. In Chapter 12 Roy Thomas looks at the results. In the 1920s Lanchester's Laboratories was set up, originally to allow Fred the time and premises for research and development. In the end it became a production unit for his loudspeaker and radio designs. Roy Thomas looks at the history and products of this phase of Fred Lanchester's career in Chapter 13.

The eldest Lanchester brother, Henry Vaughan Lanchester, had followed his father into architecture, and had made a considerable success of the practice, Lanchester and Lodge, which he founded. His younger brother, Fred, looked at some building features from an engineering standpoint, and wrote a paper, *Span*, reviewing the design of bridges and roof structures. The Lanchester and Lodge practice still thrives, and one of its current staff, Joe Hastings, has written a review of the paper as Chapter 14.

In the middle 1930s Fred Lanchester's health was deteriorating, and he began to have cataract problems. His close friend, R.D. Lockhart, was Professor of Anatomy at the University of Birmingham, and Fred asked his opinion on some ideas he had on the mechanics of the human eye, and the way in which retinal images were passed to the brain. Lockhart helped Lanchester, and the first paper was given to a meeting of the Anatomical Society at the University, and published in *Journal of Anatomy*. In Chapter 15 Dr. Fred Fitzke reviews this work, and its importance then, and now.

In the last ten years of his life Fred Lanchester felt he had been side-lined. He thought that his work for the country was insufficiently recognised, and that he could still be of use to industry and government in the approaching war. Lack of paid work, depression, and failing health led him to look at other subjects which had interested him for many years. He published a book on *The Musical scale*, reviewing the various scales in use, and their relationships to each other: Dr. Norman Dyson looks at this in Chapter 16.

Lanchester was aware of the difference between poetry and verse, and claimed he wrote the latter. He felt that such a venture would not enhance his standing as an engineer, so published under the pseudonym Paul Netherton-Herries. Barry West comments on the verse in Chapter 17.

David Thomas reviews another publication of this period, *The Theory of dimensions*, in Chapter 18. Norman Ricketts looks at Lanchester's ideas on contra-rotating propellers in Chapter 19.

The range of engineering and scientific topics which Fred Lanchester tackled is prodigious, and it would be impossible to cover all of them here. Alone or with others he made 463 patent applications, of which 236 were completed and patents granted.

For the final chapter, 20, Roy Thomas delved into the patent specifications and the sketchbooks and provides notes on some of Lanchester's other ideas, ranging from colour photography to player pianos, from a moped to a cold box.

The book finishes with a comprehensive list of the Lanchester patent applications and specifications, compiled by Mrs June Dawson, and a bibliography of Fred Lanchester's published work, including reports of contributions to discussions. There are also brief notes on the main collections of Lanchester publications, manuscripts and papers in the United Kingdom.

John Fletcher

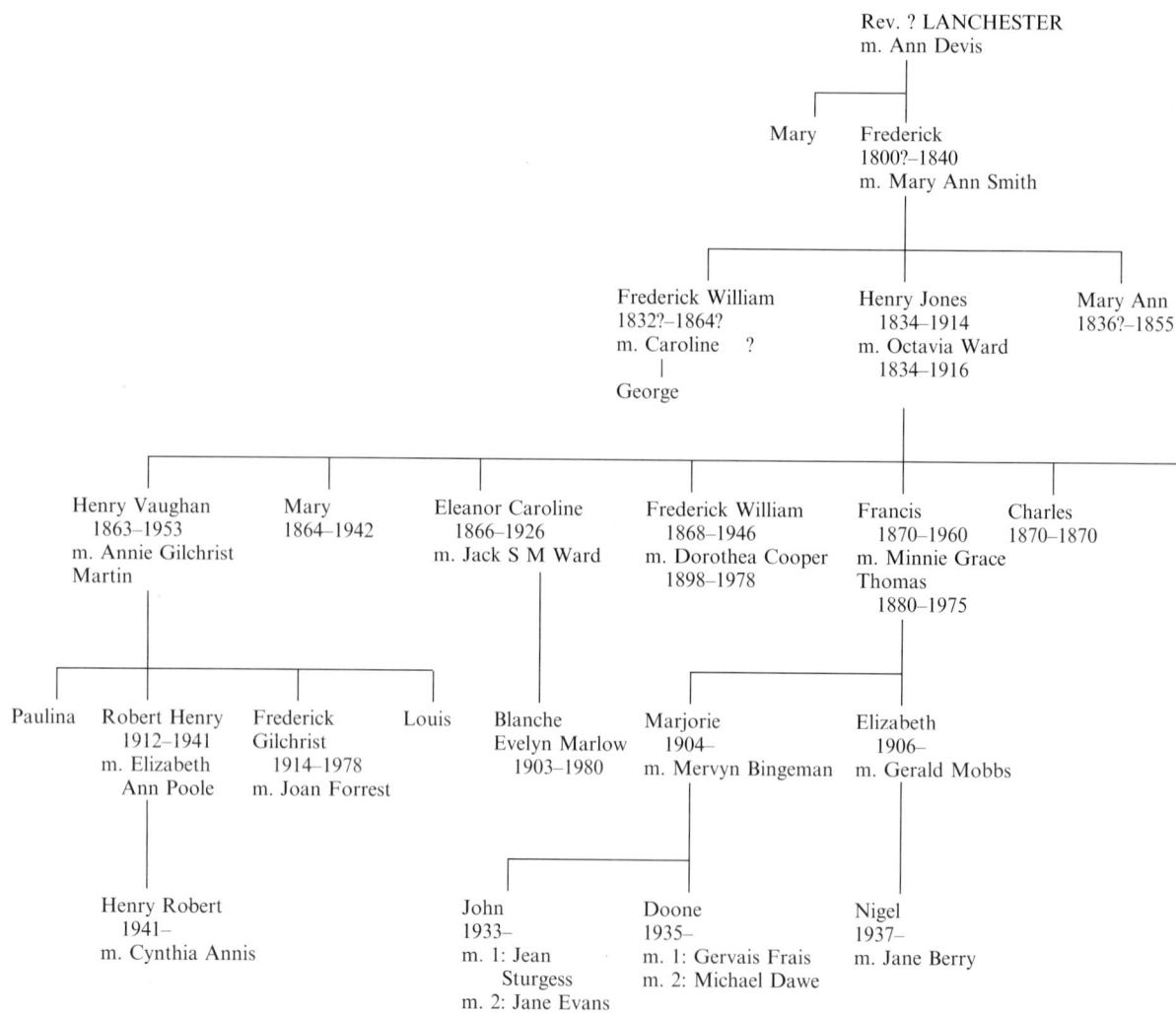

Rev. ? LANCHESTER
m. Ann Devis

Mary   Frederick
1800?–1840
m. Mary Ann Smith

Frederick William
1832?–1864?
m. Caroline   ?

George

Henry Jones
1834–1914
m. Octavia Ward
1834–1916

Mary Ann
1836?–1855

Henry Vaughan
1863–1953
m. Annie Gilchrist
Martin

Mary
1864–1942

Eleanor Caroline
1866–1926
m. Jack S M Ward

Frederick William
1868–1946
m. Dorothea Cooper
1898–1978

Francis
1870–1960
m. Minnie Grace
Thomas
1880–1975

Charles
1870–1870

Paulina

Robert Henry
1912–1941
m. Elizabeth
Ann Poole

Frederick
Gilchrist
1914–1978
m. Joan Forrest

Louis

Blanche
Evelyn Marlow
1903–1980

Marjorie
1904–
m. Mervyn Bingeman

Elizabeth
1906–
m. Gerald Mobbs

Henry Robert
1941–
m. Cynthia Annis

John
1933–
m. 1: Jean
Sturgess
m. 2: Jane Evans

Doone
1935–
m. 1: Gervais Frais
m. 2: Michael Dawe

Nigel
1937–
m. Jane Berry

FAMILY TREE

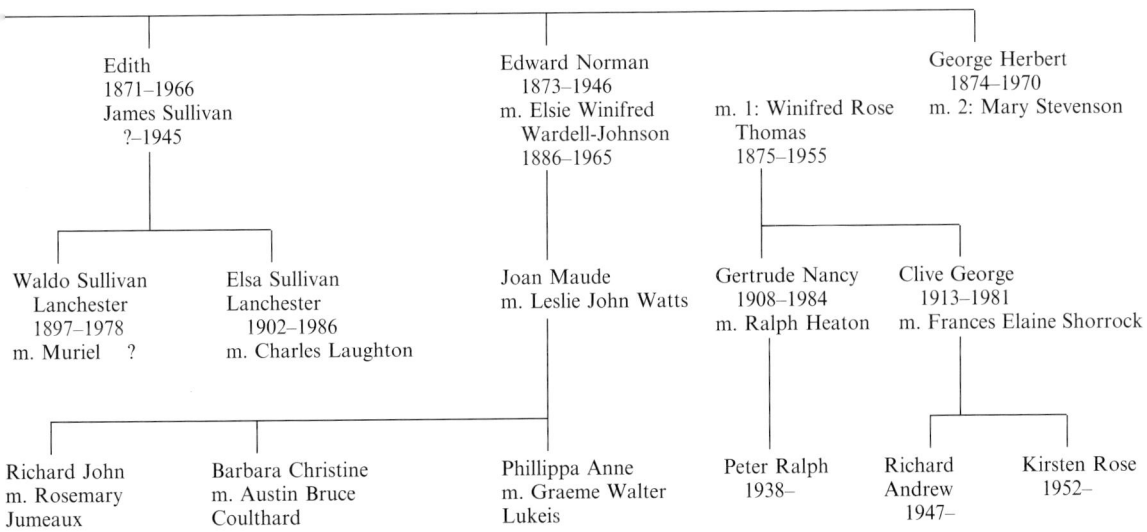

Edith
1871–1966
James Sullivan
?–1945

Edward Norman
1873–1946
m. Elsie Winifred
Wardell-Johnson
1886–1965

George Herbert
1874–1970
m. 1: Winifred Rose   m. 2: Mary Stevenson
Thomas
1875–1955

Waldo Sullivan
Lanchester
1897–1978
m. Muriel   ?

Elsa Sullivan
Lanchester
1902–1986
m. Charles Laughton

Joan Maude
m. Leslie John Watts

Gertrude Nancy
1908–1984
m. Ralph Heaton

Clive George
1913–1981
m. Frances Elaine Shorrock

Richard John
m. Rosemary
Jumeaux

Barbara Christine
m. Austin Bruce
Coulthard

Phillippa Anne
m. Graeme Walter
Lukeis

Peter Ralph
1938–

Richard
Andrew
1947–

Kirsten Rose
1952–

Chapter One

THE LANCHESTER FAMILY

by John Fletcher

The family to which Frederick William Lanchester belonged was quite extraordinary, and it seems fitting that any biography of Fred, and his brothers Frank and George, should be put into context. They were members of the family of eight children of Henry Jones Lanchester and Octavia (née Ward). Both parents were born in 1834, and Henry Jones was a successful architect in London. In order, their children were:

Henry Vaughan, born 1863

The first son followed his father's profession, being articled to him from 1879 to 1884. He went into architectural practice on his own in 1887. With two other architects Henry Vaughan entered competitions for prominent buildings, and they gained their first success in 1898. The partnership was formed and with various changes of name, the practice still survives, now Lanchester and Lodge, of London. An excellent history of the practice *The History of Lanchester and Lodge*, (Burchell, 1987) was written by Jim Burchell to celebrate their centenary.

The partnership was responsible for a large number of public buildings, both in the United Kingdom and abroad, especially in India. Among these were Cardiff Civic Hall and Law Courts, Barnsley Civic Centre, Hull School of Art, and many of the buildings of the University of Leeds. Overseas they included a theatre and cinema hall in Lucknow, and a palace for the Maharajah of Jodhpur. Henry married a musical comedy actress, Annie Gilchrist Martin in 1909, and they had a daughter and three sons. He died in 1953.

Mary, born 1864

Mary inherited the artistic talents reputed to stem from her great grandmother, Ann Devis, a sister (or cousin) of Arthur William Devis, the painter of "The Death of Nelson". George Lanchester recalled that Mary was "a painter and art teacher, and an eccentric; she had a good sense of humour" (LC. Baxter 1/4). Her painting of her parents, (frontispiece) indicates the quality of her oil painting, and she was certainly successful: her painting "Farm on the south coast" was hung in the Royal Academy of Arts in 1900. In addition to oil painting, she wrote and illustrated touring books, especially for children: *The France book, designed and coloured by M. Lanchester* was published in 1913. As a result of this she was commissioned to write and illustrate another book: *The River Severn from source to mouth ... with 58 pen and ink sketches and a map by the author* (London, T.Murby & Co. 1915). George also recalled that she carried out some research into the pigments used by the classical Dutch and Italian artists. From 1908 she lived with her parents, but after her mother's death in 1916 she lived in a wooden hut in Ashdown Forest in Sussex. She, like her sister Edith, was a vegetarian, and her family's view of her as an eccentric probably stems from these two facts: she was able to meet most of her vegetable requirements from her garden, though this would probably be regarded as less eccentric behaviour nowadays. Mary Lanchester never married, and died in 1942.

Eleanor Caroline, born 1866
Very little is known of the third Lanchester child, although she seems also to have inherited the artistic talent. She is variously referred to as "a clever caricaturist" (LC. Baxter 1/4) and "a fashion artist" (Burchell, 1987). She married her second cousin, John Sebastian Marlow Ward, and they had one daughter. Caroline died in 1926.

Frederick William, born 1868
Fred was to become undoubtedly the most famous of the eight children during and after his lifetime. He and his work are the subject of this volume, and Chris Clark's two volume history of the cars. A biography and review of his work, *F.W.Lanchester: the life of an engineer*, by P.W.Kingsford, (Kingsford, 1960) was probably the most comprehensive work to date. A more detailed biographical chapter follows. Fred married Dorothea Cooper, and died in 1946. They had no children.

Francis and Charles, born 1870
In 1870 Octavia Lanchester gave birth to twin boys, Francis and Charles. Unfortunately Charles died within a few days, but Francis, known throughout his life as Frank, joined Fred and George in their manufacturing work, though he was the commercial partner, rather than an engineer. His grandson, John Bingeman, has written the biography of Frank, which appears as Chapter Three in the present volume. Frank married Minnie Grace Thomas (the sister of his youngest brother's first wife Rose), they had two daughters, and Frank died in 1960.

Edith, born 1871
The third daughter, Edith, (known in the family as "Biddy") was "brilliant and eccentric" according to her youngest brother George. She was a militant suffragette, atheist, vegetarian, pacifist, and a member of the Social Democratic Federation and the Independent Labour Party. She studied many subjects, including higher mathematics at London Polytechnic, taking the matriculation of the universities of London and Cambridge. After a year teaching, she became secretary to Eleanor Marx, Karl Marx's daughter. When she was twenty-four she announced to her parents that she was going to live with James Sullivan, a self-educated clerk, but that they would not get married. Her father was so incensed that, with the help of his sons, Henry and Fred, and a cooperative doctor, he had her certified as "insane due to over-education". She remained in the asylum only a few days, Sullivan, their landlady, and public opinion securing her release. The "Miss Lanchester Kidnapping Case", as it became known, made national news, but the furore soon died down. She bore James Sullivan two children: their daughter, Elsa Lanchester, born in 1902, became a famous actress and married Charles Laughton. Her autobiography, *Elsa Lanchester, herself* (Lanchester, 1983), gives few insights into the family. The Sullivan/Lanchester son was Waldo Lanchester, who became well known, especially in later years in Malvern and Stratford-on-Avon, as an accomplished puppeteer: G.B.Shaw wrote a 10-minute puppet play for him: *Shakes versus Shav*, in 1949. Edith died in April 1966.

Edward Norman, born 1873
Edward seems to have been something of an individualist, like his sister Edith. He was supposed to be training for the Civil Service in the evenings, whilst studying law in a solicitors' office during the day. However, this does not seem to have been to his liking, since about 1890 he gave up all studies and went to sea, signing on as an able-bodied seaman on a windjammer. He returned from Canada to the United Kingdom at the beginning of the First World War, and served on minesweepers throughout the conflict. After the war he returned to New Zealand, met and married Elsie Winifred Wardell-Johnson, the daughter of a doctor, and they settled in Australia. They had one daughter, and Edward died in Australia in 1946.

George Herbert, born 1874

The last of Henry Jones and Octavia Lanchester's children, George was the third of the Lanchester brothers who built the first all-British, four-wheel, petrol-engined car. He was an outstanding engineer in his own right, but always seems to have been overshadowed in his career by his elder brother Fred, whom he idolised. In addition to his work on cars, George also worked for the Sterling Armament Company Ltd on machine guns, and took out many relevant patents. George was married twice: first to Winifred Rose Thomas (the sister of Frank's wife), who bore him a son (Clive) and a daughter. Clive's widow, Elaine, has written the biographical chapter on George for this volume (Chapter Four). Many years after Rose's death, and in his eighties, George married Mary Stevenson, known as "Steve", an old friend. He died in 1970.

References

Burchell, J. (1987): *The History of Lanchester and Lodge.* Macclesfield, Macmillan Martin Ltd.

Kingsford, P.W. (1960): *F.W.Lanchester: life of an engineer.* London, Edward Arnold, 1960

Lanchester, E. (1983): *Elsa Lanchester: herself.* London, Michael Joseph, 1983

Chapter Two

FREDERICK WILLIAM LANCHESTER

23rd October 1868–8th March 1946

by John Fletcher

"Hors d'oeuvre.

To write a biography, whether it be one's own, or that of another, is a serious undertaking. To justify itself it must constitute a work on which the future historian may rely at least as concerns facts. For this reason it should be documented so far as is reasonably possible. Many ignore this, often the biographer, from feelings of personal affection, or from hero-worship, is led to paint a beautiful picture, in which everything is viewed through rose-coloured glasses so that facts which might be detrimental to the reputation of their idol are glossed over or omitted altogether. Counterpart to this a perfectly honest and unbiased biography may be so distasteful to the living, friends and admirers of the subject (one might almost say of the "accused") as to be regarded by such as a social offence, the plea of de mortuis nil nisi bonum is raised, and the author is blamed."

These are the opening sentences written by Frederick William Lanchester in 1938, to be used as an introduction to his autobiography (LC. Baxter 1/22). It is thus with some trepidation that the author of this chapter undertook to write a short biography of the principal member of the "Unholy Trinity".

Fred never completed his autobiography, although there are some notes of the proposed contents, and a description of his childhood and early life remembered with great detail in his seventieth year (LC. Baxter 1/22 to 1/41). I have drawn on these, and on many other sources in trying to paint a balanced picture of the subject. The only full biography of F.W.Lanchester is P.W.Kingsford's *F.W.Lanchester: the life of an engineer* (1960), for which he used published material and some unpublished sources, probably the papers then kept by Dorothea Lanchester and later deposited with the Library of the then Lanchester College of Technology; these formed the foundation of the Lanchester Collection, now housed in the Lanchester Library of Coventry University. For a full description of the contents of the Lanchester Collection, its organisation and the indexes available, see pages 275–8 below.

Frederick William Lanchester was born in Lewisham, London, on 23rd October 1868, the fourth child and second son of Henry Jones Lanchester and Octavia Ward (see frontispiece). His elder brother, Henry Vaughan, was to follow his father's profession and become an architect. The next two children, Mary and Caroline, were to become artists.

In 1870 the family moved to a new house, 1 St. John's Terrace, in Hove, near Brighton. After attending a kindergarten for a number of years Fred became a weekly boarder at Oxford House, Eaton Place, Brighton, where he remained until the autumn of 1881. It is interesting to note that he was introduced to the French language at the kindergarten, and "advanced arithmetic and the beginnings [of] algebra and Euclid" (LC. Baxter 1/41, p.3) at Oxford House. Fred was not a strong child, though "by no means a weakling", but he had a strong appetite for learning, and picked up facts quickly. He found it difficult from an early age to write an essay, or a letter. The problem was not writing, but composition, and he noted that this remained with him all his life: he would draft and redraft manuscripts for papers and lectures many times before he was satisfied with them.

He had been accepted for the Hartley Institute in Southampton, but was unable to take it up until he was fourteen years old, so spent a year at Walkers School, "where they coached well-to-do but not very brilliant men for the army" (LC. Baxter 1/41, p.4). Fred started at the Hartley Institute in the autumn of 1882. In May 1884 he took examinations in six or eight subjects, and in fifteen the following year. He was especially keen on mathematics, but also took physics, chemistry, agriculture, metallurgy, machine drawing, and building construction. He seems to have taken some of the examinations at advanced level earlier than expected, but still passed satisfactorily.

In 1885 Fred began classes at the Normal School of Science, in South Kensington, living at home from 1886 when the family returned to Balham in London. In the first year he devoted himself entirely to chemistry, and mostly to physics in the second year. There were many practical classes, as well as lectures, and Fred derived great pleasure from making his own apparatus. Here he met Charles Vernon Boys, who encouraged him to contribute two papers (on the *Radial cursor* and the *Pendulum accelerometer* (see pp. 235–6 and 239–240 and Fig. 3 in Chapter 20) to the Physical Society of London. They were published in *Philosophical Magazine* in 1896 and 1905 respectively. It was at this time that he was introduced to the gas engine, which was to feature so much in the first stages of his career.

Surprisingly, although the Normal School ran classes in engineering, they were not to Fred's taste, so he attended evening classes at Finsbury Technical College to redress the balance. Here he derived great pleasure from the practical side, learning the techniques of using hand tools on metal, a skill which was to prove invaluable in his later life, for although he liked to know the theoretical basis for his work, he also wanted to master the practical skills it involved. In his third year he took mining and metallurgy, geology and surveying at the Normal School, but the mining course ceased to interest him, and he abandoned it without taking the examinations.

He now found himself without paper qualifications, but needing to find a source of income. His many applications for jobs having proved fruitless, he took a job as a contract draughtsman with a patent agent, doing most of the work at home. This proved to be good training, for he patented an invention of his own, the Isometrograph, an instrument for assisting draughtsmen to draw parallel lines (Patent 16432 of 1888, see p. 232–3), which he almost immediately sold for twenty-five pounds. Together with his older brother Henry, Fred also patented an automatic bookmarker (Patent 650 of 1889), but this was less successful.

In 1888 Fred's uncle, John Field Swinburn, the husband of Octavia's younger sister Jane, and a gunsmith in Birmingham, introduced him to several of his friends in that city. One of them, Tom Birkett Barker, owned the Forward Gas Engine Company, in Saltley in Birmingham, which employed about fifty men in general engineering, though the company had a contract with Crossleys to repair and maintain their Otto gas engines. He took Fred on as assistant works manager, and so began Fred's domicile in Birmingham, which was to be his home for more or less the rest of his life. He first took rooms in "Fairview", St. Bernards Road, Olton, where he remained until July 1895.

Barker's proposed contract included the clause that any of Fred's ideas which were patented should become the property of the company. Fred recounts with great relish that he insisted on the deletion of that clause from the contract, a shrewd move, and one which showed almost immediate financial benefit for him.

Fred's third invention of this date was to prove the opening into industry which he was seeking. Patent 12502 of 1889 was for a pendulum governor for gas engines, which was taken up on a royalty basis by the Forward Gas Engine Company. The following year he took out a patent for a starter method for gas engines, consisting of a low-pressure explosion, followed by a series of low-pressure impulses until the engine has acquired sufficient speed. It was much simpler than a compressed air starter, and safer than turning the flywheel by hand. He signed an agreement with Crossleys for a royalty of £3 per starter fitted by them, with a minimum payment of £1,500 a year.

Charles Linford, Forward's works manager, was an ill man, and died a year after Fred's arrival. He had taught Fred a great deal, and Barker was happy enough to promote him to the vacant position of works manager in 1890. When Linford's widow asked for money for her late husband's drawings relevant to the Company's products, Fred assured Barker that this was not necessary. He made a new set of drawings, with up-to-date additions and amendments. George Lanchester was just finishing school, and in October 1889 Fred arranged for him to come up to Birmingham, and paid for his apprenticeship to Barker.

While he was taking out the patents for his gas engine self-starter (Patents 19513, 19775 and 19846 of 1890) Fred was advised to contact Dugald Clerk, a patent agent in Birmingham. This was the beginning of a lifelong, and very fruitful friendship. Soon after, Dugald Clerk was considering taking out a patent on an improved gas engine starter which would clearly replace Fred's, so asked him if he wished to object. Fred said "No", and later purchased a half share in the new device.

The breadth of experience of a wide range of gas engines which Fred obtained at Forward was to prove invaluable in his later work on engines. One development which he made was of direct benefit to his employer: he linked a high-speed gas engine to a dynamo, providing the electricity for the works.

In 1892 Fred visited the United States, partly to see if there was a market for Barker's gas engines, and partly to negotiate the sale or use of his own gas engine patents. He visited New York, Pittsburg, Chicago and Washington, but clearly had been poorly advised about the situation in the States. Gas was so cheap, that the more economical, but more expensive British gas engines were not saleable there. He returned to England in February 1893 "wiser perhaps but without having achieved anything" (LC. Baxter 1/41 p.39).

Before leaving for America Fred had resigned his post as works manager at Forward Gas Engine Company, and his brother George, having completed his apprenticeship, took over the position. Their middle brother Frank, had worked from the age of fourteen in various banks in London, and from 1895 was also working for Forward: the company's cashier had absconded, and Frank had been offered the post. The three Lanchester brothers were together, in Birmingham, working for a company with good engineering workshops at their disposal, so the stage was set for the next major development: the design and construction of the first four-wheeled petrol-engined all-British motor car. Chris Clark, in the first volume of this trilogy, *The Lanchester legacy*, details the development of the first Lanchester cars (Clark, 1995).

On his return from the United States in February 1893, Fred had to take stock of his position. "I had little or no steady revenue from any source. I had no regular employment" (LC. Baxter 1/41, p.41). He had set up a small business in Birmingham to make dust-proof bearings for cycles to his own design, and put Frank in charge. The venture was not a commercial success, though he was at pains to free his brother from any blame for its failure. As he (uncharacteristically) wrote later, "My designs were not above criticism, I had been rather over confident" (LC. Baxter 1/41, p.41). He closed down the firm, with a loss of £800. The problem seems to have been that he underestimated the ability of the larger companies to make similar products, in larger quantities, and at a lower price.

For some time, the three brothers had been sharing a rented house in St. Bernards Road, Olton, and it was here that Fred carried out his experiments on model gliders, trying to establish some of the technical features necessary for his theory of flight. The gliders were released from the first floor windows of the house, and their line, height, and angle of glide were all meticulously noted.

In June 1894 Fred gave his first public paper, *The Soaring of birds and the possibilities of mechanical flight* to the Birmingham Natural History and Philosophical Society, based on the results he had achieved to date. He was advised to submit the paper to the Physical Society of London, with a view to publication, which he did in 1897, but they rejected it. Although a report of the meeting survives, the full text of the address was lost (see pp. 8–9 below). At the same time Fred began work in earnest on the development of his first motor

Fig. 1. Fred Lanchester with one of his model gliders, 1894

car, assisted by George and Frank, and using a rented workshop close to the Forward Gas Engine Company's works. The proximity was essential, since they were using the Company's equipment in the evenings and weekends to make parts for the car. Chris Clark gives full details of this first car in Chapter Two of the first volume of *The Lanchester legacy*.

In July 1895, Fred had moved out of the house in Olton, to take cheaper lodgings at Cobley Hill, a farmhouse in Alvechurch, but suffered a bad bout of rheumatic fever that autumn (LC. Baxter 12/76). He found living at the farmhouse inconvenient, and the three brothers rented rooms in Lincolns Inn in Corporation Street in the centre of Birmingham. This was a newly built block of what was intended to be let as offices, but there was clearly a greater demand for bachelor accommodation, and so that is what it became. Here he remained until his financial situation had improved sufficiently to allow him to move to a house, 53 Hagley Road, which was to be his home for the next twenty years.

The "scandal" of the kidnapping of Edith Lanchester, and the certification of her insanity occurred in October 1895. It is fully described by Elsa Lanchester in a Prologue to her *Elsa Lanchester, herself* (Lanchester, 1983). Fred was one of the two Lanchester brothers who helped their father in the affair, and Edith said later that he was the least kind of them. It is ironic that Fred, with good education, but lack of formal qualifications, should assist in the certification of his sister on the grounds of "over-education".

It was at this time that Fred's twin passions, aerial flight, and petrol engines, began to come together. He had realised that the possibility of heavier-than-air flight was feasible only if suitable lightweight engines could be developed, and together with his younger brother had been working on the first prototype from about 1890. In addition to the first car, they tested the engine in a launch.

Full road tests on the first car were carried out in February 1896, after short runs round local streets. The development of the car, and its engine, had required more financial resources than the Lanchester brothers were able to muster, and support had come from

Fig. 2. Fred Lanchester driving, with George in passenger seat

James and Allen Whitfield, who were to become shareholders in the Lanchester Engine Company, when it was formed in December 1899. Prior to this, in 1897, a second car, using many of the parts of the first, and a six-seater phaeton, were built. This was shown to the public in 1899 at the Automobile Club Exhibition and Trials at Richmond in Surrey. It was awarded the Gold Medal, and was subsequently named "The Gold Medal Phaeton". In the same year, a third model, based on the second, was produced.

With the formation of the Lanchester Engine Company further financial support was needed, and this came from Vernon and John Pugh, and J.S.Taylor. Fred was officially general manager and chief designer of the company, but, since they had to make most of their own parts, and cash was short, he was in effect works manager as well. Production methods needed to be worked out, and staff trained for the new product. Fred noted later "It took something like eighteen months before we were able to deliver our first cars to the public," but nevertheless, in the company report for the year ended 31st July 1902 the directors reported "some of our cars in the hands of the public had been running upwards of a year and proving in every way satisfactory" (LC. Baxter, 1/2, p.3).

Although profits improved in 1902–3, it became impossible for the company to continue without a further injection of capital. A share subscription of 42,405 shares of one pound each in December 1903 failed to attract the market and in March 1904 there was no alternative but to call in the receiver. The shortage of capital, and cash flow problems were at the heart of the bankruptcy. The Lanchester method of car production involved interchangeability of chassis and of bodywork, innovations in components, and minimal tolerances in all manufactured parts. All this entailed more expensive production methods, and greater stocks of parts, which were all expensive in capital as well as labour.

The receiver allowed the company to continue, for it was now manufacturing cars on a full production basis: the 10 h.p. and 16 h.p. models. After showing a profit in the following year, the receiver reconstructed the company in 1905 as The Lanchester Motor Company Limited, new money being found by the existing members of the board, and by issuing second debentures to the creditors.

There were some important changes in Fred's relationship to the reconstructed company. He was unable to provide any of the additional money required by the company, and resigned as general manager, becoming consulting engineer and designer, on an annual fee. This also freed him to take similar posts with other companies. In return, the Lanchester Motor Company ceased to have any rights to his later inventions, though many of his ideas were incorporated in the company's products.

This marked the beginning of a new era for Fred: he was able to combine his interests in flight, and aircraft, with those of cars and manufacturing. During 1906 and 1907 he reviewed and revised the text of the 1894 paper he had delivered to the Birmingham Natural

History and Philosophical Society on the flight of birds. He had sent it to the Physical Society in 1897, but they had rejected it for publication. He noted later that the text of this paper had been lost, but that many of his ideas were incorporated in later publications, in particular Chapter Four of *Aerodynamics*, the first of the two-volume work on aerial flight, *Aerodynamics* and *Aerodonetics; constituting the [first and second] volumes of a complete work on aerial flight*, published in 1907 and 1908. The work made an immediate impact on the new world of flight, especially in Germany, and was to prove seminal in the development of the subject in later years. In retrospect it was arguably his most important publication, and went into four editions by 1918. (For a review of the content and importance of this work see Chapters Five and Six below).

In September and October 1908 Fred visited Germany and France, firstly to try to arrange translations of his two-volume work (known jointly as *Aerial flight*), and in this he was partially successful: Professor Carl Runge of Gottingen had already read *Aerodynamics* and had offered to translate it into German. He did so, with the help of his wife Aimée, and it was published as *Aerodynamik* ... in 1909. Fred was less successful in France, and it was not until 1914 that C.Benoit's French translation of the second edition of volume 1, *Aerodynamique* ... was published by Gauthier-Villars in Paris.

The second reason for his visit was to meet other people in both countries, interested, and involved in the theory of flight and the production of aeroplanes. The Wright brothers were also to give a demonstration of their flying machine at Le Mans.

The journey began on 16th September 1908 with a meeting with Carl Runge in Gottingen, then with Lilienthal, Parsifal and Prandtl, in Hanover, Leipzig, Berlin and Potsdam. On 16th October he arrived in Paris, and the following day visited the factory of Voisin Frères the aircraft manufacturers. He watched the Wrights' machine flying demonstrations at Le Mans on 29th and 30th October 1908, "It was not a very impressive show" he noted later (LC. Baxter 1/15), and his meeting with Wilbur Wright on 29th October did not go well. They seem to have had different, and irreconcilable views on the value and importance of theory versus pragmatism in the development of powered flight. Fred had always felt that one needed to know if, and why something might work, whilst the Wrights clearly believed in the "try it and see" approach. Early in November 1908, Fred returned home.

About this time, 1908, Fred felt that he was giving more lectures, and decided to have his voice trained to improve his fluency and voice projection. He went to Edwin Stephenson, the organist of Birmingham Cathedral for lessons, which included singing lessons. He was a good pupil, and learned to love singing as well as public speaking. He became an accomplished amateur singer, with a special love of German lieder, and later wrote of his wish to have more musical evenings, but was frustrated by not having a pianist equally devoted to lieder.

On the motor vehicle side, Fred was freed from the day to day involvement with the production of Lanchester cars, though he remained responsible for the design of all models until 1910. The gradual reduction in Lanchester work enabled him to take consultancies with other companies in the same industry.

In 1909 he joined Birmingham Small Arms Ltd. as consulting engineer and technical advisor, an association with them, and with Daimler Company Ltd. which lasted for twenty years. Some of their problems, such as vibration in the crankshaft of the six-cylinder engine, were similar to those which other engine manufacturers had met, but failed to solve. The longer stroke of the Daimler engine exacerbated the problem, and Fred devised the torsional vibration damper as the answer. The solution was so effective that it was "patented, and my receipts in royalties in this country were very substantial. During the War I disposed of my U.S. patent for a substantial figure" (LC. Baxter 1/2, p.6). In 1911 he developed the harmonic balancer for car engines (small counter-balance weights linked to the crankshaft) which was designed for the Lanchester engines, but licensed to other engine manufacturers.

His work for Daimler ranged beyond the motor car: the Renard Roadtrain, the petrol-electric bus, the petrol-electric railcar, and aircraft design. The rail coach, incidentally, did

Fig. 3. Fred at his desk, about 1920

several runs between Coventry and London, faster than the contemporary London and North Western Railway express service. Equally important for Fred, the Daimler Company provided larger research resources to support his activities.

Aircraft design, and the use of aircraft were still Fred's first love. Early in 1909 Norman A. Thompson, an electrical engineer who had taken an interest in aeronautics, contacted him after reading his *Aerial flight*. Thompson wanted to go into the design and production of aircraft, but needed financial support. This he obtained from Dr. Douglas White, and together they approached Fred to become a technical advisor to their new company, White and Thompson. He worked with them on flying boats, thus tying up his theoretical work on flight, his ideas on the use of aircraft, and the practical design of them. For example, one of his papers to the Advisory Committee for Aeronautics (*Memorandum on the use of aluminium alloy sheet in place of fabric for aeroplane wings, etc.* (its Reports and Memoranda (new series) No. 359, 1917)) used the results of research he had done in conjunction with White and Thompson.

On 30th April 1909 Fred was asked to join the select Advisory Committee for Aeronautics, a new committee set up by the government to advise on, and coordinate research into aircraft design and use. He was to remain a member of the Committee until 1920, although the relationship seems at times to have been somewhat stormy. His written reports to the Committee are listed in the bibliography of publications on pp. 260–270. In 1916 he was sent to France to visit the air forces, and report to the Committee on various problems which had arisen with aircraft and their armament.

From 1908 Fred became increasingly involved in both the motor car and the aeronautical fields at a national level. Since 1907 he had been giving papers to the young Institution of Automobile Engineers, and in 1910 he became President. He was a frequent contributor to the verbal and written discussion of other members' papers, though he found writing less easy than speaking. From an early age, as we have seen above, he found writing difficult, and there have been suggestions that he suffered from dyslexia: an educational psychologist commented recently, after examining some of Fred's manuscripts and typed drafts of papers from the 1930s, that he "showed all the classic signs of dyslexia". On the other hand, he had an extremely quick mind, and revelled in debate, the more heated the more he enjoyed it, and he had had his voice trained to help his public speaking.

His work for the Advisory Committee for Aeronautics concentrated on the technical aspects of the subject and he gave up his commercial links with the aeronautical engineering industry in 1912, feeling that his involvement with White and Thompson might be seen as affecting his impartiality in advising the government. He did not, however, resign his consultancy with Daimler, for whom he also worked on aircraft and their engines.

In 1913 Fred attended the International Mathematical Congress in Cambridge, renewing his friendship with Carl Runge, who was also there. Fred invited Runge to Birmingham, to see the factory, and continue their discussions on flight. Their contact was inevitably broken by the outbreak of the war, but it is clear that the two men got on well together, sharing knowledge, interests, and above all, ways of thought and research.

Following the outbreak of the First World War, the Lanchester Motor Company had obtained contracts for building staff and armoured cars based on the 25 h.p. and 38 h.p. car designs. Meanwhile Fred's work with Daimler included designs for cars, lorries, vans and other vehicles for war use. The Daimler Company also obtained a contract to produce aircraft engines. Fred was deeply involved here, and with the production of the French Gnome aero engine. There are several pages of detailed test results on these engines in the sketchbooks. (These are artists' sketchbooks which Fred used as notebooks. He seems to have kept one on each of his desks, turning to a new page when he needed to make notes. The Lanchester Collection has thirteen of them, and there are thought to be others still in existence.)

His foresight in seeing the strategic use of aircraft became apparent in 1916 when he published his second major work, *Aircraft in warfare: the dawn of the fourth arm.* In publishing his views he undoubtedly offended many of the higher authorities on the military side of the government, who were clearly intent on pursuing the world war in the same way as previous campaigns had been carried out, ignoring the impact which air forces would have on the conduct of war. The book was the one-volume publication of sixteen articles he had written for and published in *Engineering* between September and December 1914 and which contained the now famous n-square Law, one of the foundation stones of operational research as we know it today.

In 1917, Harold A. Titcomb, had written a report on the German iron and steel industry, and Fred was asked to comment on it, especially regarding the possibility of air attack, to disable the war supplies. This he did, in 1918, in a paper to the American Committee of Engineers in London, entitled *Memorandum to the American ... concerning the Committee report on Germany's iron and steel industry and the war.* Fred later wrote that the paper was "supplied to the British War Office and [it] was in the hands of the military advisors of the Air Ministry. It was not, however, accepted [until just before the armistice] but it was served up to Lord Weir piecemeal as their own work" (LC. Baxter 12/11, p.4).

His reputation grew year by year, and the professional honours were not slow to follow. In 1910 Fred had been elected President of the Institution of Automobile Engineers, a professional body with which he had been associated since 1906. In 1915 he was appointed a member of the Air Inventions Committee of the Admiralty Board of Invention and Research. In October 1916 he became President of the Junior Institution of Engineers, and was granted Fellowship of the Royal Aeronautical Society in June 1917.

As a member of the Advisory Committee for Aeronautics, with monthly meetings, and two of its sub-committees, on Aerodynamics and on Internal Combustion Engines, with their two or three meetings per month, Fred found he was having to spend more time in London, and he hated staying in hotels. In 1917 he leased 41, Bedford Square which became his main residence until April 1924. It was a large house, but Fred felt able to spread himself, and to furnish it to his own taste. There were rooms for members of his family, notably his nieces, to stay and help him to relax with his favourite pastime, making music.

Marjorie Bingeman, one of the nieces, remembers staying in the flat at the top of 41, Bedford Square, in her 'teens, with her sister and nanny. Fred had given them a 'cello and violin, and paid for their music lessons. He was undoubtedly a favourite uncle, who was always doing interesting things. At this time he was also developing patents on both piano and player piano mechanisms (9557 and 11016 of 1912), which, although sold to Orchestrelle Ltd. do not seem to have been followed up.

Their music making was helped at one point when Edwin Stephenson, Fred's voice tutor from Birmingham and his family moved into the top floor of the house, after Stephenson had been appointed organist of a church in London. An unexpected, and happy coincidence was that Fred was able to renew his acquaintance with Edwin Stephenson's niece, Dorothea Cooper, the daughter of the vicar of Field Broughton in the Lake District.

Further honours were to follow, but for Fred probably the greatest professional accolade at this time was the granting of the honorary degree of Doctor of Laws by the University of Birmingham on 18th September 1919. He had never acquired any academic qualifications, and the honorary degree pleased him greatly: he liked to be known thereafter as "Doctor Fred". September 1919 was a momentous month, for on 3rd September he had married Dorothea at her father's church.

Sailing was another of Fred's interests, he and his brothers had had a boat in the 1890s to test out their new engines, but it was in 1910 that he purchased his first sailing boat which he kept at Bembridge on the Isle of Wight. Sailing was very much a relaxation for him, and with friends he would take the boat across the Channel for weekends and holidays. During the war he sold *Iseult*, his first boat, and purchased a Bristol Channel pilot boat, which he renamed *Troubadour*. He always intended to fit it with an engine, but never found time.

In 1920 Fred resigned from the Advisory Committee for Aeronautics, probably feeling that after the end of the Great War, his work on the military uses of aircraft was finished, that the payment offered was not commensurate with the work involved, and to allow him more time to devote to his consultancy work with the motor industry. Since his gradual withdrawal from the everyday management of the Lanchester Motor Company, during the period 1911 to 1914, Fred had undertaken consultancies for a wide range of companies, both within and outside the motor industry. His work with White and Thomson has already been noted, and in January 1924 he was appointed consultant to Wolseley Motors, at a fee of £1,000 p.a. The company were experiencing some design problems, and the possibility of patent infringements, and wanted the advice of one of the most respected car designers, who incidentally had a great deal of experience of motor vehicle patents. His work on spectacle lens production for William Gowlland of Croydon is noted on p. 241.

His resignation from the Advisory Committee reduced substantially his commitment to

Fig. 4. Fred and Dorothea, in their garden soon after their wedding in 1919

Fig. 5. Three views of Troubadour, *Fred's second sailing boat*

meetings in London, and over the next few years he decided to move back to Birmingham. George had a house in Moseley, which had a substantial piece of land attached. Fred wanted to build a house for himself and Dorothea, and asked Henry Vaughan, his oldest brother, and now an eminent architect, to design it. Unfortunately Fred did not like the design, so set about making one himself. The Lanchesters were so anxious to occupy their new house that they moved into it, in April 1924, before the workmen had finished, which may have been inconvenient and noisy, but Fred had his house, with study, garage, and workshop. It was named Dyott End, and was destined to be their last house.

At the end of 1924 Fred wrote to Carl Runge in Gottingen sending best wishes for the new year, thus breaking the ten-year enforced silence caused by the war. As Iris Runge noted in her biography of her father, "In the correspondence, which then resumed, political differences, too, obviously had to be addressed. Lanchester's views were those of an English imperialist, which, from the German point of view certainly appeared as prejudices." "... his lively correspondence with Runge, to whom he presented above all his reflections and reservations about the theory of relativity, with which he was still deeply preoccupied, showed that his disapproval of German politics did not extend to individuals." (Runge, 1949).

In order to give himself even better research facilities, and premises to carry out his consultancy work for his clients, and especially for Daimler, Fred formed, in 1925, Lanchester's Laboratories Ltd. From a commercial standpoint this was a disaster. Daimler had agreed to support the new company, and Fred was to place the control of several of his more valuable patents in the hands of the board of the company. Almost immediately, Daimler withdrew their financial support, and Fred had to rely on his own slender resources, and whatever finance other members of the board would provide.

1925 and 1926 were busy years. Fred took part in the discussions following many papers delivered to the members of the Institution of Automobile Engineers, he attended the British Association meeting in Oxford, and renewed his friendship with Carl Runge; in May 1926 the work he had done for aeronautical science was recognised by his appointment as an honorary fellow of the Royal Aeronautical Society, coupled with the award of its gold medal, and the invitation to give the Wright Lecture.

But since 1924 the great economic depression had been affecting Fred's life. Lanchester's Laboratories Ltd. was a financial drain, and he needed a regular source of income. In 1926 he signed a sole contract with the Daimler Motor Company in Coventry, but in 1927 Daimler released him from the "sole" aspect of the contract, and freed him to take on other consultancies, a situation which remained until the contract ceased on 4th September 1929.

Although the relationship between Fred and Daimler lasted for twenty years, to the mutual benefit of both parties, there were undoubtedly rough times during this period. Daimler wanted Fred to solve production problems, and also undertake "developments of a speculative kind". Fred understood, and coped with the former, but as he later wrote: "The fact impressed itself upon me that before embarking on any expensive development work the commercial position should be more thoroughly investigated, and the object in view should be very definite" (LC. vol. 26–1). The speculative projects specifically referred to are the Renard train, the petrol electric bus and the rail coach.

On the other hand, the contract with Daimler freed Fred to carry out research projects for himself, and for other companies, safe in the knowledge that Daimler were paying a consultancy fee regularly into his account. By November 1927 the imminent recession was hitting Daimler, and senior staff were being asked to take cuts in their salaries (LC. Baxter

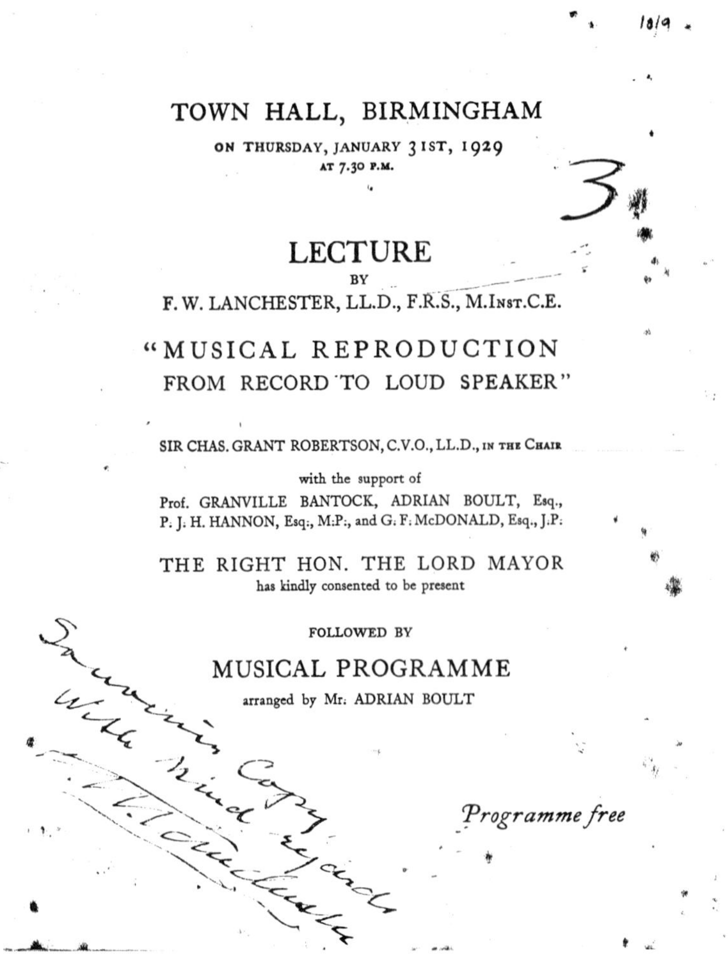

TOWN HALL, BIRMINGHAM

ON THURSDAY, JANUARY 31ST, 1929
AT 7.30 P.M.

LECTURE

BY

F. W. LANCHESTER, LL.D., F.R.S., M.INST.C.E.

"MUSICAL REPRODUCTION
FROM RECORD TO LOUD SPEAKER"

SIR CHAS. GRANT ROBERTSON, C.V.O., LL.D., IN THE CHAIR

with the support of

Prof. GRANVILLE BANTOCK, ADRIAN BOULT, Esq.,
P. J. H. HANNON, Esq., M.P., and G. F. McDONALD, Esq., J.P.

THE RIGHT HON. THE LORD MAYOR
has kindly consented to be present

FOLLOWED BY

MUSICAL PROGRAMME
arranged by Mr. ADRIAN BOULT

Programme free

Fig. 6. Cover of souvenir programme for Euterpe-phone recital (LC. Baxter 10/9)

7-30   INTRODUCTORY.

Andantino—*Lamare*—ARTHUR MEALE.   HMV B 2353

Toccata and Fugue in D minor—*Bach*—G. D. CUNNINGHAM
     (Organ of Kingsway Hall).   HMV C 1291

Wolverine March—SOUSA'S BAND.   HMV B 2869

7-45   LECTURE PROGRAMME.

NATIONAL ANTHEM.

INTRODUCTION BY THE CHAIRMAN.

LECTURE, illustrated by excerpts from records as follows, rendered by
         THE LANCHESTER EUTERPE-PHONE.

                              Illustrating in particular :

(a) Rienzi Overture—*Wagner*—
     PHILADELPHIA SYMPH. ORCH. (STOKOWSKI). HMV D 1226   Drums.

(b) "The Lost Chord"—*Sullivan*—
     GATTY SELLERS   HMV C 1237   Organ Bass.

(c) Scheherazade—*Rimsky Korsakov*—
     PHILADELPHIA SYMPH. ORCH. (STOKOWSKI). HMV D 1437   Pizzicato.

(d) Preludes 16,17—*Chopin*—ALFRED CORTOT. HMV DB 959   Piano.

(e) Hungarian Dance No. 5—*Brahms*—
     J. H. SQUIRE CELESTE TRIO.   COLUMBIA 3605 R   Strings.

(f) Serenade—*Toselli*—BENJAMINO GIGLI. HMV DB 1002   Vocal.

(g) Largo al factotum—*Rossini*—STRACCIARI. COLUMBIA L 2129   "

(h) "Cock o' the North"—
     PIPE-MAJOR D. SMITH.   HMV B 980   Volume Control.

(j) Lison—*Fragson*   ODEON 36378   Old Records.

(k) Air—*Bach*—KREISLER. GRAMOPHONE AND TYPEWRITER GC 47947   "

(l) "Praise the Lord oh my Soul"—*Wesley*—
     TEMPLE CHOIR   HMV C 1436   Choral.

(m) "Tristan and Isolde (Act 3)—*Wagner*—
     LONDON SYMPH. ORCH. (ALBERT COATES) HMV D 1413   Grand Opera.

         5 MINUTES INTERVAL.

8-40   MUSICAL PROGRAMME.
         Arranged by MR. ADRIAN BOULT.

CLASSICAL.

Scheherazade, 4th movement, allegro molto—*Rimsky Korsakov*.
PHILADELPHIA SYMPH. ORCH. (STOKOWSKI).   HMV D 1439-40

Trio in B flat Major, Op. 99 (2nd and 3rd Movements)—*Schubert*.
CASALS, CORTOT, THIBAUD.   HMV DB 948-9

Symphony No. 1 in C minor, Op. 68 (2nd and 3rd movements)—*Brahms*.
Alternatives { PHILADELPHIA SYMPH. ORCH. (STOKOWSKI) HMV D 1500-1
{ ROYAL PHILHARMONIC ORCH. (WEINGARTNER). COLUMBIA L 2146-7

         3 MINUTES INTERVAL.

POPULAR.

Selections from the following :—

"Eine Kleine Nachtmusik" (3rd movement, minuet)—*Mozart*.
STATE ORCHESTRA, BERLIN (OSCAR FRIED).   BRUNSWICK 80037A

Waltz in E flat—*Chopin*—BACKHAUS.   HMV DB 1131

"Invitation to the Waltz"—*Weber*—
PHILADELPHIA SYMPH. ORCH. (STOKOWSKI).   HMV D 1285

"The Lute Player"—*Allitsen*—PETER DAWSON.   HMV C 1313

"Goyescas," Intermezzo—*Granados* PABLO CASALS.   HMV DB 1067

"La Tosca"—*Puccini*—LOTTE LEHMANN, *soprano*,
and JAN KEPURA, *tenor*.   PARLOPHONE R 20048

"Adoree"—*West*—ALBERT SANDLER.   COLUMBIA 5070

"Only a Rose"—*Friml*—WINNIE MELVILLE and DEREK OLDHAM. HMV B 2570

"My Angel," Fox Trot—*Rapee*. "In my Bouquet of Memories,"
Fox Trot—*Akst*—PAUL WHITEMAN AND HIS ORCHESTRA.   HMV B 5510

"Under the Double Eagle"—*Sousa*.   HMV B 2361

Figs. 7 and 8. Programme of the Euterpe-phone recital

5/7). Fred complied with Percy Martin's request to take a one-third cut in his fee (from £3,000 to £2,000 p.a.), on the understanding that this was treated as a loan to the company, to be repaid when business conditions improved. At the same time, Daimler acknowledged a debt to Fred of £1,775 12s. 0d.

Less than a year later, in October 1928, the company's financial situation had clearly not improved, and Martin wrote again to Fred suggesting that in the interests of economy his contract would not be renewed when it concluded in October 1929. Instead, he offered a retaining fee of £1,000 p.a. By the April of 1929, however, things were deteriorating, and Martin wrote: "It is evident that we are drifting further and further apart" (LC. Baxter 5/26), and withdrew the proposed retaining-fee contract. Martin also made proposals about Daimler's involvement with Lanchester's Laboratories Ltd., for which he could see no future, and was advising the (Daimler) board to withdraw. He suggested that Fred should buy out the Daimler interest for either £8,000 and grant Daimler a free licence on all Lanchester's Laboratories current patents, or £10,000 with no free patent licence. After taking legal advice, Fred agreed to accept the former offer, and purchase the Daimler

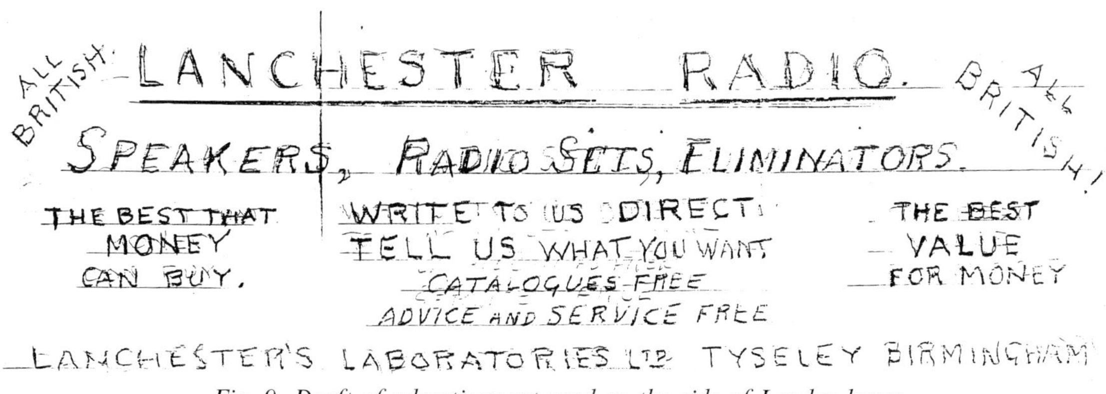

Fig. 9. Draft of advertisement used on the side of London buses

interest for £8,000. The following day, 11th April 1929 Fred's solicitor advised him to petition for the winding up of Lanchester's Laboratories Ltd. (LC. Baxter 5/31). Fred did not, and later regretted it. So ended one of the most fruitful cooperative relationships between Fred Lanchester and a client company.

In his autobiographical notes Fred says "Ultimately I acquired their [Daimler's] interest by purchase and finding the outlook for development or research work unpromising, embarked on the manufacture of high class radio products, more especially moving coil speakers and more generally specialising in musical reproduction" (LC. Baxter 1/2, pp.6–7). On 31st January 1929 Fred gave a lecture entitled "Musical reproduction from record to loud speaker" in Birmingham Town Hall, followed by a short record recital to demonstrate the quality of the sound reproduction equipment he was manufacturing see Figs 6–8. The audience was impressive, including the Vice-Chancellor and Sir Granville Bantock of the University of Birmingham, Adrian Boult (conductor of the City of Birmingham Orchestra), the Lord Mayor and local M.P.s.

For five years Lanchester's Laboratories Ltd. produced radios and loud speakers for general public sale. Roy Thomas looks at the various products in this area in Chapter Thirteen. Fred was reluctant to trade through the conventional retail outlets, who were, in any case, doubtful about the novel systems he used, and their saleability: the recession had hit them too, and capital for stockholding was short. Lanchester's Laboratories Ltd. concentrated on mail order sales, with little advertising outside the specialised press. There were exceptions: advertising posters were placed on the sides of some of the London buses (see Fig. 9).

The recession hit sales badly, and Fred was a proud man, unwilling both to admit defeat, and to lay off staff as early as he should have done. From early in 1931 he ran the business himself. and it showed a small profit in 1931/32. By January 1934 Fred was approaching sixty-six, tired from long hours of work at the Laboratories, and the worry of keeping the company viable.

Fred Lanchester was a large man, "heavily and strongly built, with a large head, yet he was quick of movement and also had something of boyish ungainliness in his appearance" wrote Iris Runge in her biography of her father (Runge, 1949). He had been a reasonably healthy man until now, but his health began to fail. He had had problems with his eyesight for some time, but now a retinal haemorrhage threatened the sight of one eye. At about the same time an abscess in his neck was operated on, found to be more serious than had been thought, and a second operation was necessary. It was not until late in the year that Fred recovered, by which time the Laboratories had gone beyond rescue. Like so many other companies at this time, it was forced to close down, and its assets sold in 1934, whilst Fred was in hospital.

Fred Lanchester was now beginning what turned out to be the final, and very depressing phase of his life. His commercial ventures had failed, partly because of the economic state of the country, partly as a result of bad advice, but also because sometimes he had refused to act on good advice. There was no work available to him, and he began to feel unwanted, and short of money. He turned to writing, knowing that it would not be income-earning, but might increase the world's awareness of his contributions to industry. He wrote many articles in *Engineering*, and gave papers to the Institution of Automobile Engineers. He also took on further non-commercial ventures to occupy his declining years. He became increasingly bitter that the important contributions he had made to the development of so many subjects were not being recognised in official quarters.

In the early 1930s Fred had struck up a friendship with Robert Lockhart, Professor of Anatomy at the University of Birmingham, and began some experiments on eyesight. The research which he carried out, and the papers which he produced are dealt with by Dr. Fred Fitzke in Chapter Fifteen below. The friendship between the two men continued after Lockhart moved to Aberdeen to become Head of that University's Department of Anatomy, and the Lanchester Collection contains an almost complete record of their frequent

Fig. 10. Fred Lanchester, about 1933

correspondence throughout this period, and the Second World War, until Fred's death in 1946.

The second re-discovered interest was in music. He had been trained as a singer, was interested in German lieder, and had been an accomplished performer. Now he turned his attention to the musical scale, or more correctly, the variety of musical scales which were, or had been, in use throughout the world. He examined them as a physicist, rather than as a musician. Dr. Norman Dyson reviews the product of this work, *The Musical scale* in Chapter Sixteen below.

Fred entered a new area in his *The Centenarian: a Lakeland story told in verse* published in 1935. For many years he had written verse, limericks and doggerel, but never before for publication. But the long ballad had more in it than a release from boredom during a period of ill health. It was semi-autobiographical, with many of the actions, and especially the philosophies mirroring those of the author. He was obviously embarrassed by what his professional colleagues might say of an eminent engineer who wrote verse, so *The Centenarian*, and the later volume of verse *The King's prayer*, were published under the pseudonym Paul Netherton-Herries. Barry West reviews these forays into verse writing in Chapter Seventeen below.

Yet another subject occupied him at this time – relativity. What Fred tried to do was present the complex problems of space, time and gravity in such a way that students and the general public might be able to understand them. He eschewed the approach of Einstein, and preferred to adopt the methods used by Minkowski, but the work, *Relativity ...* although published conventionally by Constable in 1935 made little impact on the public for whom it was intended. Kingsford (op. cit. p.208) noted that only 260 copies were sold in the first year, and the much needed income was not forthcoming.

Whilst all this writing activity was progressing, Fred seems also to have been working on a revised edition of his *Aerial flight* again intended for the general public. There are notes for this new book, variously entitled "Fundamentals of flight", or "Basic principles of flight", but it was never completed. In the same field, aeronautics, he had been working for some time on a new book on skin friction, and in November 1936 he gave a paper, "The Part played by skin friction in aeronautics" to the Royal Aeronautical Society. It was well reviewed in the engineering press, and was published in the Society's *Journal* in February of the following year.

Another new occupation came to Fred in 1935, which clearly gave him great pleasure, though little income: he was asked to be an external examiner in mechanical engineering at Birmingham University, a position which he held for three years. He certainly took the duties seriously: he seems to have interviewed all final year students, not merely those on the border line.

During this period of build-up to the Second World War Fred was against the appeasement and disarmament lobby. He had shown himself very strongly patriotic during the First World War, and this showed again in the 1930s. His protests took many (written) forms. *A King's prayer, and other poems*, was a protest at the disarmament movement, and in February 1936 he had two thousand copies printed, and sent copies to all members of both Houses of Parliament. It was reprinted in the coronation issue of *Engineering* in May 1937.

A further blow to his health came in 1937, when he was diagnosed as having Parkinson's disease. This affects the physical movement of the patient, and is degenerative, but the brain remains active, hence Fred's ability to continue with the projects noted above. The threat of another war made him think of other new ideas, including patent applications for bombs (Applications 2458 of 1937, and 6509 of 1939) though he did not follow these through. There is also a considerable volume of correspondence between Fred and various colleagues from his past industrial activities, and it is clear that he was working on an autobiography.

By 1938 Fred and Dorothea Lanchester were in difficult financial circumstances. He wrote to Admiral Sir Reginald Bacon in 1937: "I am out of a job so fill up my time writing books and technical papers ... I cannot afford to run a car now" (LC. Baxter 12/11, pp.2, 4). He had no regular source of income, and although he continued to carry out research and follow up his other interests, they did not bring income. He wrote to many old friends and colleagues with whom he had worked in an attempt to secure not only acknowledgement of the debt he felt the country owed to him, but also some financial reward. Some of his friends, probably led by Frank Lanchester, approached the Society of Motor Manufacturers and Traders, who, through their trust funds took over the Lanchester' bank debt by providing an interest-free mortgage on their house and gave them an annual income of £200.

Fred had held very strong views, which he was not averse to committing to paper, on the conduct of the First World War, and the build-up to, and outbreak of the Second World War saw him equally vehement in his criticism of affairs political and military. There are copies of letters to the Prime Minister, both before and during the war, on both strategic and technical matters. One particular set of documents in the Lanchester Collection is especially interesting. At the outbreak of the Second World War, Fred wrote an article showing that, in his opinion, Gibraltar was undefendable, since the airfield could not be used, and Spain could not be relied on to remain neutral, or provide bases for its defence. Although he offered the article for publication, it was refused, and he concluded that the Government wished the public to believe that Gibraltar was impregnable. This was intended to be a chapter in a book on the national situation in 1939, but the book was never completed. A summary of the chapter on Gibraltar is now published as Chapter Eleven of the present volume.

Fred was a fervent patriot, and in the build-up to the Second World War he felt that his experience and expertise should be used. He did not want a government post, for in April 1939 he wrote to C.C.Walker of the De Havilland Aircraft Company: "It is impossible to work for or with the Government unless you are a politician and aspire to cabinet rank, or become a civil servant for which the retirement age is 65" (Fred was 71). But he wanted to

Fig. 11. "In agricultural battle dress and battle axe" (October 1943, age 75)

help the war effort: "... it seems to me that there is scarcely an aircraft concern in the country that could not profitably make use of my collaboration" (LC. vol. 5–72).

It was not until June 1941 that the Society of British Aircraft Constructors offered Fred an appointment as a consultant, with an annual fee of £200. He needed the money, but was unwilling to accept charity. The following month he wrote to Captain Pritchard, Secretary of the Royal Aeronautical Society asking for his opinion of the offer, and was reassured that the S.B.A.C. wanted to be able to call on his services at any time. He accepted the appointment.

By 1942 Fred's health had deteriorated even further. The cataract had become worse, and he was almost blind in one eye. Dorothea had learned to type, so his work continued, with him dictating to Dorothea. In December 1942 he published *Relativity and radiation*, adding the appendices in 1943 and a note on the atomic bomb in 1945. In that same year, 1945, he issued a second edition of his *Musical scale*, incorporating amendments resulting from comments on the first edition.

Frederick William Lanchester died on 8th March 1946, following two strokes. His ashes were buried in his parents' grave at Lindfield, near Haywards Heath in Sussex.

So what are we now to make of the life and work of Frederick William Lanchester fifty years after his death?

On the technical and engineering aspects it is clear that he was a genius, in the full sense

Fig. 12. Fred Lanchester, about 1945

of the word. He has been called "A British Leonardo" (Gold, 1984), and with justification. Others, more capable than the present author, are assessing his place in the development of the motor car, aeronautics and engineering in this and the other two volumes celebrating the centenary of his production of the first motor car.

There can be few men who have made their mark in such a wide range of subjects. The seat of his genius is not hard to find: an extremely agile brain, and despite requiring great effort on his part, the ability to communicate his ideas to others. He was also able to transfer ideas or methods from one area of study to another. Two examples will suffice. His study of the flight of birds led to explanations of the reason for them flying in V formation, which he was able to transfer to the advisability of war aircraft flying in that formation too. His work on pianolas, with their use of pneumatic tubes and holes in the player roll, was transferred to the design of an arithmetic calculator using the same principle (see p. 243 below).

For various reasons his contemporaries were slow, or unwilling, to appreciate the quality and foresight of his thinking. Even when his ideas were recognised and accepted, there was a long delay before some of the accolades which were undoubtedly his due materialised. He never received the national honour which in later life he felt was due to him for the work he had done. He was very disappointed to receive a cheque for only £250 in 1933 as recognition for his work for the Air Ministry. His aeronautical theories were better known, and his work more appreciated in Germany than in Britain, which must have disappointed him, for he was, above all, a patriot.

Throughout his life, and even in his old age, the Establishment never quite knew how to handle Fred. On so many things his thinking was far ahead of his time: he carried out experiments on rocket propulsion at the farm in Alvechurch in 1895; he filed a patent application in 1897 for an aerial torpedo. But his reputation as a maverick lasted all his life. As late as March 1941, a lecturer at Aberdeen University wrote to Fred asking his advice, and added that he would not want "... it to be known that I am taking the action of asking your advice [as it] would be as fatal to my career as immorality or communism." (LC. vol. 23).

It is difficult now, fifty years on, to assess the man himself. It is rather like having an incomplete jig-saw puzzle: the pieces are the letters and manuscripts which survive, and the recollections of the few people alive who remember working with him. Some of the pieces fit together well, and one has a reasonably clear idea of aspects of the personality. Other pieces do not seem to fit in anywhere. Inevitably a biographer must arrive at his own conclusions from the evidence which is extant. What follows are, therefore, the views of the present author, having read a large proportion of the documents in the various collections. Eric Baxter and I indexed the papers in the Lanchester Collection, (see pp. 275–8) and these are available to researchers. Others, reading the same material, may see the man differently. It must also be remembered that Fred was an inveterate disposer of "rubbish" (he notes that on one occasion he threw away a ton of paper from his office), so much of his personal correspondence must have been lost.

He was a big man, both physically and intellectually. He dominated those working with him, many of whom seem to have been in awe, or in fear of him. Only a few friends seem to have been able to tolerate his undoubted quick temper, and accept that this was one of the prices one pays for rubbing shoulders with genius. Again Iris Runge noted he was "a very clever, very determined man, full of bright ideas and funny expressions, not exactly gifted as far as social contact was concerned, indeed, as Runge learned from others, occasionally rather trying for those who had to deal with him, but for Runge himself he had nothing but a charming affection and friendliness" (Runge, 1949).

Fred was egoistical: the draft pages of his autobiography are almost entirely devoted to listing his achievements, and although there are a few references to good teachers, most are to their failings. He seems to have been very intolerant of anyone who disagreed with his views, on any subject, and particularly objected to having his writings assessed by a referee before being accepted for publication. In February 1939 he wrote to his friend Robert Lockhart "I do not like the system of referees at all. If I or you were to devote a few months to the study of some particular phenomenon there is no-one on God's earth competent to criticise our conclusions" (LC. vol. 21). There are many examples of extended correspondence with editors and referees who criticised his work, and he was not averse to entering into verbal battles in discussions following papers he presented to professional meetings. As Eric Baxter noted in the introduction to his *Catalogue* . . . "At times Lanchester seems almost to relish the war of words, even perhaps to exaggerate the extent and importance of the engagements in which he became involved. In writing to Sir Charles Boys in 1937, following a controversy on dimensions, he refers to 'a regular dog fight' in the columns of *Engineering*, and continues 'I think you will agree that . . . I have laid my opponents in the dust'" (Baxter, 1966) (LC. Baxter 12/21).

He was very suspicious of his academically qualified and successful colleagues. In particular he disliked those who had achieved senior positions in the universities if he considered that they had not achieved the standard of research which he thought was appropriate to their position. Again, from his letter to Robert Lockhart in February 1939, "There are no super-men in the world of science, mostly poor sticks these days, universities have increased and multiplied at such a rate that there are not enough good men to go round. And yet these are the men who are set up in judgement of the work of others" (LC. vol. 21). In his defence, it must be said that the same worries were being expressed by many academics in the 1960s and the 1990s.

As a family, the Lanchesters were generally not close, and seem to have found personal relationships difficult. Caroline's daughter Blanche wrote to Elsa Lanchester of ". . . the accustomed Glacification of the normal human instincts so inherent in the Lanchesters . . ." (quoted in *Elsa Lanchester, herself.* p.313). Henry Vaughan married late in life, Mary and Edith never married, Fred was fifty when he married Dorothea. His contacts with other members of the Lanchester family seem to have been rare (non-existent in the case of Edith, but that is not surprising in view of the "kidnapping" incident). There are references to his nieces, who visited him in Bedford Square, and played music with him. They have memories of his generosity to them, and his ability to make things to amuse them. Iris Runge notes in her biography of her father that during his visit to Gottingen in 1907 Fred made himself popular with the two sons of the family by making little gliders from sheets of mica. Needless to say, they flew, beautifully stable, in a room or on the stairs.

Fred's relationship with George and Frank was different. They seem to have shared interests, and leisure activities, from cycling in the early days in Birmingham, to sailing on the Solent. Nevertheless, those letters between Fred and his brothers which survive are almost always limited to the business in hand, and the personal notes are formal.

The other outstanding exception to this family coolness was his relationship with his wife Dorothea. The Lanchester Collection includes a few of the personal hand-written letters he wrote to Dorothea, and there are indications that he wrote frequently, even daily, when they were apart. He wrote to Dorothea from a hotel in Southampton where he was staying whilst sailing in the Solent, "To the most wonderful of God's creations, greetings . . ." (LC. vol. 9–64), and signed another letter "With all my love to you, dearest. Your affectionate (rare and precious) antique, Fred" (LC. vol. 9–58).

There can be no doubt about the importance of Frederick William Lanchester in the development of the motor car, the theory of flight, operational research and general engineering in the United Kingdom and throughout the world. Alone, or with others, he made 426 applications for patents, and 236 of them were completed and granted. The range of subjects covered is enormous, from player pianos to aircraft landing gear, from colour photography to draughting instruments, from gear cutting machines to loud speakers. His legacy remains far beyond his lifetime.

References

Baxter, E.G. (1966): *Catalogue of the private papers of F.W.Lanchester in the Library of Lanchester College of Technology, Coventry.* Thesis submitted for Fellowship of the Library Association, 1966

Clark, C.S. (1995): *The Lanchester legacy, Vol. 1: 1895–1931.* Coventry, Coventry University, 1995

Gold, K. (1984): A British Leonardo, in *Times Higher Education Supplement*, 17th February 1984

Lanchester, E. (1983): *Elsa Lanchester: herself.* London, Michael Joseph, 1983

Kingsford, P.W. (1960): *F.W.Lanchester: the life of an engineer.* London, Arnold, 1960

Runge, I. (1949): *Carl Runge und sein wissenschaftliches Werk.* Abhandlungen der Akademie der Wissenscaften in Gottingen: Mathematish-Physikalische Klasse, Dritte Folge, Heft 23, 1949 (selected translation by Janet Osborn, Lanchester Library, Coventry University)

Chapter Three

FRANK LANCHESTER

(22 July 1870–28 August 1960)

by John Bingeman

Frank Lanchester was born at Hove, Sussex on 22nd July 1870. He was actually christened Francis though throughout his life he was known as Frank, and even his closest friends never knew otherwise. The fifth child in a family of eight, his parents were Henry Jones and Octavia Lanchester (see frontispiece) living at No. 1 St. John's Terrace, Hove near Brighton; the house was later to be re-numbered and renamed 49 Church Road. His father had recently moved from Lewisham following his appointment as architect to the Stanford Estates at Hove.

The family tree starts with Frank's great grandfather, the Reverend Lanchester who was chaplain to the Earl of Onslow. He married Ann, sister of the painter Arthur William Devis (1762–1822), who should not be confused with their father, another Arthur (1711–87), who was also a well known portrait painter of the period. Arthur William, a product of the Royal Academy, where he received a silver medal from the hand of Sir Joshua Reynolds, joined the East India Company as a draughtsman and spent his formative years painting in the East. On his return he exhibited at the Royal Academy, painting mainly large historical and romantic scenes, the most famous being the "Death of Vice-Admiral Nelson in the cockpit of HMS Victory" which now hangs in the National Maritime Museum. His second wife was Margaret, sister of the Reverend Lanchester, so there is a double tie between the two families. Not too much is known about Frank's grandfather, Frederick, son of the Reverend, except that he was thought to have been an owner-captain of a merchant vessel, and that he married Mary Ann Smith, the daughter of a map publisher.

Frank's father, Henry Jones Lanchester, A.R.I.B.A., was a successful architect and surveyor; his work for the Stanford Estates included the Grand Drive from the sea front and the four avenues on each side, as well as many other developments around Hove. He was involved in local government work, from sanitary committees following the cholera outbreak of 1865–6, to advising on foreshore preservation from sea erosion. In later life, he gave expert advice to select committees on such things as widening London Bridge.

Frank's mother, Octavia was highly educated, an unusual accomplishment for a woman in those days. From 1860, she taught Latin and mathematics at a school in Westbourne Grove, Paddington, prior to her marriage on 23rd April 1862 to Henry Jones Lanchester. Her education may have been due to the fact that she was an eighth child, hence Octavia, and needed to make a living. Her father was a coach builder and her mother, Jane Marlow, came from a family of gold and silver smiths.

It is therefore not surprising that Henry Jones and Octavia should bring forth an unusually gifted family. Their first born, Henry Vaughan (Harry) became an architect and, in partnership with Rickards, was exceptionally successful in architectural competitions. They designed the Cardiff City Hall and Law Courts, Central Hall Westminster, and many other university and public buildings. He became a pioneer in town planning and adviser in this field to the Governments of India and Burma. Among his work he included a maharajah's palace and the government buildings in Lucknow.

Their second child, Mary, became a painter, and her flower pictures both in oils and watercolour were exhibited at the Royal Academy and other galleries between 1896 and 1910. She taught painting at Roedean School and had several books to her name.

The third child was Eleanor Caroline, known as Carrie, who was unusually tall for that era, and designed ladies' hats. She married her cousin and had one daughter, Blanche.

The fourth and most eminent member of the family was Doctor Frederick William, the pioneer motor and aircraft engineer. Sir Harry Ricardo wrote for the Royal Society: "In Lanchester we had one of the very rare combinations of a great scientist, a great engineer, a mathematician, an inventor and a true artist in mechanical design. He was a man, too, of remarkable versatility, for his interests and his inventions ranged over a very wide field, including musical instruments (for he was a great lover of music) and all forms of sound reproduction."

Then came Francis (Frank), my grandfather, and the subject of this chapter. He was actually a twin and thought to be the weakling of the two. Frank, who was always a great story teller, claimed that he had all the attention so Charles, his twin, promptly died!

Edith, known as Biddy, was the sixth child, and was highly educated and intelligent like her mother. She had socialist ideas, was secretary to Karl Marx's daughter Eleanor, and a member of the Independent Labour Party. If that was not enough, she had very advanced ideas for her time and lived with James Sullivan as his common law wife, an outrageous thing to be doing in Victorian England. Her father kidnapped her, with the help of her brothers, with an Urgency Order signed by a Doctor G.Fielding Blandford, a well-known mental specialist, naming the cause of insanity as "over-education"; she was conveyed to Roehampton Asylum. It became a cause célèbre and on 30th October 1895 *The Times* reported that the Lunacy Commissioners had ordered her release four days later. Her case helped to change the mental health laws. After that she devoted her energies to the women suffragette movement. She was the mother of Waldo who ran the Lanchester Marionettes at Malvern; and Elsa Lanchester, the well known actress of the London stage in the 1930s and the screen in the 1940s and 1950s. Elsa married the famous actor, Charles Laughton.

The seventh was Edward Norman, who emigrated to Australia, married and had a daughter Joan Maude who married Leslie John Watts (Jack). They in their turn had three children and there are now a number of grandchildren.

The eighth and final child was George Herbert, the third family member of the Lanchester Motor Company and his biography is ably covered by Elaine Lanchester, his daughter-in-law, in Chapter Four, below.

Frank's upbringing must surely have been influenced by his brothers and sisters in this most lively and creative household.

While his illustrious elder brother Fred enjoyed a full education ending as a national scholar at the Normal School of Science (later called the Royal College of Science) and the Royal School of Mines, Frank was not so fortunate. The Hove building boom ended. As his father's finances were always under strain with so many children, the family moved back to London in 1886. Earlier, Frank had gone to Brighton Preparatory School and then on to Dr. Walker's College, Hove which specialised in coaching for the services; he left school at the age of sixteen.

George Lanchester records how he, Frank and Edward not only needed to find work but had to help their parents' finances. Edward took off to the antipodes (Australia) to start a new life and Frank got a job in a City tea importing firm. Frank found his workmates uncongenial and shortly afterwards left to join the Union Bank of London as a messenger boy and later qualified to become a cashier.

In the 1950s, I exchanged many letters with my grandfather, and still have over forty of his letters today. A 1954 letter encourages me to take up rowing:

"I indulged in this fine aquatic sport at Putney when I was in a London Bank (a 'beastly' sedentary job from aged 18 to 22 before coming to the Midlands & becoming a 'quasi' Engineer) and did not do badly getting stronger every day, till promoted from No. 2 to

Stroke in our Eights Racer, with a win or two, in Skiff Races thrown in and out when Bargee on the Thames got in our way."

In a second letter he tells how, after working at the Bank until 8 p.m. "We were often afloat practising even in Winter till 10 o'clock p.m. freezing, actually, our perspiring and shivering bodies, good for you."

By 1890, Frank was spending weekends with Fred and George at Fred's house at St. Bernard's Road, Olton near Birmingham. Fred had become works manager at the Forward Gas Engine Company the previous year, and George was an apprentice with the company.

Frank and George were members of the Speedwell Bicycle Club based in Birmingham and they appear in the club's group photograph of 38 members. In the Speedwell's report for the year ending 1894, Frank is the club's honorary secretary and he is listed as attending five of the six committee meetings. By 1898, Frank was the club's captain; he won a prize for coming second in a 25 mile handicap race and was one of a 'pioneer' party of eight who made a pilgrimage to Harrogate for the North of England Cyclist Camp. The team was led by Charles Vernon Pugh, the club's president. Pugh is better known for his two companies Rudge-Whitworth and ATCO Motor Mowers, and remained a lifelong friend of the family. Frank must have enjoyed the first Harrogate meeting for despite living under canvas, he attended their next three annual meetings up to 1901. In 1898, the A.G.M. report records Frank as attending ten out of ten committee meetings, as one would expect from the club's captain.

Fred also shared his younger brothers' interest in cycling, and true to character, re-designed the bicycle hub. As a side-line to his other activities he set up a little factory for making the patent hub, in rented premises in St. Paul's Square, Birmingham; a first-class toolmaker was taken on as foreman, who later took charge of the jig and tool shop when the Lanchester Engine Company was formed. 1893 was the year that Fred prevailed upon Frank to give up his job with the Union Bank of London and take charge of the cycle

Fig. 1. Speedwell Cycle Club about 1894

business as general manager. Other work from Fred involved commercial transactions with the sale of patents and an infringement action demanding patient detective work. It was during this time that Frank joined his two brothers with the Forward Gas Engine Company at a salary of thirty shillings a week plus a small commission on sales. In fact, through his flair for salesmanship, and much to the surprise of the firm's owner, his sales commission often exceeded his salary.

In 1893–4, Fred built a single cylinder $3\frac{3}{4}$ inches × $4\frac{1}{2}$ inches stroke vertical high-speed benzoline motor of 2 b.h.p. to propel a stern-wheel paddle punt. "No more than 800 r.p.m. – daring enough at that date" to quote Doctor Fred giving the 24th Thomas Hawksley Lecture to the Institution of Mechanical Engineers in 1937: r.p.m.s of 150 to 200 were the norm at the time. The purpose of the boat was to demonstrate and prove the engine since the use of road vehicles was difficult before the Emancipation Act. This 1896 Act officially permitted the use of such engines on the road up to 12 m.p.h.

Frank, when working on the punt, realised that his rate of work was so much faster than the professional boat builder. When challenged, the professional replied:

"if you were a boatbuilder all your life, you would not work so damn fast!"

His two brothers were only free on Sundays and the respectable residents of Olton were suitably shocked to see them working as they walked to and from church, and coined the nickname, "The Unholy Trinity"!

The completed boat had the first Lanchester experimental engine that ran on benzoline through a wick carburettor. This carburettor was to feature on all early Lanchester cars. The engine's ignition used a hot tube heated by a blow lamp. The punt was taken by a horse drawn lorry and launched on the Thames at Salter's Yard, Oxford. The *Motor Boat & Yachting Magazine* in 1955 credited the boat to be:

"the first British Motor Boat in 1894"

In 1898, a second launch was built by Messrs Bathurst of Tewkesbury using an 8 h.p. two cylinder opposed engine identical to the experimental cars, but water cooled with a reversible propeller designed by Fred. The launch was for James Whitfield, one of the financial backers of the experimental cars, and ran on the river from his home at Bredon to Tewkesbury for over 40 years.

In Anthony Bird's book *Lanchester motor cars*, Bird records that:

"The (cycle hub) side-line business was short lived as Fred Lanchester's designs were too expensive to compete with the established cycle-component manufacturers, but that it started the association of the 'Unholy Trinity'. Frank Lanchester's great gift was a spontaneous natural warmth, which gained him many friends and made it possible for him to persuade hard-headed, and mostly reluctant, business men to put up funds to develop his elder brother's inventions. He played a very big part in the flotation of the Lanchester Engine Company in 1899, and an even bigger part in helping to keep it afloat in the first difficult years."

Certainly, Frank's commercial knowledge and his social skills were an indispensable counterpart to his brothers' technical skills.

The Lanchester Engine Company was registered on 13th December 1899, with Frank appointed company secretary at a salary of £200 per annum. The five directors were: the Speedwell Cycle Club's president, Charles Vernon Pugh, as chairman, his brother John V.Pugh, Joseph S.Taylor, Hamilton Barnsley and James Whitfield. The assets of the company at its flotation were Fred's many patents, the 1895 car rebuilt in 1897, the Gold Medal Phaeton of 1898, a third similar car not yet finished and the punt valued at £300.

There had been an earlier syndicate to finance the experimental cars that included Tom Barker, the owner of the Forward Gas Engine Company, and money put up by the brothers, James and Allan Whitfield. The latter received shares in the new company in recognition of their earlier contributions. The directors all knew each other and were closely connected in business. A few years later, John Pugh was to design the first detachable and interchangeable car wheels for Rudge-Whitworth, and acknowledge in an *Autocar* article that the original idea had come from Lanchester's tangential spoked wheels and splined hubs used on the 1895 experimental car.

The year 1900 was a year of intense activity for the three brothers. The first task was to organise and equip the new works at Armourer Mills in Montgomery Street, Sparkbrook, Birmingham which was to remain their main production shop until 1931. The first 10 h.p. production model was on the road by August 1901, only twenty months after the company had been launched: no mean achievement. Besides being company secretary, Frank was responsible for sales and publicity. The 10 h.p. car was a technical masterpiece by the standards of the day, as well as a great sales success and between August 1901 and 1905, over 300 were sold, much to Frank's credit. The production rate reached three to four cars a week, and prices were: 10 h.p. £525, 12 h.p. £550, 16 h.p. £650 and 18 h.p. £725. These were complete cars: many makes quoted chassis-only prices. Lanchesters came with tool kit, and a complete range of consumable spares.

Frank, throughout his life, was meticulous in answering customers' letters and this was one of his fundamental rules from which he never deviated. In letters, customers praised the promptitude and courtesy with which letters were answered and spares despatched. Even when Frank was in his eighties, I invariably received replies by return of post.

To promote sales in April 1902, an "invitation drive" to the press for an outing to Worthing was a great success. Nine air cooled 10 h.p. cars were assembled at the Carlton Hotel for the press to sample the delights of "Lanchester luxury". The *Autocar* reporter was so impressed that he referred to "this successful and enjoyable run" in two editions and went on to acquaint readers in glowing terms of the general arrangement of the Lanchester. The *Autocar*'s photographer recorded the start at St. James's Park and Frank driving the first car into Worthing; it is recorded that the convoy made quite an impression as it passed through Ewell, Epsom and Dorking.

Frank did everything to promote the Lanchester name. He was elected a member of the Automobile Club, now the Royal Automobile Club, in April 1900, was a founder member of the Society of Motor Manufacturers and Traders, and a year earlier had founded the Midland Car Club, now famous for the Shelsley Walsh Hill Climb, and was their first honorary secretary.

At the end of the non-stop trial from Glasgow to London, Frank was there to promote publicity and welcome the driver, Mr. Hartenfield. The photograph shows Frank, dressed immaculately, sitting in the passenger seat in the mud spattered car that had specially fitted auxiliary fuel tanks built on the wings. This was one of the many Lanchester successes in reliability events. Frank was on the Automobile Club's Committee for Reliability Trials from 1902 which must have been helpful.

Frank went over to Ireland to steward for the Automobile Club (R.A.C.) Gordon Bennett race in 1903 and used a 10 h.p. Lanchester as a course car. The race was run over a distance of 327½ miles and was won by Jenatzy in a 60 h.p. Mercedes for Germany. I have Frank's rather nice steward's badge, a miniature replica of the Gordon Bennett trophy; the club's "Rules and Regulations covering all Automobile Competitions dated 1st March 1900 – price 6d"; and his snapshot album.

The Lanchester company made no pretence of building racing cars but contented themselves with taking part in reliability trials, fuel consumption trials and smoke emission trials; in these they did well. Fuel consumption trials were usually a walk-over for Lanchester, as the wick carburettor was much more economical than most early spray carburettors. With the introduction of the 16 h.p. cars, Lanchester did begin to feature in speed events.

Fig. 2. Frank Lanchester in 1904

In July 1903, George and Archie Millership shipped two 16 h.p. tourers to Dublin for the Phoenix Park Speed Trials which followed the Gordon Bennett race. Frank, driving the 10 h.p. course car, was beaten over the standing start mile and flying kilometre course by a 10 h.p. Renault. George, in one of the 16 h.p. cars, retrieved the Lanchester honour with a 2.5 second win against Mr. Bucham's 20 h.p. "Special Design" car.

Today, it is hard to appreciate what it was like to be a pioneer motorist. George wrote that in towns, the public was for the most part hostile to the motorist; some people were tolerant and curious, and a small minority were interested, but those who came from their suburban homes into town daily in their carriages, and particularly J.P.s who dispensed law in the police courts, were most active in demonstrating their hatred, and often expressed it in no uncertain terms in spite of their oaths to administer law without bias. The law prescribed a speed limit of 12 m.p.h. for motor vehicles, and as nearly all cars were able to travel safely and under control at 20 m.p.h. or more, it was a foregone conclusion that the motorist was a law-breaker to be punished, often viciously, when his presence was demanded in the police court. It was generally of no avail to bring witnesses, and to many magistrates it only aggravated the crime to engage a solicitor for defence.

On the way to Ireland, George recorded the following account of Frank's quick wittedness in saving the day with the police and the inevitable court appearance.

"Archie Millership and I took two 16 h.p. cars to Holyhead for transportation to Dublin for the Phoenix Park Speed trials, in which we were competing. With our usual friendly rivalry we made a sporting race of the journey as soon as we got clear of the town. In crossing Anglesey, passing through the village of Caerwen, Archie was leading by about six yards, a dog ran out just missed it and promptly committed suicide under my wheels. Having learnt by hard experience that when a dog meets its death by hard collision with a motor vehicle, the veriest mongrel is transformed into a valuable pedigree animal, we did not stop. The dog was dead and there was nothing to be done about it, so we drove on to Holyhead where we had to embark the cars. The loading up was organised; Archie draining off the petrol and manoeuvring on the quay side; I was in the hold, manoeuvring the cars into place, and Frank who had arrived by train was attending to the transport formalities.

The Police at Caerwen sent an officer to track us down at Holyhead and whilst our respective parts were being played, he arrived at the Quay. Frank being the most prominent of the three of us, in his capacity of directing operations, the Officer charged him with killing the dog at Caerwen. Frank was duly apologetic and said that if he had done so, it was the fault of the London and N.E. Railway, producing his return ticket. The Officer then turned to Archie who was manhandling my car into place ready to be lifted and lowered into the hold, and said: 'I want your name and address'. Archie was about to give it, but Frank intervened and said: 'Don't give it, the Officer has already accused me of an offence of which I have no knowledge and he cannot accuse two people of one offence'. Whilst the Officer was thinking that one out, Frank went on board and warned me not to give my name. I immediately went up to the Captain and asked him if a policeman had any authority on his ship. He replied: 'Emphatically not', and proceeded to turn the Officer off.

However, the Officer doggedly refused to leave us and followed us to the Station Hotel, where we very leisurely had lunch, leaving him to cool his heels in the Hotel entrance. Whilst we were at lunch, he telephoned the Chief Constable for advice, and after lunch we were confronted with the pair of them. After some cajoling, the Chief said: 'If you were gentlemen you would give the Officer the information he wants'. We replied that as he had accused each of us for the same offence without a shred of evidence, we preferred not to be gentlemen.

The Chief said to the Officer: 'There is nothing you can do, you had better return to Caerwen'. Having a few hours on our hands before sailing, we escorted the Officer to the platform and gave him a rousing send off!''

This was the second time that Frank had come to George's rescue from the law. Earlier the police had accused George of driving from Cardiff to Newport, Monmouthshire a distance of fourteen miles in thirty-four minutes, an average speed of twenty-five m.p.h. when the legal limit was twelve. In fact it had been George's fiancée, Rose Thomas who had been driving under his tuition. George recounts what followed:

"It so happened that on the crucial date I was laid up with a bad attack of flu, so I rang Frank who was doing a sales engagement in South Wales and happened to be staying at my fiancée's house. [Frank was to marry Rose's sister, Minnie later that year.] I asked him to attend the Court and apply for an adjournment. Rose decided that she wished to go with him to the Caerleon Police Court.

On arrival at the Court, the usher mistook Frank (who always dressed immaculately) for George's solicitor and conducted him to the solicitor's bench. The ever resourceful Frank rose to the occasion and conducted the case admirably. After the police and their witnesses had sworn on oath that I was the driver, Frank put Rose into the witness box, and in a professional manner asked her to tell the Bench who drove the car on the occasion and route which the police had described. It was obvious that they could not have mistaken an attractive petite girl, as she was, for me so the case for the prosecution collapsed. I was duly acquitted thanks largely to Frank's able impersonation.''

In the *Autocar* for 13th April 1901, Frank contributed a lengthy article on "Autocars and motor cycles at the Easter manoeuvres" and in his typical witty style, describes these exercises under the direction of Major Cardell, Royal Artillery. Everyone appears to have been soaked and the motor cycles were used to call back the troops to dry out for the morrow; while Frank's car, "the old Lanchester that had years before distinguished itself on the field of Richmond [the faithful 1897 Gold Medal Phaeton which in 1900 had completed the Thousand Mile Trial, and today is on view at the Science Museum], ran sixty or seventy miles, placing the commanding officer in personal touch with his assistants, on advantage of this Major Cardell gave most generous praise."

In this exercise, Frank had sown the seeds that led to the forming of the Motor Volunteer Corps (M.V.C.), the Army realising the usefulness of cars. George wrote in the autumn 1959 *Veteran Car Club Gazette*:

"It was not until 1902 that the Army began to be interested in motor transport. In that year a band of motorists formed a Motor Volunteer Corps. Its object was to place the services of their cars and drivers at the disposal of officers in the services in the conduct of military training exercises. It was initiated by Mark Mayhew (owner and director of a corn milling company) and was supported by Sir J.H.A.Macdonald, the Hon. C.S.Rolls, Colonel Crompton and W.C.Bersey amongst others, all pioneer motorists. My brother Frank and I joined. Members bought their own uniforms to a design issued from Headquarters. They were of grey-green cloth with green cuffs, and tunic facings, and we looked rather like characters from a musical comedy. I recall being sent to York to report for duty to Colonel X. I spent one day conveying him to various units engaged in an exercise; the rest of the week I was detailed to drive his wife and daughter round sight-seeing. The Corps was short lived and in 1904 it was disbanded."

I was able to check George's *Gazette* story when going through the (Royal) Automobile Club's journals. The R.A.C. took a great interest in the M.V.C. Their journal records that George took part on the very first occasion when the M.V.C. was on duty in the north of

Fig. 3. Frank Lanchester in MVC uniform in 1904

England from 27th to 31st March 1904. The exercise was based at York with the M.V.C. commanded by Captain S.F.Hammond; Major General Sir Leslie Rundle K.C.B., directing the exercises, "expressed his appreciation of the services rendered".

Later that year, Frank took part in exercises starting at Droitwich on 7th August. It was a four day exercise and they were quartered at Worcester for the night of the 8th and at Tewkesbury on the 9th, ending the exercise at Windsor the next day. Despite George ridiculing the dress, Frank looks very smart in his M.V.C. uniform that had cost him £7!

Lieutenant Colonel Mark Mayhew, referred to by George in his article, was the commanding officer of the M.V.C. In fact it lasted a little longer than 1904. By 1905 it had an establishment of 159 consisting of forty-five officers with the two Lanchester brothers among its 114 members. While the Corps had provided excellent service, it was now found necessary to reorganise and augment its numerical strength. Not surprisingly, there was an apparent disinclination on the part of owners of suitable cars to be enrolled as "privates" instead of members! The M.V.C. was disbanded in 1906 and the officers were transferred to the Army Motor reserve. The M.V.C. statistics for 1905, the last full year, were impressive:

10.18 days duty per person,
814.26 miles for each member's car
A daily average of 75.42 miles
Cost to the Crown was 5d per mile

In fact, it must have been useful remuneration for Frank and George when the company was going through its financial problems. The rates were:

30/– daily allowance
12/6 for each night away
¾d per mile petrol allowance

Frank had a close liaison with the Army and was on the council of experts to advise the Mechanical Transport Committee at the War Office. Not inappropriately, the committee bought a 10 h.p. Lanchester, painted khaki and with the royal cypher, for their use and convenience at Whitehall. A magazine photograph dated 1903 shows Frank and George being chauffeured in some style by a sergeant with a private acting as footman; the caption reads:

"The New Car is going very well. – Major Lindsay Lloyd, R.E."

The compliment comes from Major Lindsay Lloyd, Royal Engineers, who was the secretary to the Mechanical Transport Committee.

More Lanchester cars were sold following Archie Millership's success in trials, when a further War Office contract was won to supply six 10 h.p. air cooled tourers.

Besides all these activities, Frank found the time to court and marry Minnie Grace, youngest daughter of William Thomas of Oakridge, Llandaff. Mr Thomas was the contractor who, with Frank's eldest brother the architect Harry, was building the new civic centre at Cardiff. The wedding took place at St. Marks Church, Gabalfa near Cardiff, on Tuesday, 10th November 1903 and was conducted by the Reverend John Davis, M.A. The Lanchester Engine Company presented Frank with a splendid solid silver tray suitably inscribed to mark the event.

Frank and Minnie set up home at 48 Sanden Road, Edgbaston, and appropriately, our only photograph has a 10–12 h.p. tourer outside!

An admirer of Lanchester cars was Sir Alfred Watkins who commissioned two gold brooches depicting a Lanchester 20 h.p. tourer with tiller steering. The brooch is most unusual and always makes a conversation piece with its diamond hub caps, white enamelled

Fig. 4. Frank and Minnie Lanchester's first home, 48 Sanden Road, Edgbaston, Birmingham

body panels and green upholstery. Sir Alfred gave one to his wife and the second to Frank for Minnie to wear on suitable occasions.

Despite making a profit in 1903, on 4th March 1904 the company went into voluntary liquidation. A public subscription to raise £42,405 had failed, and the directors lacked the necessary faith to carry on the business. The receivers appointed Arthur Gibson as managing director, and he was able to carry on the firm without any change in personnel and within a couple of months was making steady profits. Gibson was able to terminate the receivership and appointed Hamilton Barnsley, one of the five directors of the former company, as the new managing director. This post was previously held by Fred, who now became general manager. Outwardly there was little apparent change, apart from the firm's new name: The Lanchester Motor Company.

Following the reorganisation, Frank ceased to be the company secretary; his sales position was clarified as chief of sales which was always his main forte. Even so, it must have been a worrying time, being newly married and with a daughter Marjorie arriving on 17th October. Fortunately, Frank seems to have been on good terms with Hamilton Barnsley, who continued to be a director, as his wife consented to be one of Marjorie's two godmothers with Rose Thomas (Mrs George Lanchester), and Edgar Thomas (Frank's brother-in-law), as godfather. Frank did have a little bonus in the form of a hundred shares in the new Lanchester Company as a 'thank you' from Fred, who had received shares for the goodwill and patents that were passed on.

Once through these financial problems, the company was to prosper for over twenty years. A new four cylinder 20 h.p. model, with a choice of aluminium-panelled touring or landaulet bodies, was launched in early 1905 and it proved to be an immediate success. Frank expanded the sales organisation and from Lanchester's London office in Oxford Street, he opened showrooms at 95 New Bond Street in 1908; a Manchester depot and repair department was also opened to cover the north. A few years later their old friend, Archie Millership was to return from demonstrating Wolseley cars for Herbert Austin, and took over responsibility for sales under Frank for the Midlands and the North.

To be nearer to London, Frank and Minnie set up home in Surbiton and their second daughter, Elizabeth (Betty), was born shortly afterwards on 14th October 1906. By this time all Frank's work was in London representing the firm and devoting his energies to the political development of the motor industry in general. This was to be Frank's life for the next fifty years.

Frank was one of the twenty-five founder members of the Society of Motor Manufacturers and Traders (S.M.M.&T.). At their inaugural meeting on 22nd July 1902, he was elected a

Fig. 5. The Lanchester Motor Company showrooms in Bond Street, London

council member and played an active role in their affairs for nearly sixty years. The society was responsible for the now famous annual Motor Show originally held at Olympia, before the expanding needs of the industry took them to Earls Court, and after Frank's time to the Birmingham National Exhibition Centre on a bi-annual basis. He was vice-president 1917–19 and president a year later, and was in the centre of the phenomenal change brought about by the motor industry in the first half of this century. Aged eighty-two, in a remarkable speech at the Society's golden jubilee luncheon at the Royal Automobile Club on 17th July 1952, Frank likened the leading motor organisations in the early days, to the three cardinal virtues of faith, hope and charity.

"Firstly, the Automobile Club (afterwards deservedly Royal) truly represented faith. Faith in the future of motoring and tireless in promoting its developments and popularity.

Secondly, the Automobile Association which decidedly represents hope. Hope that our road journeys should be happy and successfully accomplished and hope also that we should be freed from police interference and restrictions.

Thirdly, the Society of Motor Manufacturers and Traders can well claim the remaining virtue of charity, because (vide their articles) all the energies in their Society – as members – are voluntarily devoted to the good of motoring causes, whether by exhibitions or official or governmental contracts. Profits from the shows are largely devoted to the support of the British motor car interests throughout the world."

Another article in the motoring press is a well deserved tribute to Frank when elected president of the S.M.M.&T. in 1919:

"A man of many qualities, Mr. Lanchester brings to the new office a cultured mind, a finished style of address and the saving grace of a dry and effective humour. Mr.

Lanchester in his quiet way, has always been deeply interested in all movements affecting the trade. On the executive of the Benevolent Fund his counsels have been marked by common sense and humanity, and in the more controversial discussions in the governing bodies of the Industry he has always been found on the side of sane and progressive legislation. In every way, therefore, we think the Society has made an excellent choice in its new President, and we look forward to Mr. Lanchester's period of office as being likely to produce measures of advantage to the continued prosperity of the industry."

While president of the S.M.M.&T., Frank had the privilege of welcoming King George V and Queen Mary to the Olympia Motor Show, the first time that a reigning monarch had attended. Frank's younger daughter Betty presented a bouquet of pink carnations to the Queen, much to the chagrin of Marjorie, the elder daughter. Before their Majesties left, Minnie, Marjorie and Betty were presented to the King and Queen. Frank escorted the King and Queen around the stands for over an hour exceeding the planned time and causing the public to have to wait outside from 10 to 11 o'clock! *The Motor* dated 19th November 1919 records that:

"the King expressed to the President [Frank] the pleasure the visit had given him and his surprise and delight at the progress which had been made in the motor industry since the Armistice".

A less complimentary remark is recorded by Anthony Bird in *Lanchester motor cars* (op. cit). On the Lanchester stand, while inspecting one of the most expensive and talked-about cars in the show, whose interior had marquetry panelling and other lavish fittings, the King was provoked to say:

"Very fine Mr. Lanchester, but more suited to a prostitute than a prince, don't you think?"

History does not relate how Frank handled the remark! The car is now owned by Mr L. Aas in Scandinavia and was present at the 1995 Lanchester Centenary Rally. Its interior is still original, a most impressive example of English coachbuilding craftsmanship, and in my opinion showed taste and quality.

While president, he had to deal with an increasingly hostile press. To resolve the issues, sixty members of the press were invited to meet the S.M.M.&T. Council at a luncheon at the Savoy in April 1920. The president's address refuted charges against the motor industry that by slow and uncertain deliveries, they had brought about profiteering in second-hand cars above the regular list prices of similar cars. Labour and steel costs had increased two and a half fold since 1914; yet the cost of cars had only risen by ninety-five per cent. Two passages from Frank's address are worth repeating:

"The fact that car prices have not gone higher (and for this the buyer should be grateful) is due to the greatly improved machines and plant now in wide use, and also to the greatly increased output thereon. It is interesting to note that the prices of motor cars have been increased less in proportion than the prices of most other commodities."

"In explaining and justifying the causes for late deliveries of cars is the moulders' strike, but this strike, although most disastrous in holding up car manufacture and output was only one cause. The vital fact was at the conclusion of the war entirely unknown conditions had to be faced, and a period gone through in which there were difficulties that the most experienced business brains could not anticipate or appreciate. We had for years been working under pressure of the needs of war. The workman was glad to do his utmost. This atmosphere of energy and the resulting output, naturally, had its influence on manufacturers' minds when estimating the possibilities for the future. Events have proved that all those estimates were too optimistic. The war ended, everybody thought that hard work was over, the entire atmosphere of the workshops changed. Hours were reduced, overtime was prohibited by the trade unions. Strikes in various important industries began and all the original forecasts as to output and progress were consequently disappointed."

At the seventh annual luncheon for the Agents Section Limited, with representative motor agents from all over the country, there were two burning topics: importing foreign cars and the question of car prices. Sir Percival Perry, in his speech made it quite clear that in view of the political economy, he had undergone a complete volte-face and ceased importing Fords. He earnestly warned agents of the danger to this country in buying and continuing to buy foreign cars. The Agents Section chairman, Mr. A.Noel Mobbs, (later Sir Noel) was not to be drawn into the controversy and contented himself by saying that he was in favour of getting everything possible made in this country. [Sir Noel's youngest son, Gerald, was to marry Betty Lanchester in 1933.] Frank, the president of the S.M.M.&T., had the difficult task of justifying increases in car prices to the agents. When referring to the success of the Motor Show, he said that they had reached veritable Olympian heights and (with his eye humorously surveying the members of the press) had managed to obtain what was necessary to the success of every enterprise, namely "a good press". On prices, Frank said he was convinced that every increase could be justified and he saw no hope of reduction.

Shortly after being president, Frank took the chairmanship of the car manufacturing section of the S.M.M.&T. and held this office from 1923 to 1927.

The R.A.C. had close ties with the Royal Motor Yacht Club at Poole. The S.M.M.&T. also ran aero and motor boat exhibitions in addition to the motor shows and these two might explain why Frank remained a member of the Royal Motor Yacht Club until his death.

Frank was twice president of the British Motor Trade Association; from 1931–4 and a second time between 1940–6, following the death of Lord Austin. On ceasing to be president on 9th May 1934, the Association presented him with a large silver salver with all the council members' signatures; and in 1946, they presented him with a gold Rolex watch, inscribed:

Presented by the Council of
The Motor Trade Association
To
Frank Lanchester Esquire
President 1940–1946
"A Token of Esteem"

And finally in the 1950s, a beautiful embellished parchment inscribed:

Presented to
Frank Lanchester Esq
in appreciation, affection and gratitude
for over 50 years of untiring service to the
BRITISH MOTOR INDUSTRY
and, in particular, to mark the occasion
of his election as an Honorary Member of
the Council of The British Motor Trade Association,
with which, as Councillor and
twice President, he has been
continuously associated since
its foundation in 1911.

Fig. 6. Frank Lanchester in 1920

Many other honours followed. The Institution of Mechanical Engineers elected him a Companion of the Institution. The Fellowship of the Motor Industry elected him president in 1949; and at the Fellowship's request, Frank defied the rules by carrying on as president year after year until he decided after five years: "it was high time some other member was given the honour of the presidency".

At the annual luncheon, in the Connaught Rooms, described as a stupendous success, a new president was elected, in the presence of a large proportion of the 500 membership, when tribute was paid to: "Frank, the revered president, renowned for his wise and witty speeches, such being demanded of his office."

The Nineteenth Century Motorists was another elite dining club formed for the purpose of "eliciting the greatest possible amount of reminiscence" between motorists who had driven cars before 15th April 1900, the date when the 1,000 Mile Trial finished. Both Frank and George were members and attended the inaugural dinner on 14th December, 1927 at the Piccadilly Hotel costing the sum of twelve shillings and six pence!

An early member of the Automobile Club of Great Britain and Ireland (as the R.A.C. then was), Frank was elected a member of the Club in April 1900. He served on the Trials Organisation Committee from 1902, and was elected on to the committee of the Club on 8th November 1919 and remained on the committee until he resigned in May 1957. Following his election, he and Brigadier-General Lord Montagu of Beaulieu (father of the present Lord Montagu) and two others were appointed to be members of a panel of experts to advise the Ministry of Transport on matters relating to traffic and roads. For the next twenty years he continued to play a very active part on a number of sub-committees and in appreciation of the tremendous work he had done since joining in 1900, he was elected a vice-president of the Royal Automobile Club in 1938. This honour was all the more unusual as he had the distinction of being the very first vice-president elected from industry, no mean accolade from such a prestigious club.

It may have been Frank's experiences with the Speedwell Cycle Club that prompted him to start the now famous Midland Automobile Club on 12th August 1901, and become their first honorary secretary. Frank was supported by Charles Vernon Pugh, president of the Cycle Club, who became a vice-president of the new club. Among the committee were Herbert Austin and J.D.Siddeley both famous for their cars. Brother George recalls the club's first hill climb contest on Gorcott Hill some twelve miles from Birmingham, but it was the steeper Sun Rising Hill which proved the better test and was the origin of the club's badge. After a most successful event in 1903, the following year was not so successful as the locals rose up in wrath, supported by the authorities and the police, and Frank recalled their very words, still fresh in his memory some fifty years later in the Golden Jubilee Hill Climb souvenir programme: "For God's sake, move on"!

This led to the committee's decision: "We will have our own Hill". Following this decision, the rights to Shelsley Walsh were acquired and the inaugural meeting on "our own hill" took place on 12th August 1905. Frank was subsequently elected an honorary life member.

Fifty years later, I had the honour to be present at the Golden Jubilee meeting at Shelsley, with both Frank and George present. Interestingly, a 35 h.p. Daimler recorded 77.6 seconds in 1905 and fifty years later Ken Wharton in an E.R.A. held the record at 35.80 seconds: it says a lot for the veteran cars that they were able to put up such creditable performances. In 1905, George drove the new 20 h.p. Lanchester and recorded a time of 165.4 seconds, achieving eighth position on handicap. Cars varied from 6 to 50 h.p., and thirty-five completed the climb out of a total of forty-one entries.

Besides hill climbs, there were other events. The club had an outing to Wroxall Abbey in May 1902, the home of the M.A.C. president, Mr. J.Broughton Dugdale and spent the afternoon viewing the old chapel. Others with more modern yearnings visited Mr. Dugdale's workshop, which was equipped with up-to-date plant and tools and where an embryo car was discovered! The *Autocar* photographer recorded the car line up which included two

Lanchesters. Their annual dinner at the Grand Hotel, Birmingham on 10th January 1903, is the earliest record of Frank as an after dinner speaker; the menu shows him replying to the toast, "The Club". It was at the same dinner two years later that George received a silver bowl for winning a "balloon chase"; my mind boggles at the thought of such an event! I have some sympathy for a "Mr. Jefferys C.Allen-Jefferys" writing to the Automobile Club, who described them as:

"road hogs, who care for no one but themselves".

In 1953, Frank was one of the selected few to write greetings in the *Golden Jubilee Number, The Motor, 1903–1953*. He recalls his close association with the Dangerfield family over three generations. He also remembers the early days before *The Motor* magazine when the Dangerfield family published *Cycling*, recalling the nostalgia of his youth and the Speedwell Cycle Club in the 1890s.

The Government's Roads and Traffic Committee was formed in 1935. Frank and his old friend Major, now Colonel, Lindsay Lloyd from the Mechanical Transport Committee some thirty years before, were advising on a wide range of topics. These varied from reviewing the Home Office 1934 figure of 39,332 prosecutions for obstruction; to the designing of new roads, roundabouts and bridges; while Frank had the additional task of submitting a memorandum on amenities. Typical of Lanchester thinking, Frank came up with the idea of underground garages and was asked by the committee to write a memorandum on the subject.

Not only was Frank contributing to the British car industry, he was also on the international scene representing our country from 1912 to 1936. He was thrice president, a rare honour for an English-speaking person, of the Bureau d'Automobile Internationale, in Paris. Across the Atlantic, Frank and Mr. H.G.Burford represented the S.M.M.&T. with Roland Wian for the Motor Agents Association, at the Second World Motor Congress in New York from the 11th to 13th January 1926. Afterwards they went on to inspect the Canadian motor industries at Ottawa and Toronto. One report records Frank speaking for himself and his colleagues, as saying:

"We are very glad to see such extensive manufacturing of cars in Canada. It is also gratifying to see such a high percentage of this manufacturing is done in Canada. The activity shown in the firms we visited indicates to our mind the importance of the industry in this country. We are sure it will continue to grow and expand with much attendant advantage to the Dominion as a whole."

Frank went on to represent Great Britain at World Congresses at Rome in 1928 and at Berlin in 1933.

In 1938, the Coventry authorities chose Frank to welcome King George VI at Tile Hill railway station and escort him round the various "shadow" factories being built around Coventry, as part of the preparations for war. Frank had known the King from the 1920s when he had been Duke of York and had purchased a number of Lanchester cars. On coming to the throne in 1936, the King ordered three Lanchesters, one of which was a sports model which he always drove himself. In this car he used to drive the Queen, with Princesses Elizabeth and Margaret, then children, from London to his country house at Royal Lodge, Windsor, nearly every weekend. Frank held the Royal Warrant that allowed the Lanchester Motor Company the privilege of using the Royal Coat of Arms on their letter-head and publicity.

In the family archives, there are signed letters from King George when he was Duke of York. The oldest letter is typed, and dated 24th April 1923. It is signed personally by "Albert", and thanked Frank for a motor clock on the occasion of his wedding to the present Queen Mother. The clock was actually an antique clock from a sedan chair.

More impressive is a letter all in manuscript by the Duke, only three days after his daughter's birth, which read:

April 24th 1926
Dear Mr. Lanchester

Thank you so much for your kind congratulations to us on the birth of our daughter. Also for the lovely flowers which we also received.

I am
Yours sincerely

Albert

Our "daughter" being referred to is our present Queen Elizabeth II.

Little has been said about Frank's family. After the birth of their second daughter, Elizabeth (Betty), they moved to a rented house at Belmont, Surrey in 1908 to see if the area suited them. This must have been so as they built their own house "Newstead" which backed on to Banstead Down Golf Course, and moved there in 1911. Holidays were a great feature for the family and they usually went to the seaside in the spring and summer. The earliest photographs are at Littlehampton and by 1914 they were holidaying at Bembridge, Isle of Wight. On the outbreak of war, the family curtailed their holiday to return to Belmont and were stopped twice by the Army who wished to commandeer their car. Frank had to remonstrate with the soldiers who had little idea what they were meant to be doing, and agreed that he could not possibly hand over the car with two small girls when miles from home.

Frank was forty-four at the start of the First World War. Initially, the Lanchester Motor Company was, rather short sightedly, switched to manufacturing munitions. There was a shortage of motor transport and through Frank's efforts, orders for armoured cars, aero-engines and even paravanes for naval minesweepers were obtained. There was one order for forty-two armoured cars, of which thirty-six went to the Russian front and six were sunk on their way to join Lawrence of Arabia in the Middle East. The Royal Naval Air Service were among others who used Lanchester armoured cars. Orders followed for lorries, high speed searchlight tenders, field kitchens, and lumber wagons carrying winches for observation balloons; these were all based on the Lanchester 38 h.p. chassis.

Frank enrolled as a Metropolitan Special Constable; the only knowledge of his duties comes from my mother, corroborated by sister Betty, who remember him in uniform, and taking him food on foot escorted by Nanny while he guarded the local reservoir! A duty hardly befitting a person of his ability but no doubt typical of many things that occurred when the country was plunged into war.

After the war, the family returned to Bembridge on holiday in 1919, and were joined by Fred, who had recently married Dorothea Cooper (Dorothy) and was spending the second half of their honeymoon in his yacht at Bembridge. There is a lovely story of how Dorothy arrived in the late evening drenched, feeling seasick and wanting a bed! There are photographs of Minnie steering *Iseult*, Fred's first sailing yacht (no engine!), together with Frank (with his arm in a sling), Fred (looking bored) and Dorothy. Today, *Iseult* (with an engine!) can be seen in Yarmouth harbour, Isle of Wight, and is now owned by Joe Lester. Frank's two daughters were taught to sail in a pram dinghy under Fred's coxswain's tuition in the waters around Bembridge.

Fred liked to treat his nieces as if they were his own children, and was a generous uncle. Both Marjorie and Betty were given identical Sheraton vertical bureaux of an unusual design which they both have to this day. We have Octavia Lanchester's answer to Marjorie's excited letter in 1916 telling her grandmother about all Uncle Fred's presents including the desk. Fred loved music and had a small yacht-piano in *Iseult*; to encourage his nieces he gave Marjorie a violin and Betty a 'cello.

Subsequent holidays were spent close by at Seaview where Frank rented the same house, "Little Glen" in Fairey Road, for a number of years.

Frank was a gifted writer as well as an orator. Rudyard Kipling wrote on 3rd January 1921, to congratulate Frank on his article in *The Motor*, "The Rape of the Policeman from Arcady", and it must have given him particular pleasure to have received the compliment "It is very well told", from such a famous writer. Kipling was a great supporter of the firm; he bought car Number 16 in 1902 and subsequently owned a number of Lanchester cars, including a notorious and rare 18 h.p. car which he named "Jane Cakebread". In the same letter, Kipling teases Frank: "Do you remember . . . had a knack of blowing up . . . and was taken to pieces once on a vacant plot at Worthing which she completely filled with her associated entrails". Kipling concludes: "Good luck to you and the Firm for this Year."

A few years later, Frank became chairman of the Bond-street Association Limited. He arranged their first annual dinner which took place at the Hotel Victoria, and invited Colonel Ashley the Minister of Transport, whom he knew as a member of the panel of experts, to be the principal guest. The Minister explained the Government's approach to London traffic congestion, and he congratulated the Association saying their policy should be widely copied. He went on to say he knew:

"nothing nicer than on a summer morning to walk up Bond-street and see its charming shops and the beautiful women who promenaded there".

The Association had introduced a one way traffic system in conjunction with Woodstock and Albemarle Streets. The Minister wanted the idea copied, and in answer to the many other transport problems, his appeal: "not to shoot the pianist, as he was doing his best", was greeted with laughter!

The idea of the one way system came from Frank's elder brother Harry Lanchester the architect, whose paper to The Royal Institute of British Architects had proposed such a scheme. At the time in 1909, Harry's revolutionary ideas had been received by his architectural peers with derisory comments:

"however brilliantly in its conception, from somewhere in the clouds", and, "Mr. Lanchester's Paper as a flight of fancy".

Thanks to Frank, one of Harry's proposals actually came to fruition. Frank was chairman of the Association, renamed Bond Street (London) Traders Association, from 1924 to 1928.

The financial depression of the late 1920s meant that the company could not sell their usual quota of high class cars. The directors, who now included Frank and George, failed to persuade the senior directors to follow the trend of other firms by building a smaller car. George had actually designed a small car, but their plans were rejected year after year. Frank obtained War Office orders for armoured cars, and despite prestigious orders from the Duke of York (King George VI) and Indian princes, trading became very difficult. To make matters worse, the chairman and managing director, Hamilton Barnsley, became ill and Frank was appointed to deputise as managing director. While Frank obtained further accommodation from the Midland Bank to cope with new armoured car orders, the company was still in deep trouble. In desperation, Barnsley signed over the firm to their neighbour the Birmingham Small Arms Company (B.S.A.) on 28th December 1930 subject to the agreement of the Lanchester shareholders. He died the next day from a heart attack. Frank personally interviewed each of the shareholders and obtained their agreement for the sale of their £64,321 shares to the B.S.A. for £55,000 which, after paying off the overdraft, worked out at eight shillings and three pence in the pound. B.S.A. had owned the Daimler Company since 1909, so the acquisition of the Lanchester Motor Company brought in urgently needed new designs. George became involved in the design of a new range of Lanchester 18 and 10 h.p. cars, using an overhead poppet-valve engine.

Frank became London director of the Lanchester Motor Company and joined the board of Stratton-Instone Limited, the Daimler, B.S.A. and now Lanchester agents, who took over the Lanchester showrooms at 95 New Bond Street. After Stratton-Instone was sold, Frank held responsible positions in the sales organisation of the Daimler-Lanchester-B.S.A. Group, and was elected a director of the Lanchester Motor Company in February 1936.

The Lanchester car had been popular with Indian princes. Probably the most famous was Prince Ranjitsinhji, the Jam Sahib of Nawanagar, better known as Ranji the cricketer, who had been a Lanchester owner since 1904, and whose 12 h.p. had done well in the Delhi–Bombay trial in 1905. In 1936, Frank set off on a tour of India and Ceylon on behalf of the Daimler and Lanchester companies, to promote sales and to visit other Indian princes. Frank recounts how he spent a night in a tree with Ranji's A.D.C., Colonel Geoffrey Clark, waiting over a kill to shoot a tiger; Frank failed to shoot a tiger, but was presented with a skin.

Sadly, most of Frank's records and library of motoring books were lost in the Coventry blitz; he wrote on 3rd August 1953:

"Alas my old valued Library Books of the past were lost, when my Office went up in Smoke (Damn Hitler) in 1941."

Sadly the first Lanchester, the 1895 car, was also destroyed in this raid.

In another letter of November 1956, when I was taking part in Anthony Eden's ill-fated Suez Operation, he recalls:

"In 1936 on my way to India (on Business & Pleasure combined) I passed through the 'classical' parts that you are now so closely linked with; and was invited to climb the 'Pyramids' – up on to camels and other lofty athletics which 'scared' me stiff because having broken my arm in England a month or so before, I had to content myself with a mule for transport at the finish".

Frank brought back impressive photographs of the Prince's line-up of twenty-four Lanchesters dating back to 1904. Also on display was a special miniature car that Frank had organised for the young Prince. The Prince ordered a special straight-eight Lanchester tourer similar to other one-off cars that were supplied to special customers like King George VI. The Prince's grandson, the present Maharaja has recently exported the straight-eight tourer to New Zealand.

Daughter Betty recalls:

"When Daddy [Frank] was in India, Mother came to stay with us, and I introduced her to 'gin & orange' and when he came home, I said 'Will you have your usual 'gin & orange' Mother?' Daddy said 'Oh no dear, Mother will have a sherry'!! He thought 'no lady' should drink gin! But when I told him it was the Queen's tipple (the Queen Mother today) – he was quite satisfied, and Mother was allowed to drink her 'gin & orange' ever after!!"

Following the amalgamation with Daimler, Frank and Minnie sold their home at Banstead in 1931 and moved to the Kenilworth Hall Hotel, in Kenilworth, Warwickshire a large house standing in spacious grounds. With furniture in store, their intention had been to buy a new home in the Midlands but this never happened and they remained at the Hotel until driven out by bomb damage in 1941. Betty took them into her home, Arden

Cottage, Wootton Wawen for the next four years, until they were able to move back into the repaired Kenilworth Hall Hotel. They remained there until late 1947 when they moved to a more residential hotel, The Shieling Hotel, Beauchamp Avenue, Leamington Spa.

I have happy memories of my grandparents at Kenilworth in August 1939. My mother left my sister and myself with nannie in the care of Minnie, while she went off house hunting. We had hurriedly left Malta when war was threatened. Grandfather was great. He had to suffer us each evening; firstly to see his latest car (it was invariably different each night), followed by persuading him to charge through Kenilworth ford near the Castle, known to us as the "car-splash", at high speed; no doubt to the consternation of any other road users! My sister, aged just four, always started the day over breakfast with "Hallo Gramps" at the top of her voice, to the amusement of Hotel guests and Frank's embarrassment!

Some years later, Frank wrote what he called, "nonsense verses" to my sister his only granddaughter. We used to have two cats and one was partial to warm car bonnets which prompted the following:

> Oh "Sandy Pet" we will "have you" yet
> For scrambling on my Cars Bonnet
> With muddy paws and outstretched claws
> You've left your foul marks on it
> Instead, confine your antics to the House
> Catching more usefully a lively mouse
> Or if you seek more sporting habits
> Go right away and kill some Rabbits.
> If not we Motorists plans are laid,
> We'll "Collar" you, "Cat Marmalade".

Another little thing grandfather did for my mother was to address "luggage labels" for our many moves; father was in the Royal Navy and we used to follow him around. Frank's writing and printing were of professional standard; re-reading his many letters one is struck by the clarity and straightness of his writing which never changed even in his late eighties. He knew exactly how to address our luggage with, "via Tilbury", Inverness or wherever, perhaps as a result of experience gained from sending off spares to Lanchester customers throughout the world.

From 1936 to 1958, Frank remained a director of the Lanchester Motor Company at Coventry and their sales consultant. He led a busy life, spending one or two days each week in London using the R.A.C. Club in Pall Mall as his second home. Now the doyen of the motor industry, he was in great demand as an after dinner speaker. Few, if any, men in the British motor industry had such a wide circle of friends. His keen sense of humour and his pleasant and friendly manner were qualities which were largely responsible for his popularity. Occasionally he invited me to lunch at the R.A.C. Club in Pall Mall; the hors d'oeuvres on a revolving trolley with so many choices will never be forgotten!

Towards the end of the war, Frank designed and built a model of his mobile builders' platform, mainly out of Meccano, and took out patent number 579219 which would have been registered around 1945. The patent was to speed up building houses; the bricklayer was able to hoist himself, while a second platform could be raised and lowered to keep him supplied with materials. Only one of these platforms was ever built and was used within the Daimler/Lanchester Works at Radford, Coventry. The idea never caught on; it was just too sophisticated for the average builder. This was Frank's second patent. Some years earlier he had designed and made a small model of a revolving fireplace. The principle depended on two back-to-back rooms sharing a common chimney, so that when the occupants moved from one room to another, the rear fire shield could be turned reversing the fire. Today, the idea may seem rather bizarre, but eighty years ago with no central heating, the occupants of a bungalow when going to bed in a freezing bedroom, would have found the idea most attractive.

Frank used the shed at Betty's Arden Cottage as a workshop, where his grandchildren were always welcome to see what he was up to. Later when I was making a bureau at the age of twelve, Frank brought me large pieces of mahogany from the Works, slightly scorched from the Coventry blitz, after the insurance company had written them off. He was full of encouragement and supplied antique style handles for the drawer.

On 14th December 1949, the fiftieth anniversary of the Lanchester Company was celebrated at a luncheon at the Dorchester Hotel, with Frank and George as the honoured guests. The gathering was a particularly happy occasion and it was attended by leading personalities in the motoring industry; it fell to Sir Patrick Hannon to propose the toast: "The Lanchester Brothers". The *Autocar* reported:

"There are few firms in the World's motoring industry that have been able to claim fifty years of motoring existence, and fewer still that have reached that honoured stage in their history, when founder members still enjoy an activity some younger men might envy."

In his eighties, Frank's driving can best be described as enthusiastic! He never dawdled and fog never seemed to worry him as he would appear out of the gloom at high speed! I can remember seeing close on eighty on the speedometer, (pre-war speedometers were slightly optimistic), when driving his 1939 14 h.p. Roadrider De Luxe Saloon. He was bringing me home after delivering my Lanchester Eleven for repairs at the Works. After thanking him and enclosing my cheque for the repairs, he writes:

Fig. 7. Frank Lanchester about 1948

"You must firmly tell your Car to behave like a Lanchester, ie Be Quick off the mark, and not to behave extravagantly, and above all wear well or else it will get the sack!"

The *Motor Industry* magazine in December 1953, reporting Minnie and Frank's golden wedding said:

"... our Grand Old Man of the Industry ... at 83 years of age, is as evergreen as ever, quietly efficient in his daily routine and as energetic as many men twenty years his junior."

Sadly, 1954 was the year he saw his last car model, the Lanchester Fourteen, go out of production, and its replacement, the new Lanchester Sprite, hardly got off the ground. There were only six prototypes, and at the age of eighty-six, following the resignation of Sir Bernard Docker the managing director, his fellow directors finally defeated him. No further cars bearing his illustrious name were manufactured.

"AJAX" in *The Policy-Holder* on 8th September 1960, wrote at the end of Frank's obituary:

"At this particular show, the name of Lanchester was absent, and I could tell by his manner and his voice that Frank Lanchester felt the situation acutely. To him, a motor show without a Lanchester car was like Hamlet without a ghost.

Even if the car bearing his name had to cease production, there was not the slightest reason why the name of Lanchester should have been dropped. Gottlieb Daimler had no more to do with the design or production of the British-built Daimler car than I (AJAX) had; he ceased all association therewith in 1897, but the magic name of Daimler was retained.

We have the Rolls-Royce, the Armstrong-Siddeley and others, so why not the Lanchester-Daimler?

Dr. Frederick William Lanchester, L.L.D., F.R.S., M.I.C.E., M.I.Mech.E., M.I.A.E. was one of the most brilliant scientists of the century and one of the founders of the British Motor Industry. He was the first man to introduce scientific principles into motor car design, and for this time-honoured name to be dropped and run the risk of being well-nigh forgotten in the course of time is a tragedy of the third largest industry of England which his brother Frank felt severely during the last year or two of his life. The name of Daimler – a German – was retained but the name of Lanchester – an Englishman – was abandoned. Why?"

In the evenings, Frank and Minnie played cards; it was a form of "two people bridge" and Minnie always won. She used to keep a little purse with her winnings! By nature, Frank always had to be doing something, but he did enjoy reading and had a taste for biographies and histories, particularly Winston Churchill's books. They both enjoyed a glass of whisky each evening and Frank smoked all his life. Fortunately, the cigarette seldom went back in his mouth after a couple of puffs, as he never stopped talking! His repertoire of stories was without equal.

Frank was not religious but would attend on special festivals like Christmas and Easter. Minnie attended Holy Trinity Church, Leamington Spa regularly.

Frank died on 28th August 1960 at Breton Lodge Nursing Home, Leamington Spa, after nearly two years of gradual decline through old age. Taking my grandmother, then living with my parents, to see Frank was my first experience of witnessing an aged relative declining. There were times when he could recall his wonderful fund of amusing stories and anecdotes from the past. The funeral service was held at the Church of St. Giles, Stoke Poges. Frank's ashes were laid to rest in the Stoke Poges Gardens of Remembrance.

The Queen Mother did remember him, the purveyor of Lanchester cars to her family; the Comptroller at Clarence House wrote on 5th October 1960:

Dear Mrs. Lanchester,
    Queen Elizabeth The Queen Mother has just learned of the death of your husband, and Her Majesty commands me to write to send to you and your family a message of her deep sympathy in your great loss.

                        Yours sincerely

                        signed: Adam Goulon
                        Comptroller to
                        Queen Elizabeth The Queen Mother.

Minnie lived on for many years to the ripe old age of 95 and died on 15th February 1975. Her ashes joined Frank's at Stoke Poges on Friday 21st February. At the time of writing their two daughters Marjorie (my mother) and Betty are aged 91 and 89 respectively.

For the first half of this century, Frank Lanchester was the front man well known to the media. Today, it is the reverse. It is Frank's brothers, Fred and George, whose many engineering achievements rightly earn the Lanchester name an honoured place in our country's roll of famous engineers. In reality, it was the strength of the "Unholy Trinity" team, that made the name Lanchester so famous.

Reference

Bird, A. and Hutton-Scott, F.: *Lanchester motor cars*. London, Cassell, 1965

Acknowledgments

It was some years ago that Chris Clark of the Lanchester Car Register asked me to write a biography of Frank, as so little was known about him compared with Dr. Fred and George. I hope my efforts will go some way to rectify the situation.

I would like to record my thanks for all their help in clarifying information to: Frank's daughters Marjorie Bingeman and Betty Mobbs, his grand-daughter Doone Frais, and niece-in-law Elaine Lanchester.

Thanks are also due to Chris Clark, author of "The Lanchester legacy, Volumes One & Two"; the librarians at the R.A.C. in London and the National Motor Museum, Beaulieu, for their helpful research.

And finally, thanks to my wife Jane for her great help and patience with editing.

Chapter Four

GEORGE LANCHESTER

11th December 1874–13 February 1970

by Mrs Elaine Lanchester

George Herbert Lanchester was the youngest member of a family of five brothers and three sisters born to Henry J. and Octavia Lanchester on 11 December 1874. His father was an architect living in Lewisham, a village in those days, but practising in London, who moved to the Hove area in 1870 on his appointment as architect to the new development of Stamford Estates. These were being built on what had originally been farmlands between Brighton and Cliftonville and which were to become the new suburb of Hove. The house built for them, originally 1 St. John's Terrace, and the birthplace of George, was subsequently renamed 49 Church Road.

My father-in-law told me of the many happy memories he retained of those early childhood days spent in that unspoiled area of countryside with their own home only a couple of fields from the sea itself. To the bonus of fresh air and the freedom and enjoyment of country pursuits were later added, as development proceeded in the area, the delights of living adjacent to the building sites where all manner of fun and games could be invented. One childish game that he and his nearest brother, who was only one year older, devised, was called "fishing" and was intended to add a little excitement to the days which could not be spent in the open air. This entailed the making of a long fishing line which consisted of lengths of twine with a hook on the end. With this they would repair to an upstairs window, or preferably, on the roof of the house around which ran a low parapet, where they would hide. When sounds from below indicated that a visitor or perhaps a tradesman was approaching the door they would carefully lower the line and attempt to catch the unwary caller's hat or bonnet – impish delights of anticipation were sadly met with only a limited rate of success, but the game was well worth the chastisement that followed.

Early schooldays at an adjacent kindergarten held no terrors and following days at the next stage in Brighton were stimulating and enjoyable. However this phase of George's boyhood came to an abrupt end in 1886, and it was a great wrench when the family had to return to the London area. A severe trade slump had ended the building boom in Hove, and it had become financially impossible to continue living there. Inevitably, this crisis had a serious effect on George's schooling, and to help his parents through their difficulties, he left school at the age of fourteen and went into industry, as had his older brother Frank and the intervening brother, Edward, who was soon to emigrate to the Antipodes. When he was fifteen, George moved to Birmingham to join his second oldest brother, Frederick, five years his senior, who had arranged for him to be apprenticed to the Forward Gas Engine Company of which Fred was the designer and works manager.

Four years later in 1893, George took over the job as works manager, Fred having resigned so that he could turn his attention to the development of the motor car, and the aeronautical research which was becoming a major interest. Within two years, George was to contribute to the making and assembly in 1895/6 of the first British built four-wheel petrol-driven motor car, in a workshop adjoining the Forward Gas Engine Company's premises. This car, and the next two experimental Lanchester cars, were the joint efforts of Fred and George. Eventually in 1897 George resigned from the Forward Gas Engine

Company to work full time with his brother, becoming his technical assistant on motor car projects, at the experimental workshop Fred had set up in Ladywood Road.

After the formation of the Lanchester Engine Company in 1899 George was appointed production engineer. He was responsible for making the jigs and special equipment needed for the production models which followed the experimental prototypes. This work kept him fully occupied and in addition he supervised the repair shop, and in the evenings would sometimes deliver cars. At weekends, he and his brothers and Archie Millership, took part, in Lanchester cars, in a great variety of hill climbs, speed trials and reliability runs.

During those early days, George was not averse to using other forms of transport, nor were his two brothers, as all three were keen cyclists. A photograph shows George and Frank posing with the bicycles at a meeting of the Speedwell Cycling Club (see Fig. 1, Chap. 3). Other photographs show them enjoying various forms of river sports – punting, rowing, boating. In 1901, the Midland Automobile Club was created, of which George was a founder member and vice president, and as a result of the club's involvement with trials and hill climbs, a pioneer of this sport. Their first hill climbing contest took place in 1901 on Gorcott Hill, a few miles from Birmingham. Three other climbs took place, but as the site did not prove satisfactory, the club committee instituted a search, which led in 1902 to Middle Hill, Broadway. As this was not sufficiently testing, further sites were sought in the Cotswolds area, Cleeve Hill and Birdlip and ultimately Sunrising Hill, where two events were staged in 1903 and 1904. The club crest in fact was devised from this site. Eventually the final choice was found on an estate belonging to a club member, Mr Montague Taylor, at Shelsley Walsh in North Worcestershire. The first meeting there was held in 1905, in which George drove a 20 h.p. Lanchester, and gained eighth place on handicap. The Midland Automobile Club is believed to be the only club in the country to have held meetings at the same site over such a long period of years. It was a great thrill, for George, many years later, to be present at the meeting in 1965 which marked the Diamond Jubilee of the Shelsley Walsh hill climbs, and to be driven to the event in a 1910 Lanchester Landaulette.

During these early years and bachelor days, George lived with his brothers, Fred and Frank, at St. Bernard's Road, Olton. When he left his earliest quarters in Saltley, Fred rented the Olton house, which was comfortable and convenient, and they were looked after by an old family servant who had nursed them in their infancy. It was in the yard of this house where in 1893 the three brothers collaborated in building the hull of the first river launch, for which the gas engine and hull had been designed by Fred. Eventually, for financial reasons, Fred left the house in 1895, to move into the country, where he rented rooms at Cobley Hill, a farmhouse near Alvechurch.

George and Frank chose to find new quarters in central Birmingham, where they rented a large room in Lincoln's Inn Buildings, opposite the Victoria Law Courts in Corporation Street. These buildings were originally intended to be used for housing legal chambers and offices, but as this proved to be unsuccessful, the floors above the ground floor were utilised as bachelor lets. The brothers divided their room with curtains and furniture into a bedroom and living area. By 1897 the three brothers were together again as Fred decided to leave Cobley Hill and join them. This was because he needed to be nearer the Ladywood Road workshop he had established. He rented two rooms and stayed at Lincoln's Inn for several months whilst he searched for a permanent base which he found eventually at 53 Hagley Road, Edgbaston, Birmingham.

The three brothers had become known as "The Unholy Trinity" whilst living at St. Bernard's Road, Olton, and maintained a close relationship throughout their working lives. At Lincoln's Inn they were joined by friends and had soon established a close and industrious little community of congenial companions who enjoyed a great deal of socialising. Prominent amongst these was Archie Millership who arrived in 1897, and took two rooms next door. He was an amateur motorist and already a first class driver who owned a variety of foreign vehicles which he used to take to the Lanchester brothers for maintenance, adjustment and overhaul. After being given a brief mechanical training upon the formation of the Lanchester Engine Company, he joined it as a tester of parts, and sales demonstrator, and eventually he became a member of Frank's sales staff.

In 1895/6 Britain's first four wheeled petrol engine motor car was built as a joint effort by Fred and George, as were the next two experimental mechanically identical cars, in a workshop adjoining the Forward Gas Engine Company's premises. In 1897 George left the Forward Gas Engine Company to become Fred's technical assistant and share with him the experimental work and trials carried out at the Ladywood Road workshop.

After the conversion of the first car to a two cylinder 8 h.p., the second car was built, an 8 h.p. two seater Phaeton. This was completed and put on the road in 1898 and in 1899 it was entered and driven by Fred in the Richmond Show and Motor Trials. It was awarded the Gold Medal for its design and performance, and eventually placed for safe keeping in the Science Museum in South Kensington. In 1898 a third 8 h.p. motor was made, identical in every way with the car motor except that it was water cooled. This was for a privately commissioned water boat. Subsequently this boat was fitted with a modern four cylinder in-line engine in 1923 and in 1939 was still in commission on the Avon at Evesham, with its original propeller.

In 1899 the Lanchester Engine Company produced six 10 h.p. cars (the actual prototype for this car had been designed by Fred in 1898 and built in the Ladywood Road workshop). These cars were in the nature of a trial trip and intended for the use of the directors and as demonstration models. This initial production had to be carried out by relying very largely on the resources of the Ladywood Road workshop. Fred had to organise and equip the new factory whilst George and Frank had to get together the executive team. The commercial, sales and publicity duties were dealt with by Frank, who also doubled up as company secretary. George's responsibilities were assisting Fred in the shop layouts, design and supervision of jigs, tools and gauges and any problems relating to such work. Ready made gauges were not then available, and had to be made in their own tool room, as well as other special kinds of equipment. Men already trained in the Ladywood Road workshop formed the nucleus of the works personnel at Sparkbrook. Fred was chief designer and general manager, and George held the position of chief production engineer. His range of duties covered such tasks as jig and tool designer, repairs supervisor, advertising copywriter, apprentice master, trouble shooter and chief tester. After hours, and at weekends and special occasions, he drove in speed trials, reliability runs, hill climbs, carried out tests and delivered cars to customers.

In one of the early Brooklands meetings which started just after the turn of the century, Archie's success in 1906 led to the first War Office contract for Lanchester cars – the War Office had an M.T.V. section. The harmony between the brothers contributed very greatly to the success and efficiency of those early years. George's respect and admiration for Fred was profound and sincere, and he never shirked the responsibility he felt to "keep the peace and pour oil on troubled waters" when Fred in later life was frequently irascible, and abrasive in his business relationships.

Frank was not an engineer, but had his own formidable skills to add to the partnership, to which George was to pay tribute in a paper he read years later on 21st February 1948, before the Veteran Car Club.

"After 1899, Frank's first contribution was in the nature of propaganda – he had to convince sceptics of the possibilities of the new form of locomotion and spread the gospel of motoring among his many friends and acquaintances. This was not an easy job in those days when most people of means were horse owners or horse minded. Frank was ideally cast for this role for he had a ready made facility for making friends. He entered into the business enthusiastically, and played his part with consummate skill. It was to him more than to any of us that credit is due for the formation of the Lanchester Engine Company in 1899, when a factory was acquired at the Armourer Mills in Montgomery

Street, Sparkbrook, for the manufacture of six 10 h.p. air-cooled Lanchester cars. This factory remained the production shop of a succession of Lanchesters until 1931. Frank's flair for making friends brought both social and economic distinction to the Company."

Many years later, whilst he was living in Itchenor, and at the time of Frank's death in 1960, my father-in-law said to me "If Frank had not chosen to throw in his lot with us, he would have been a great success in a completely different career as a diplomat. He had all the intelligence, charm and skills to shine in this field".

George and Frank continued to live at Lincoln's Inn Buildings, but the end of their bachelor days was approaching. The separation came in 1903 when Frank married Minnie Thomas, the younger daughter of William Thomas, of Gabalfa, near Cardiff, and during the same year, George became engaged to Rose, the elder daughter. The Lanchester and Thomas families had become acquainted as a result of the success of the oldest brother, Henry V.Lanchester, an architect, elected A.R.I.B.A. in 1889. He, with his partner, A.E.Rickards, had won an open competition in 1898 for the designs for Cardiff City Hall and Law Courts. The building contractor for this tremendous undertaking was the master builder, William Thomas.

After their wedding, Frank and Minnie moved to a house which was not too far away in Birmingham, and George and Rose, who married in 1907 in Gabalfa, found their own first home at Annesley, Dyott Road, in Moseley. Thus a close contact continued between the three brothers. Annesley was to remain the home of George for the next twenty four years, until the move was made to the Warwick area. It was on a plot of land adjoining the house at Annesley, that Fred and his wife Dorothy, eventually lived when they left

Fig. 1. Rose Lanchester, about 1953

Fig. 2. George Lanchester, about 1910

London and built a house and small workshop to Fred's design. Here they remained until Fred's death in 1946.

In 1904, the Lanchester Engine Company went through a difficult financial patch and was reconstructed and renamed the Lanchester Motor Company in 1905. Fred ceased to act as general manager, though remaining as chief designer until 1909 when he finally left the company to fulfil his appointment as a consultant to Daimler Motor Company, as well as pursuing some other interests. George took over as chief designer, retaining Fred's services in a consultant capacity. They collaborated on equal terms, and in 1910–11 the 38 h.p. six cylinder, and the 25 h.p. four cylinder model which followed it, were both introduced under this arrangement. These however, followed pretty closely the earlier 20 h.p. and 28 h.p. models which dated from 1904 and 1906. Subsequently, Fred became too fully occupied with his consultancy work, aircraft engineering and the Advisory Committee for Aeronautics to devote any more time to the Lanchester Company. Thus the six cylinder sporting 40 h.p. model of 1914 was the first car built entirely to George's sole design, and in which his talent for bodywork design was also given full expression, producing one of the best looking and most stylish cars of all time. The responsibility for the bodywork design of all future models, in addition to the car engines and other technical activities, added to the zest and stimulation of his work and contributed to "the immense satisfaction" George said he felt "in maintaining the high traditions Fred has established".

In 1914, the onset of the First World War ended the production of motor vehicles. Frank always maintained a close contact with the armed forces, and after great persistence, the Ministries realised that the greatest contribution to their war effort lay not in the production of shells at the Lanchester factory, to which the company had been put, under George's direction, but in the provision of armoured cars and 1A aero-engines, and contracts were awarded. Subsequently, many other instruments of war were produced, including minesweepers, paravanes, and kite balloon mobile winches, high speed searchlight tenders, field kitchens and lumber wagons – all based on the 38 h.p. vehicle chassis.

After the Armistice, the Lanchester Motor Company was without any new car designs, and the pre-war designs were obsolescent. George commenced work on the design of a 40 h.p. six cylinder car which adopted many very advanced features from aero-engine practice combined with well established Lanchester features. The prototype was given its first run in 1919 and was in production from 1920 to 1929. At this time too, Lanchester racing cars appeared at Brooklands. One nicknamed "Softly-catch-monkey" won an event in 1921. This car was based on a modified 1911 38 h.p. model, and it continued to race for several more years. Another model based on a 1913 38 h.p. achieved 90 miles per hour beating more modern racers, such as Bentleys and Vauxhalls. A Lanchester 40 h.p. fitted with a two seater racing body on a standard chassis took part in 1922. Another Forty was prepared by George and A.W.Bird for Lionel Rapson to test his company's tyres to destruction. Driven by J.G.Parry-Thomas, the car broke thirty records in fifteen hours and achieved an average of 104 m.p.h. for the last hundred miles in 1924. In 1922, based on the 40 h.p. engine and gearbox, George designed an experimental armoured car of a new type – six wheeled with four wheel drive, and utilising deep armour plate girders. This was the first example of an armoured car, specially designed as a fighting vehicle. After various teething troubles had been sorted out, the vehicle was accepted by the Mechanised Warfare Department, two cavalry regiments were mechanised, and the armoured car saw service in the early part of the 1939–45 war.

In 1923 a design was produced for a 21 h.p. six cylinder car using cheaper sliding gears for economic reasons; this was followed in 1924 by a 21 h.p. model with a slightly increased diameter for the cylinder, becoming the 23 h.p. The 30 h.p. Straight Eight came in 1928, being identical to the 23 h.p. model but with increased length due to the fact that it had two extra cylinders. This was the last new Lanchester car designed by George: introduced as a replacement for the 40 h.p. model, it was described as "the smoothest Straight Eight ever". This car produced outstanding performances and was capable of great speeds. During this period, George's designs for the bodywork of his cars was being perfected, and he was involved in new techniques and materials for use in the bodywork.

Following on the Wall Street Crash of October 1929, the financial crisis of 1930–31 had very serious repercussions on both the Lanchester and Daimler concerns. Enforced economies had to be made and the Lanchester Company was constrained by its range of cars which catered very largely for the "carriage trade, and the richest owners". The B.S.A. Group, Birmingham, already owned the Daimler Company and was looking for a place where their capacity for the production of smaller and cheaper cars could be increased. In January 1931, B.S.A. Group therefore acquired the Lanchester Company, and both brothers, Frank, at that time managing director, and George, chief engineer and designer, transferred their services to B.S.A. so that they could assist the new owners in Coventry. The resulting production of the Lanchester 18 h.p., the 10 h.p. and the 14 h.p. cars, smaller and more economical vehicles, but carrying with the name Lanchester the reputation of quality and prestige, was successful in selling well until the 1939–45 war. The last purely Lanchester design was the 15/18 model, but from the 1930s onwards, the designs were wholly of those of the Daimler Company. The period of five years following the takeover of their company was not a very happy or fulfilling time in George's career, and in 1936 he resigned to join the Alvis Motor Company, on his appointment as chief engineer of their Automobile Division.

In 1931 George and Rose left Annesley, Moseley, the home which they had loved, and moved away from Birmingham. They rented a house at 28A Warwick New Road, Leamington Spa where they remained whilst a new house was built for them at Fulbrook Edge, Sherbourne Hill, Warwick, to which they finally moved in early 1933.

At Alvis, George was responsible for the design and development of the 17 h.p. Alvis Silver Crest and followed this with the 12/70 Sports Tourer, the last private car that he ever

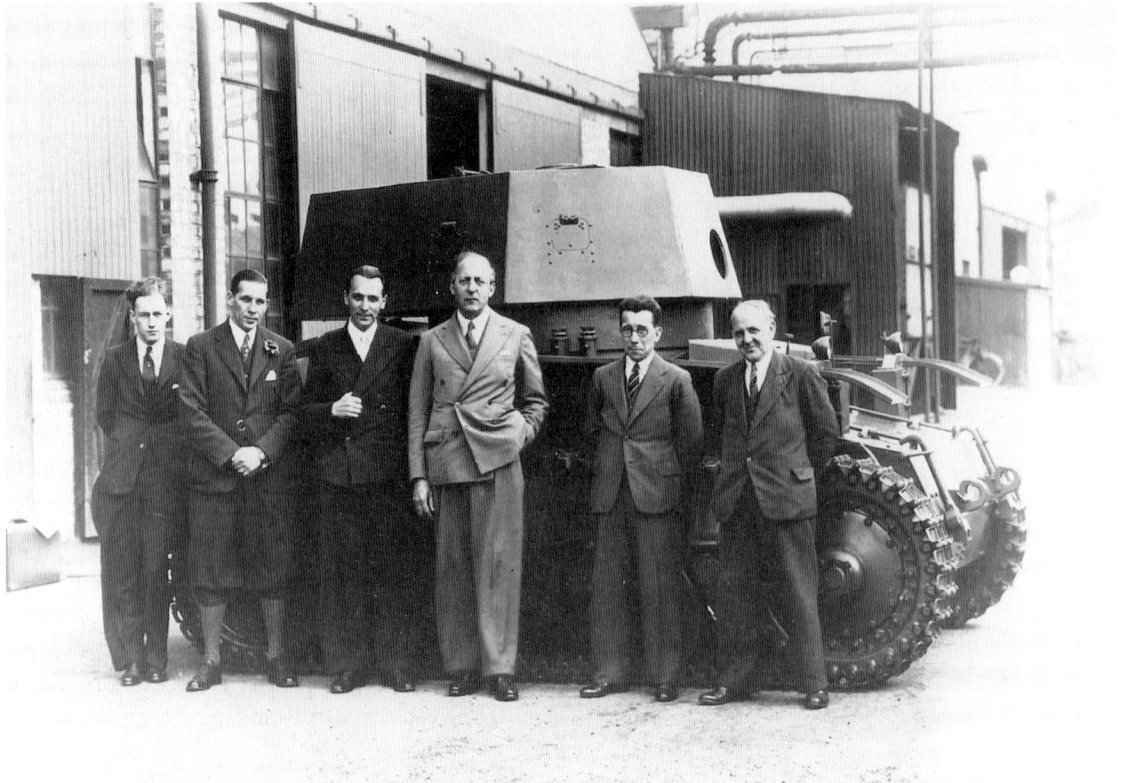

Fig. 3. George (in centre) with other Alvis staff, and an Alvis armoured car, in 1939

designed. He then transferred to the Alvis-Straussler Mechanical Warfare Department as chief engineer. He was responsible for the design and development of several light tanks for various foreign governments and the Alvis four wheeled independently sprung armoured cars for the R.A.F. which were successfully used in the Middle east by the British, and by the Dutch in the Far East before the outbreak of the last war. Later he designed and developed the Alvis "Hefty" tractor. With the expiration of the Alvis contract in mid 1939, George's career in the Midlands came to an end. He and Rose faced the upheaval of a move to London with their customary optimism and courage.

The final phase of George's working career came on the outbreak of the 1939–45 war with his appointment to the Sterling Armament Company. Their much loved home at Fulbrook Edge was put up for sale, and it was while they were still in occupation there that my own introduction to the Lanchester family came about. In the early summer of 1940 I met my future husband, their son Clive, whilst I was stationed at Warwick, during the first stage of my wartime service with the Auxiliary Territorial Service. By Christmas George and Rose had left Warwick and moved to the Star and Garter Hotel in Richmond, Surrey, where I was able to join them for a family gathering during my short period of Christmas leave.

At Sterling, George's work commenced straight away with the supervision of a contract already in hand, requiring completion. This was in connection with sound ranging vehicles used in the detection of hidden gun emplacements. He then redesigned the German Schmeisser 9 mm gun to make it easier to put on a rapid production. The redesigned gun was a great improvement on the original and the new model was renamed the Lanchester sub-machine-gun and was used by both the British and Australian navies. It was during the period of the necessary testing of the gun, which was being carried out in the Northampton area, that George and Rose moved up to Northampton in mid-1941. On their eventual return to London, they took a flat at 12 Old Manor Court, St. John's Wood, north London, which was to remain their home until January 1949.

At the end of 1943 George was appointed technical advisor and consultant to Sterling Engineering Company, a subsidiary. His work here (the company being sub-contractors) included organising the manufacture of breech blocks for the Oerliken 20 mm cannon, and bomb-release gear for Vickers. He was also responsible for designing and making a hydraulic flap damper for the Barracuda aircraft which had been experiencing a very serious wing vibration.

At the end of the war, George moved on to another subsidiary of the Sterling organisation when he became technical advisor to Russel Newberry Diesel Engine Company, and worked on design products in industrial diesel engines. Another aspect of his work for this firm was to act as general consultant on various domestic engineering projects. I remember being involved myself in a small way by trying out in my own home several items including a floor polisher and a carpet cleaning machine. By 1952, my mother-in-law's deteriorating health was causing great concern, and George decided to reduce his business commitments to a part-time arrangement on a basis of two days a week, which continued for a further nine years when his contract with the firm ended in 1961.

Since their marriage in 1907, George and Rose had lived a very fulfilling and active life. The first of their children, a daughter, Nancy, was born in 1908 and the second, a son, Clive, followed in 1913. To George's work with its heavy commitments were added the domestic duties of home life, and in 1914 came the extra strains and trials of the First World War, bringing with it hugely increased responsibilities, when the Lanchester Motor Company was taken over completely for the production of armaments. George found time nonetheless to become a member of the War Volunteer Reserve Corps and several photographs in the family possession show him undergoing training at the Corps head-quarters at Ragley Park.

He and Rose soon made a wide circle of friends and were the centre of a busy and hospitable community. Early in their married life, they started renting a series of country cottages, a tradition which they maintained until the mid-1930s. During these weekend retreats from city life they were able to relax and provide the hospitality which had always been such a feature of their home in Moseley. Many years later, my father-in-law told me that family members and friends flocked to these weekend visits where they were always assured of a warm welcome, and chores were cheerfully shared. On some occasions as many as forty to sixty people turned up, and girls from the village were "roped in" willingly to help provide extra scones, small cakes and sandwiches.

Longer holidays, especially when the two children were small, were usually spent in South Wales, Devon and Cornwall, and other places on the south coast. The cottages rented for weekends had to be within easy reach of Birmingham, and favourites places were Stratford-on-Avon, Evesham, Worcester, the Malvern Hills and the Cotswolds. They had always loved country life and open air activities and whenever possible, proximity to the river was chosen so that punting and boating could be included.

During these rural interludes, George became a keen gardener and was able to indulge in simple carpentry which he had always enjoyed, so he was frequently occupied with making items of cottage furnishings: book cases, shelves, tables and chairs for both inside and outside use, garden sheds and so on. He undertook the never ending little jobs of household maintenance and repairs, and during the early 1920s he started the stick carving hobby that he enjoyed for the greater part of his life. He wrote an article in which he explained that:

"This hobby started accidentally when my wife and children and I were walking in Witcomb Woods in Gloucestershire. I happened to see a straight slender branch of ash suitable for a walking stick, and having cut it off and after the wood had been seasoned, I carved a small head of Mephistopheles. When completed however it was too light and slim for a walking stick so I gave it to my wife as a 'swank stick'. Of course, others followed by request of my two children and nieces. From these developed the habit of looking for suitable sticks in woods and hedgerows, and from the various shapes of the head ends I drew inspiration for a figure, head or animal that the shape suggested. The majority were made of ash, but I found many other woods suitable – hazel, sycamore and holly being the best media, but acacia, cherry and sweet chestnut were also found useable. The only tools I used in carving were ordinary pocket knives. The painted decoration was by water colour and the finish provided by clear cellulose lacquer. One of the 'rules of the game' was that the whole figure had to be carved from the natural wood and nothing was to be added or stuck on. The one exception was the Elephant's Head, in which the tusks were inserted".

At some stage during the 1939–45 war my father-in-law told me that he found this hobby to be a very satisfying form of relaxation. Altogether he carved about sixty sticks, and they provide a remarkable demonstration of his inherent artistic ability and attention to detail.

Annesley, their home in Moseley, was a centre for fund raising events in aid of various charities, and they loved entertaining their friends, and organising the amateur theatricals and fancy dress parties which were particularly popular. Several photographs have survived which show Rose and George in some very striking costumes. They were keen supporters of the professional theatre in Birmingham, Stratford-on-Avon and the Malvern Festival.

The enjoyment of music had always occupied a very special place in both their lives and their entertaining included musical soirées. This interest and delight in musical matters was further supported in a more practical manner by the help they gave to the organisation of Midland musical festivals in the early 1920s, and particularly by their connections with the Imperial League of Opera. This was founded in the early 1920s by Sir Thomas Beecham and the first opera season under the auspices of the League took place in 1928. The original purpose of this Imperial League, as stated by Sir Thomas, was "to build and endow an Opera House in London, and to assist in the building and endowment of Opera Houses in some of the leading provincial cities – allowance must be made for the fact that public

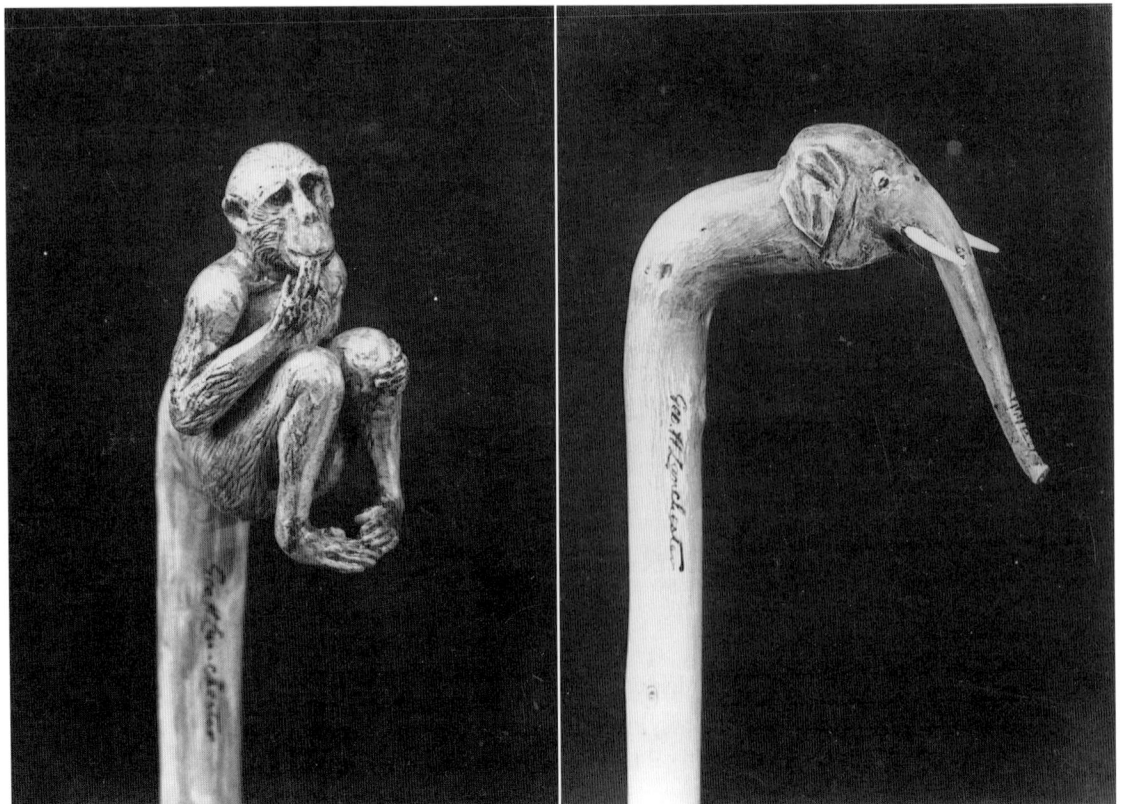

Figs. 4 and 5. Two of George Lanchester's "swank sticks"

Fig. 6. George and Rose in fancy dress

opinion for such a large enterprise, is at this stage, not fully ripe". Birmingham was one of the cities selected, and Rose and George were very active members of the Moseley and District Committee, of which Rose served as chairman in 1928, and George provided secretarial and publicity support throughout the period of their connection with the Imperial League of Opera.

It was in 1932 that he first became involved in searching for the solution to the mystery of the Yellow Emperor's South-Pointing Chariot, to which his attention had been drawn by a friend, Dr. J.B.Kramer. Though this engineering problem was not connected to any aspect of his normal engineering projects, it excited him, and as he always loved a challenge, this one proved irresistible, and so he pursued it until he found the solution. He made two working models of the Chariot – one of which I believe is in the Birmingham Museum of Science and Industry. He became actively involved in a number of organisations. He was a founder member of the Institution of Automobile Engineers at its inception in 1907, and was elected Chairman of the Birmingham Centre in the 1920s, and Member of the Council in 1927. He became a member of the Institution of Mechanical Engineers in 1927, having been elected some years later than his brother Fred. He became a member of the Veteran Car Club of Great Britain, though he was not one of its founders, in 1930. Other organisations which he joined included the Fellowship of the Motor Industry and The Circle of 19th Century Motorists.

Their flat in St. John's Wood became my own home base during the war years until late 1945, when I demobilised and waited for my husband's return from Germany on his own demobilisation in 1947. It was here that I met Miss Mary Stevenson, who lived in the flat above on the next floor. She became a firm and valued family friend and was a great help to my parents-in-law. "Steve" as she became known to all who met her, eventually became my stepmother-in-law on George's remarriage in 1961.

George and Rose would not have lived in London from choice, but they cheerfully made the best of things and coped with the hazards and difficulties of the war years. They were a devoted couple and kindness itself to me and still managed to extend hospitality to family and friends who had to visit London at that time. Though heavily involved throughout the war years in a variety of armament and engineering projects, George still found time to keep up with other interests and organisations. In 1943–4 he served as President of the Institution of Automobile Engineers and during his term of office he actively furthered the negotiations which led to its amalgamation with the Institution of Mechanical Engineers. Subsequently he was a member of the Council of the Automobile Division of the Institution's activities in 1952. Some years later, on 25th September 1961, he derived great satisfaction at being able to attend the official opening of the Lanchester Room at the Institution of Mechanical Engineers.

In 1945 Frederick W.Lanchester was awarded the honour which gave him the greatest pleasure, the James Watt Gold Medal. Sadly, his health was too frail for him to attend any ceremony, though he was able to dictate a message which George read for him, adding his own tribute, when he attended the presentation, and received the medal on Fred's behalf. This afforded George the greatest happiness as he had always fought for proper recognition of his brother's achievements. In 1947, another interest was celebrated when on 3rd February, George delivered a lecture at the China Society, on the Yellow Emperor's South-Pointing Chariot.

George and Rose, had, in 1948, already realised the need to move out of London, and in January 1949, they left St. John's Wood, and took a residence at the first of what was to become a series of apartments in country hotels in the Home Counties. All had to be within easy travelling distance of central London, because of George's business commitments and continuing involvement with a variety of activities and organisations, and social functions he was required to attend. By the summer of 1953, my parents-in-law had moved to Slindon, West Sussex.

Their son, having been appointed to a firm in Chichester, found it necessary to move his own small family south from Warwickshire. Two adjacent flats were rented for the winter months, into one of which George and Rose moved, whilst a search was made for a home where the two families could live under one roof, yet remain in separate quarters, thus bringing an end to their travels. A house was found in Itchenor, south of Chichester, which provided the ideal solution and which became our joint home in the spring of 1954. Though providing the two required units, these were joined at a large communal kitchen, but the sharing of this presented no problems and the arrangements worked harmoniously.

My mother-in-law's health had deteriorated to the extent that she was virtually bed-ridden, but we had been able to find a domestic "treasure" who relieved my father-in-law of the housework. George did all the shopping and cooking and devotedly nursed Rose throughout the fifteen months they had together at Itchenor until Rose's death in the late summer of 1955 after nearly fifty years of marriage throughout which George said "She had put up with life as a workshop-widow like an angel".

After the move to Itchenor, George had continued with his consultancy work at Russel Newberry, on its reduced basis of two days a week, and this he continued to do until June 1961, when, to quote him "I was given the sack as being too old". The firm had been taken over, and it was a considerable shock when the new owners discovered they were actually employing a man of eighty-seven. This was hard for George to swallow because although this was actually his age, he still remained so mentally alert and active, and enjoyed his work so much, that he had not even considered retiring. Nevertheless an increasing problem with failing eyesight meant that he had to reduce his long distance driving trips, and travel to London by buses and trains, and this had become irksome.

During 1955 he had acquired a "bubble car", a B.M.W. Isetta, a completely different type of vehicle to any that he ever had before, and which he enjoyed using. He was not a man given to useless nostalgia and instead of looking backwards he preferred to look forwards, and anticipate the future. The little car had become a source of great amusement throughout the family and friends, and soon became known as Lanchester's Motor Pram: its registration plate was LMP 133. He used this for as long as possible for occasional short distance visits, local shopping and the social visiting he so much enjoyed, having many friends and acquaintances in the Sussex area. Nevertheless, the very sad day eventually arrived when he had to realise that it was no longer safe for him to continue driving, and that he had to adjust his life accordingly. He did not allow this to inhibit him, and continued to carry out work commitments, and support functions, and keep a variety of appointments in London as well as a number of other places.

He had been a keen and active supporter of the Veteran Car Club of Great Britain, founded in 1930, since its early days, and was frequently seen as a driver, judge or spectator at rallies and trials organised by the Club. He loved, and eagerly anticipated the London to Brighton Run in which he took part whenever possible (and whatever the weather) as a passenger, or at the tiller of the 1901 twin cylinder Lanchester – usually accompanied or driven by George Upton. In 1958 and 1959 he served as president of the Club, an honour which he gave the fullest attention, and which led to a great deal of extra travelling and activity, all of which he enjoyed with much enthusiasm. As the years rolled by and the older members became reduced in number, he told me with some sadness that he regretted "the changing atmosphere of the London to Brighton Runs – an element of showmanship, of theatricality was creeping in – a hint of the circus".

Two events occurred which gave George great satisfaction as they commemorated the achievements of Dr. Fred. The creation of the Lanchester Hall at the College of Aeronautics (later incorporated in the Cranfield Institute of Technology) and the Lanchester Building at the University of Southampton, at the opening ceremony of which, in May 1960, he and Dorothy, Dr. Fred's widow, were special guests.

He was actively connected with the Daimler and Lanchester Owners Club, and the Lanchester Register. He contributed a number of papers to various organisations and gave talks to the Institution of Mechanical Engineers. I have a copy of the talk he contributed in 1957 to the Newcomen Society at the Science Museum, on the life and work of Dr. F.W.Lanchester. A photograph in the *Tatler and Bystander* of 8th May 1957 shows him demonstrating his model of the 1895 Lanchester car at the Diamond Jubilee of the Royal Automobile Club, founded in 1897. Another photograph shows him at a Roadfarers' Club luncheon in 1957. Several photographs were taken on the occasion of the twenty first anniversary meeting of the Vintage Car Club, at Goodwood, Chichester, in which he is shown at the wheel of Francis Hutton-Stott's 1913 Lanchester, taking part in the parade of veteran and vintage cars. A very important occasion for him was the opening ceremony of the new building of the Lanchester College of Technology in Coventry, the name chosen to honour and celebrate the memory of Dr. F.W.Lanchester. By 1980 the name had been altered to that of Coventry Lanchester Polytechnic, and four years later this included the Lanchester Library, which was opened on 29th April 1984 by Mrs. George Lanchester, and contained its own Lanchester Collection. (See colour plates 6 and 7)

When not fully occupied with matters which took him away from Itchenor, George was involved in a variety of activities. He had always enjoyed making models and in the middle of the 1950s he constructed one of the 1895 Lanchester car which had sadly been destroyed during the Coventry "blitz" in the 1939–45 war. He had a phenomenal memory, which together with Fred's drawings enabled him to make a complete replica, all by hand, down to the smallest detail. The model was subsequently donated by his widow, Steve, to the Lanchester Library. Models of several other items were made for his daughter's family, and one of the Yellow Emperor's South-Pointing Chariot, for his son, Clive. He carved the last two walking sticks he was able to do, for his grandson, Richard, a blue tit, and for his

Fig. 7. George, with the model of the first Lanchester car (see also Colour plate 8)

granddaughter, Kirsten, a robin, bringing to an end the hobby which he first started in the early 1920s. When his daughter Nancy was a tiny girl, he made her a little car, big enough for her to drive about the garden of the home in Moseley. In the early 1950s he made toys for his son's children. I remember a trolley, a doll's pram, a fort, a piggy money-box. There were small household maintenance repairs which, even with failing eyesight, he managed to execute. Having mastered basic cooking skills, he turned his attention to cake-making, and soon we were all enjoying a range of small cakes, rock cakes being a favourite speciality. Eventually he progressed to Christmas cakes. For several years he made these for his close family and for several friends, though he jibbed at the icing and embellishments which the recipients were expected to add.

A major project arose which occupied him very fully in the latter years of the 1950s. A biography of his brother, Dr. F.W.Lanchester, had been planned and collaboration with the author, Dr. P.W.Kingsford, entailed a huge amount of work in the gathering together, collating, checking, and preparation of records, papers, photographs and Lanchester memorabilia which was required. The work was published by Edward Arnold in 1960 and entitled *F.W. Lanchester: the life of an engineer*. This collaboration gave him enormous satisfaction and pleasure – he had always regarded Fred as a genius, and held him in the greatest respect, and was tireless in his efforts to ensure that his brother's achievements received the proper recognition to which they were entitled. Although George was a brilliant engineer and designer himself, he never sought to bask in any reflected glory from his brother's eminence. He loved his work at all levels, and carrying this out thoroughly, and to the best of his ability, gave him all the satisfaction he needed.

Nor did he waste any efforts in seeking worldly acclaim or favours for his own achievements. He was a man of high principles. I remember him telling me on one occasion in 1961, that he had been very pleased to receive a good letter from our son, Richard, then a schoolboy of fourteen, which contained the information that "he (Richard) had managed to outmanoeuvre a school friend to gain his objective". The manner in which this had been achieved led his grandfather to make the observation to me that "it showed the boy had an eye to the main chance", and it made him wonder where the characteristic had come from. My father-in-law said that "it must either have come from his mother's side (mine) or from Granny Lanchester's (his wife, Rose Thomas's) or both – it certainly had not come from the Lanchester forbears".

This lack of self seeking was demonstrated by a later comment when he told me that years earlier he had been recommended for a knighthood in recognition of the Lanchester Company's contribution to the war effort in 1914–18. His reply at the time "that he would prefer a crate or two of good whisky" had apparently not been received with enthusiasm, nor was it ever followed up. His generosity of spirit and willingness to help others, and his wonderful sense of humour had always helped to endear him to people and he had a gift for retaining friends. He maintained a huge correspondence with acquaintances as well as friends over the years of his long life. George, too, was the member of the family, who with his wife Rose, maintained contact with the second youngest brother, Edward, who emigrated to the Antipodes in his youth. A devoted family man, George's close relationship with his daughter Nancy, especially after Rose's death, was the source of much comfort and happiness. He frequently stayed with her and her husband Ralph Heaton, in Birmingham maintaining his links with the Midlands and subsequently at All Stretton, Salop and sharing their holidays in Wales and Scotland and Orkney where his eldest grandson Peter Heaton, had embarked on a farming career – a venture in which George took an enthusiastic interest.

In November 1961, George married Mary Stevenson in London, his son Clive, "giving him away" in his eighty-seventh year, a profoundly satisfactory solution to what was becoming a life restricted by failing vision and a lack of mobility. With Steve, he embarked

Fig. 8. George and "Steve" at the time of their marriage

on a new lease, and after a short period in London, and an urge to move to the country, they found a home at The Hill, Chulmleigh, North Devon, where they were able to spend the rest of George's life together, and which brought them great happiness. The cottage and the garden had to be licked into shape, and even with his increasingly poor sight, George still managed to carry out minor repairs and adjustments. They found a new circle of friends in a small and congenial community, and after the arrival of Steve's little D A F car, they were able to venture further afield. Local trips and shopping in Exeter, and trips to Shropshire to his daughter, long journeys by road to Glasgow to meet his wife's relatives and friends, and tours of the Highlands of Scotland and other favourite areas further south.

There were commitments to fulfil in London. All these journeys were made possible with Steve as chauffeur. Several letters I have refer to engagements they attended: in October 1962, a luncheon in London given by the Fellowship of the Motor Industry; on 4th May 1965, the twenty-first anniversary dinner of the Guild of Motoring Writers, organised by a group of old timers in motor journalism, with many old friends present; on 13th May 1965, the dinner given by the Royal Aeronautical Society for the Lanchester Memorial Lecture.

By the end of 1966 deteriorating health led to a serious illness which lasted for two years, from mid 1967 to mid 1969, during which George's life was made bearable by Steve's devoted ministrations. By the early autumn of 1969, a great and welcome improvement in his health had come, although he was by now almost completely blind. He wrote in a letter dated 14th December "I am glad to say that after two long years of illness, I have made an excellent recovery and am almost back to my old form. In fact Steve and I have just returned from a long journey to Scotland, by road, for the purpose of attending the special invitation to the seventieth anniversary of the Royal Scottish Automobile Club in Glasgow on December 10th. We were guests at the Club for three nights and it proved to be a quite wonderful and unforgettable occasion – some Very Important People among the 300 to 400 guests". A few days later, Christmas greetings arrived which was the last communication we received.

On 13th February 1970, we received news that George had died suddenly, and peacefully, after a good meal, and whilst enjoying one of his favourite forms of relaxation, that of listening to good music – Beethoven, I heard later from Steve. A friend who met him a month earlier on 11th January in Chulmleigh told us that he was "in wonderful form and looking remarkably fit for his age". We heard subsequently that only a couple of days before his death, he had been making the arrangements for his participation in the forthcoming re-enactment of the 1900 1000 Miles Trial by the Veteran Car Club, in the original of which, George had taken part in the three seater 8 h.p. Phaeton Lanchester, which became known as the Gold Medal Phaeton, and which is now in the Science Museum, Kensington. How typical it was that in his ninety-sixth year he had maintained his zest for life, and was still looking forward to the future.

Chapter Five

LANCHESTER'S *AERODYNAMICS*

by Eur. Ing. J.A.D. Ackroyd

1 Introduction

On 17th December 1903, our view of the world became changed irrevocably. The Wrights' proving of their new technology on this momentous day ensured that, from then on, our world would be viewed from the air, explored from the air, invaded from the air. The air itself would become subject to mankind's invasion.

That is not to say, however, that the world changed overnight. The Wrights' invention neither took to the air fully-fledged, nor was the world at large ready to accept what had happened at Kitty Hawk that day. Indeed, only one garbled newspaper account had announced the event to a world by then inured to claims for mechanical flight. Yet much of value had already been achieved before the Wrights' first powered flight. With the benefit of hindsight, we can applaud the successes of the earlier gliding pioneers, Lilienthal, Pilcher, Herring, the Wrights themselves, even acknowledge the positive steps taken by the far less successful powered flight pioneers such as Ader, Maxim and Langley. At the time, however, the world saw only that these escapades had ended in disaster, the sole exception being the Wrights, of whom the world knew little. Scepticism was, by then, rife. And since the Wrights wished to see financial reward from their invention, they determined to guard it carefully throughout their negotiations, negotiations which embroiled them in protracted but depressingly fruitless correspondence with various governments who, for example, persisted in the belief that what was being offered was merely another useless design, a "paper aeroplane" dreamt up to elicit funding for another pair of crack-brained inventors, certainly not a machine which existed and which had flown.

Nonetheless, the Wrights used this dispiriting period well, particularly during the years 1904 and 1905 in which they developed significantly their invention. With successive aeroplanes built during those years, they evolved the modern system of aerodynamic control about their aeroplanes' three mutually perpendicular axes so as to control pitch, yaw and roll independently. Moreover, because of their choice of centre of gravity position relative to the aerodynamic surfaces, their aeroplanes were inherently unstable in pitch (although I doubt the Wrights realised this), yet the Wrights learned to modify the aerodynamic geometry so that at least the instability became more controllable.

In the end, Wilbur Wright cut the Gordian knot of entangled negotiations by giving a series of very public demonstrations of their latest machine near Paris in August 1908, thereby at last winning world-wide recognition for their achievements. The effect on this hitherto secure, "sceptred isle", in particular, was to be profound.

Yet it was an Englishman, Frederick William Lanchester, who had already predicted, explained and provided many of the solutions to the problems of mechanical flight. In the year preceding that of Wilbur's public flying demonstrations, and therefore before the world saw that powered flight was with us, Lanchester had published his *Aerodynamics* (Lanchester, 1907). This term had probably been borrowed from an earlier publication (Langley, 1891) by that other great seeker after powered flight, the Smithsonian Institution's august secretary and originator of this term, Samuel Pierpont Langley. Lanchester had met Langley, however briefly, in Washington in 1892, Lanchester's interest in powered flight having emerged in the previous year.

Yet when Lanchester attempted to discuss the problems of flight with Wilbur Wright in 1908, so as to assess the Wright aeroplane's performance, he was rebuffed with the phrase (Lanchester, 1926),

"the bird that talked most (the parrot) was a bad flier."

This was a phrase often on the lips of the dour, laconic Wilbur that year, I suspect. Certainly, he made a similar comment in declining to give a speech at that time (Gibbs-Smith, 1970). Perhaps Wilbur felt that, for financial reasons, he must remain wary of this Englishman who was, as other contributions to these volumes make clear, already eminent as an automobile engineer. Perhaps, also, something of the reputation of Lanchester's *Aerodynamics* had preceded him, so that Wilbur felt disinclined to discuss matters with someone whose views were so evidently difficult to understand, so esoteric. Certainly, during what little discussion they did achieve, Wilbur flatly disagreed with one of the major practical consequences of Lanchester's deep thinking.

Both the Wrights and Lanchester were eminently practical men, ingenious both in design and application. Yet there was an essential difference in their approaches. The Wrights had achieved powered flight, careful step by careful step, through an extensive programme of rationally conceived model experiments backed by full scale tests. Such investigations were conducted mainly so as to measure the air resistance experienced by certain shapes of wings selected in a somewhat ad hoc manner. But the Wrights could progress beyond this and into the realms of theoretical speculation. By a series of fairly involved calculations and analyses, they developed one of the earliest successful propeller theories based on their air resistance data. Yet, as far as I am aware, the Wrights never sought to enquire as to the origins of air resistance, probably because they realised that the scientific community could provide no answer. Lanchester, in contrast, was much inclined to think at a very fundamental level: his treatment of engine balancing for his first two-cylinder petrol engine provides a case in point. Consequently, his approach to this new subject of aerodynamics was to try to begin at the beginning, to ask the fundamental question: what is the basic cause of air resistance? The answer he gave plunged him headlong into a then profoundly difficult subject, long debated within the scientific community. And since his analyses based on that answer were often relatively crude, it was easy for that community to criticise his approach. Clearly, his application of his ideas was directed, moreover, to the subject of powered flight, about the outcome of which the scientific community was much divided. Despite the successes of earlier gliding pioneers, Otto Lilienthal in particular, in 1896 Lord Kelvin had refused an invitation to join the Aeronautical Society in London with the comment (Gibbs-Smith, 1970),

"... I have not the smallest molecule of faith in aerial navigation other than ballooning or of expectation of good results from any of the trials we hear of."

Others, such as Lord Rayleigh, perhaps kept a more open mind on the subject. Indeed, a mere handful of scientists scattered about Europe were on the point of conceiving the ideas necessary to an explanation of air resistance and, from it, discover the secret of flight. Although it is from these men that the science of aerodynamics is usually seen to have had its genesis, nonetheless Lanchester will be found to have been paralleling their ideas and, in certain respects, pre-dating them. Thus the aim of this chapter is to attempt to set Lanchester's thinking in context, within these early steps in the development of aerodynamic theory.

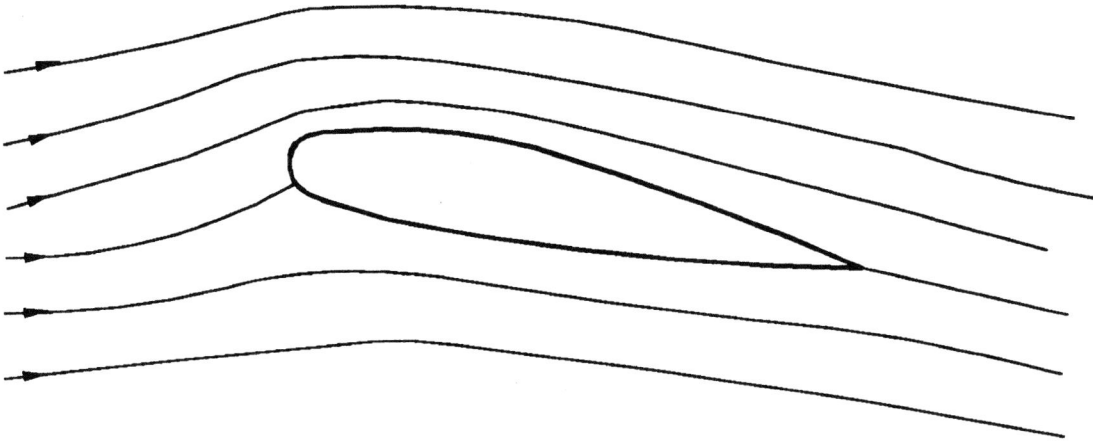

Fig. 1. The flow field about a wing at incidence

2 A Survey of Modern Aerodynamics

Considering the level at which Lanchester's thinking must be explained, it is as well that a few basic ideas in modern aerodynamics are understood before I begin my historical survey.

How a wing produces lift can be deduced from the situation shown in Fig. 1, in which we see a wing section* set at a fixed incidence in a uniform stream of approaching air. Evidently, the airflow passes smoothly around the wing and departs cleanly at the wing's sharp trailing edge. Routes taken by the airflow are indicated by the streamlines shown, and these could be revealed in practice by injecting smoke into the flow. Yet whilst streamlines reveal at all points the direction of the air's motion, it is more important to understand that the air must be passing between the streamlines, not crossing them. Looking at the streamtubes formed between the streamlines, and tracing those over the wing's upper surface back to their upstream condition, you notice that they become wider at this upstream limit. Since I have introduced the term "streamtube", I will now ask you to think of these as pipes, which must pass the same amount of air per second, say, at every section (we call this the continuity principle). And this suggests the idea that this relative narrowing of the upper surface streamtubes must be accompanied by an increase in airspeed there. Decreasing streamtube width can be likened to the river flow into a narrow gorge, in which the water speed increases. Although I shall avoid mathematical theory, I have to introduce a mathematical theorem here (called the Bernoulli equation) which demands that an increase in airspeed must be accompanied by a decrease in air pressure. Regarding the narrower upper surface streamtubes, we see, then, that the air pressure here must be below that at the upstream part of these streamtubes where the pressure is the local atmospheric pressure. Lift is therefore generated on the wing by this partial vacuum at the wing's upper surface, the wing being sucked upward by this effect.

In contrast, we can think of this lifting process in a quite different way. Notice that the air leaving the wing's trailing edge is moving slightly downward. Gradually, the contours of the wing have pushed the air downward. Action and reaction being equal but opposite in sense, the airflow must therefore be pushing the wing upward. We have here another explanation for the lifting effect already described. On this basis, however, we can take the argument a step further. Notice that, somehow, the airflow is induced to move slightly upward as it approaches the wing's leading edge. Demonstrably, though, the wing then turns the airflow to a near-horizontal direction before finally deflecting it downward. Evidently the wing must be pushing downward on this upwardly approaching airflow to

---

* The reasons for this particular choice of shape will emerge later in this section.

achieve this. Repeating my earlier action and reaction argument, we see that the wing experiences a further lifting effect because of this. For the wing, then, its high aerodynamic efficiency lies in its ability to cause this *combination* of an upward and then a downward motion in the air flowing around it. Useful as the latter argument is, on occasions the streamtube-pressure argument has advantages and we can use either, or both, of these arguments, depending on our needs.

Let me delve now into mathematics – not, I hasten to add, the mathematical theory itself but rather the problem of how we might attempt to construct a mathematical model of our wing's airflow. With the mathematical solution following from this model taken for granted, I will then describe the results which emerge from that solution.

Our choice of wing shape and incidence remains that depicted in Fig. 1. My main problem then concerns the modelling of the flow and of the air itself. As to the air, I am going to assume that this always has the same density everywhere. Now, as to the flow, I shall assume that changes in its motion, both speed and direction changes, can only be brought about by changes in pressure within the air. In addition, I must insist that such changes in motion obey the continuity principle introduced in the earlier streamtube argument. Concerning the pressure model adopted, an important consequence of this is that each and every element of air can never be made to change its rotation. As this is a vital point in later arguments, let me explain this. Regarding the cylindrical element of air shown in Fig. 2a (ignore Fig. 2b for the moment), we see its circular section and the air pressures acting around it. Every pressure arrow shown acts perpendicularly to this surface. And since the arrows' lines of action all pass through the centre of the circle, this pressure field can never persuade the element to spin about its centre, or to change whatever rotation it already possesses. Because, initially, the airstream is in uniform motion with no rotation of the air elements about their own centres, the air elements can never acquire rotation (or, to use the correct technical term, vorticity). Obviously, the elements can, and do, distort in shape as they move around the wing. Under the action of this pressure field alone, however, all air elements remain at zero vorticity and the whole flow field is said to bc irrotational.

To acquire vorticity, then, the elements must experience tangential, or shearing, forces around their peripheries. Here we can turn to Fig. 2b, where we see such tangential forces in operation. Eventually, we will return to this additional consideration. Regarding the present flow model, however, the possibility of such shearing forces is ignored.

There are standard mathematical techniques available for solving the equations of motion describing the above airflow model. When applied to our wing problem, they produce the streamline pattern shown in Fig. 3a. Clearly, this does not look like the real situation shown in Fig. 1. Indeed, were we to calculate the net effect of the pressure variations, or the upflow/downflow, around our wing, we would find that the wing experiences no lift force

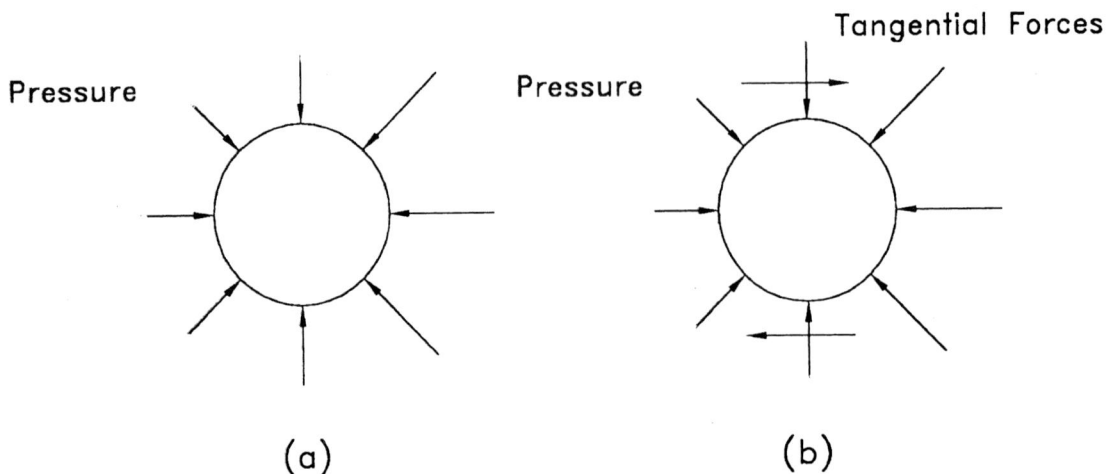

Fig. 2. The forces on a cylindrical air element

Circulation

Circuit ► ►

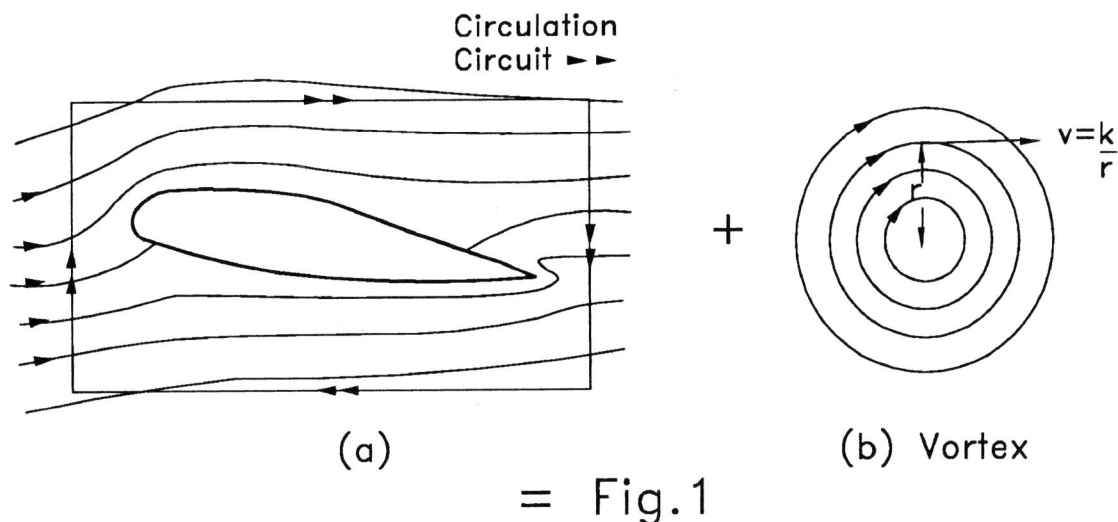

(a)

(b) Vortex

= Fig.1

Fig. 3. The effect of adding flow fields (a) and (b)

(the air resistance component directed perpendicularly to the initial airstream's direction) and, even more surprisingly, no drag force (the air resistance component in the direction of the initial airstream). Looking a little more closely at this streamline pattern, we see that our required upflow/downflow is not evident. Moreover, at the sharp trailing edge the flow whips around that edge, from the lower surface to the upper surface, before departing at some point on the upper surface. In fact, the airspeed at the sharp edge itself turns out to be infinite, a highly unlikely circumstance in reality.

In an attempt to improve our picture of the airflow, I am now going to add to it the spinning vortex flow shown in Fig. 3b. This flow has circular streamlines and is also irrotational, except at one point. The exception is the vortex centre, at which the flow has infinite rotation, or vorticity, and this is concentrated at the zero area of the centre (what is called, mathematically, a singular point). I fix this vortex centre at some point within the wing section's shape. I do not worry about the mathematical singularity itself because this now lies within the wing and we are interested only in what happens outside of the wing. I then add the vortex speeds to the airflow speeds of Fig. 3a. The vortex speed must augment the airflow speed over the wing's upper surface and therefore, by the Bernoulli equation, reduce the pressure there. At the lower surface, in contrast, the vortex speed is in opposition to the airflow speed of Fig. 3a, so that the net flow speed is less and the pressure becomes higher. In terms of our streamtube-pressure argument, the addition of the vortex produces the effects we require. Indeed, if I choose precisely the right vortex strength, in terms of its spinning speeds, I find that the net effect on the airflow is now very close indeed to our original picture, Fig. 1 (see the "equation" implied in Fig. 3). Calculating the pressure variations, I find that there is now a lift force but, surprisingly, no drag. Clearly, though, the effect of this particular vortex has been to produce the required upflow/downflow. Moreover, the infinite speed at the sharp trailing edge has been suppressed so that the flow departs smoothly, and at finite speed, from that edge.

Clearly, then, the addition of this particular vortex strength, the one which produces smooth flow at the trailing edge, has given us a flow field very close to the one we want. Indeed, if we use this model of the flow to calculate the lift force, we find that the calculated lift is very close indeed to what is measured in reality. A lifting wing, then, behaves as if it generates around itself a swirling vortex flow which is carried along with the wing as the wing moves through the air: the vortex is bound to the wing and is referred to as the bound vortex.

If, by some means, we could increase the amounts of upflow and downflow, in effect by increasing the strength of the vortex whilst still retaining smooth trailing edge flow, then we would expect to increase the lift of the wing. This turns out to be the case, and our means of doing this lie in our choice of one feature of the wing section's shape. You might have noticed that the wing shown in Fig. 1 has a symmetric shape top-to-bottom. Re-shaping this section so that it becomes a more upwardly-arched shape clearly has the effect of increasing the downflow at the tail. As it turns out, a similar increase in the nose's upflow is also achieved by this. Consequently, increasing the arch of the wing, or, to use the correct term, the camber of the wing, increases the upflow/downflow, the strength of the bound vortex and hence the lift. As to the reasons for the more detailed contours of the wing, particularly the blunt nose and the tapered tail, these will emerge presently.

So far, I have talked about vortex strength in rather vague terms. I can be a little more precise, by introducing the term "circulation". I can find the circulation of any flow field by drawing a complete circuit in the flow (see Fig. 3a), calculate the flow speed in the direction of the circuit for each and every air element on that circuit, then add up the multiple of this speed and the element length of circuit for every element on the circuit. If I carry out this calculation for the vortex, I obtain a finite answer which is the circulation, or strength, of the vortex.* However, if I draw any circuit around the wing shown in Fig. 3a and repeat the calculation, then the result obtained is zero. There is no circulation here, because there is no vortex present in this particular flow. If, on the other hand, I repeat this procedure for the flow shown in Fig. 1, then the answer I obtain is that the circulation is now not only finite, but equal to the vortex circulation used to generate this particular flow. Therefore, one particular value of vortex circulation is required in order to produce this flow field exhibiting smooth flow at the trailing edge.

To complicate the argument a little further, there is a mathematical theorem in aero-dynamics which states that, if a finite circulation is calculated for a circuit in a flow field, then this must be equal to the multiple of the vorticity and the area over which it occurs within that circuit. Where is the vorticity in our model? I stated that the flow field of Fig. 3a is irrotational and, indeed, so is the vortex flow everywhere, except the centre. At that singular point, you recall, there is an infinite vorticity spread over zero area, a curious mathematical limit which, when vorticity and area are multiplied together, gives the finite answer $2\pi k$ obtained in the footnote for the vortex (below). In fact, our circulation circuit must always include the vortex centre if a finite circulation is to be obtained, because it is at this point that the vorticity is concentrated.

My argument, no doubt you will think, is now drifting a long way from reality. How does this relate to the real flow depicted in Fig. 1? Where is the vorticity situated in the real flow field of this figure? Clearly, there must be vorticity somewhere since the flow field possesses circulation, and hence the wing experiences lift. Also, although this vortex, or circulation, theory of lift predicts lift accurately, so far I have been unable to predict anything except zero drag force. The answer to all of these conundrums is that, although I have produced a mathematical model which describes most of the flow field, this model does not describe all of it. Indeed, there is one crucial area of the flow field which is so thin that it is often difficult to see it. Also, in my model I have neglected one small, but extremely important, physical property of air.

Gases, such as air, and liquids, such as water, can be classified under the general term "fluids". And all fluids are viscous. Admittedly, the coefficient of viscosity for both air and water has an extremely small value (typically about 0.00001 Pa s in S.I. units). Nonetheless, this extremely small coefficient colludes with a further physical constraint so as to make a fluid flow in certain regions behave in a highly viscous manner. This physical constraint is

---

* In the case of the vortex, this calculation is relatively simple. For the irrotational vortex used here, the airspeed v on any circular streamline of radius r from the vortex centre is v = k/r, k having a *constant* value throughout the vortex. Thus the air speed in the direction of a circular circuit which lies on this circular streamline is always v = k/r, and is constant around this circuit. To find the circulation, we merely multiply this constant speed by the length of the circuit, which is $2\pi r$, and obtain the answer $2\pi k$. This, also, has a constant value throughout the vortex. Note that our circuit has included the vortex centre.

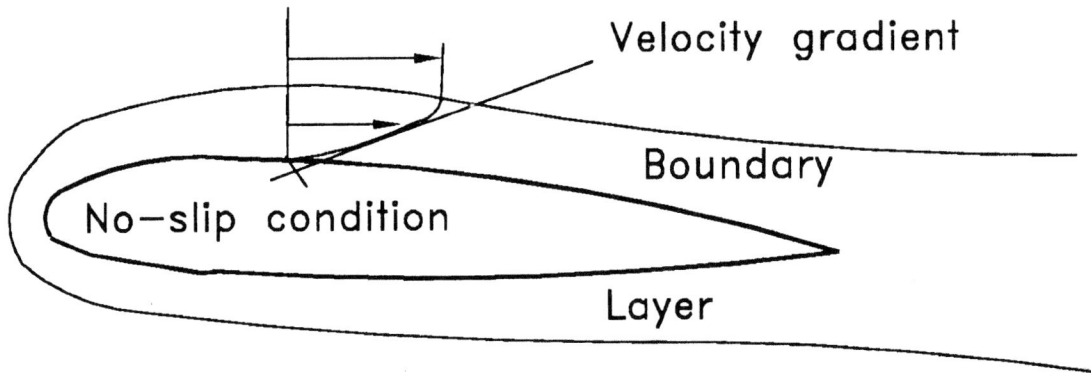

Fig. 4. Boundary layer flow about a wing

that the fluid cannot slip over any solid surface with which it is in contact; the fluid at a surface must have the same speed as that of the surface. This constraint arises from interactions at the molecular scale between the fluid and the solid.

Fig. 4 shows the consequences of the combined effect of small viscosity and what is called the no-slip condition. We see our wing as before, but now we concentrate on the flow region very close to the wing's surfaces. The flow speed increases very rapidly over an extremely thin region (I have grossly exaggerated its thickness, so as to illustrate the effect), from zero at the surface, as demanded by the no-slip condition, to the much higher speeds of the air passing around the wing and predicted in my earlier model shown in Fig. 3. Consequently, the values of the velocity gradients (a typical gradient being shown in Fig. 4) within this region are huge. It turns out that the shearing forces in a viscous fluid are proportional to the multiple of the viscosity coefficient (here, very small) and the local velocity gradient (in this region, very large). For this region, this multiple of these two quantities is then very significant, so that this region, subjected to these viscous shearing forces, behaves in a highly viscous manner. Fluid elements passing through this region are slowed down by the action of these high viscous forces, and to a greater extent very close to the surface itself: hence the higher velocity gradients here. This very thin, slower moving region is called the boundary layer, and its properties and behaviour will be seen to be crucial to the generation of lift as well as to the production of drag.

In the flow field exterior to the boundary layer, the fluid has the same small viscosity and the flow experiences changes in speed from streamline to streamline, so that velocity gradients exist here, too. Yet such gradients are far smaller than those in the boundary layer, so that this part of the whole flow field behaves as if it is virtually inviscid. Indeed, it behaves as our model fluid flow depicted in Fig. 3, changes in its motion being caused almost entirely by pressure changes. And, of course, this vast majority of the whole flow field also behaves as an irrotational flow possessing zero vorticity. Fluid elements in the boundary layer, on the other hand, are subjected to substantial viscous shearing forces, like those shown in Fig. 2b, and therefore acquire vorticity. So it is within the boundary layer that vorticity is created. This generates the circulation, the appearance of a vortex being bound to the wing, and hence the lift. Without the viscous boundary layer, then, flight would be impossible and other artifacts which rely on the same basic principle, propellers, the sails of ships, etc, would be useless. It is a staggering fact that a quantity as small as the viscosity of air is the root cause of the lift force which sustains a Jumbo Jet weighing two hundred tons in the air.

Much of what I have said so far depends on the boundary layer behaving itself so as to remain attached to the wing surface. One thing, however, a boundary layer cannot do is to negotiate a sharp edge. The boundary layer flow always separates from the body surface at such edges. Indeed, a simpler view of the lifting problem is to say that it is the boundary layer's insistence on separating at our wing's sharp trailing edge which produces the smooth flow departure there, shown in Fig. 1. This, in turn, produces the appearance of a bound vortex being generated around the wing so as to generate the lift via the inviscid, irrotational flow model depicted in Fig. 3. Indeed, provided the boundary layer's behaviour can be guaranteed so as to ensure that separation occurs only at the sharp trailing edge, the role of the boundary layer in lift generation can be ignored and lift can be predicted according to the model proposed in Fig. 3. This is one of the great simplifications of aerodynamic theory: lift can be predicted by a relatively simple inviscid flow model and the action of viscosity can be ignored.

Unfortunately, however, the good behaviour of the boundary layer cannot always be guaranteed. Our choice of a smoothly blunted nose for the wing section of Fig. 1, as opposed to a sharp nose, avoids the possibility of boundary layer separation occurring there. Even so, further problems arise. Boundary layers experience difficulty in remaining attached to a surface in circumstances in which there is a progressive rise in pressure along that surface. And this pressure rise is inevitable on all body shapes since pressure is at its lowest around the fattest part of the body, after which it must rise so as to approach the local atmospheric pressure near the tail of the body. Hence we shape the aft parts of bodies so as to have a very gentle taper toward the sharp tail (we make the body "streamlined") so as to make this pressure rise as gradual as possible in an effort to control the boundary layer as it approaches, say, the sharp tail of our wing in Fig. 1. But put that wing at a higher incidence and the upper surface low pressure deepens in value so that the subsequent pressure rise toward the tail becomes more rapid. The wing's lift increases almost in proportion to the incidence angle but eventually the consequence is that, at a sufficiently high incidence, the boundary layer separates at some point on the upper surface other than the tail*. Part of the lifting vorticity is lost, a wide wake appears which destroys in part our lifting pressure field, lift is lost and drag becomes much larger. In fact, the wing has stalled.

I have just mentioned drag and you have probably guessed already how some, at least, of this drag arises. The viscous shearing forces within the boundary layer act at the surface itself, producing a form of drag force which is usually referred to as skin friction. But the boundary layer produces a further form of drag. Being a slower moving layer of fluid, and since continuity of flow must be maintained, this thin layer acts as if elbowing aside slightly the exterior inviscid flow field. The effect is to change slightly the pressure field around the body so that there is a net pressure imbalance fore and aft: in fact, a pressure drag. If the flow field was *entirely* inviscid, this pressure drag would disappear, so, too, would skin friction, and the body would experience zero drag force, but zero lift as well. The secret of flight in this practical, viscous world of ours, then, lies in *careful choice of shape* by streamlining our wing shapes so as to minimise skin friction and pressure drag whilst maximising lift.

Poorly streamlined bodies, bodies with blunt rear surfaces (the circular cylinder is a good example), also experience the boundary layer drag described above over those parts of the surface on which the boundary layer remains attached. But the rear of a cylinder, for example, is so badly shaped that the boundary layer promptly separates around the cylinder's maximum thickness. The wide wake thus created is accompanied by a large fore and aft pressure imbalance which completely swamps the boundary layer drag.

Two further comments must be made about the effects of viscosity on flows before I leave this aspect of aerodynamics. The first concerns comparisons of flows about bodies having the same shape but experiencing differing circumstances (the bodies may differ in size, the initial flow speeds may differ, the fluid viscosities might have different values). To do this, we should compare the sizes of the ratio of those forces in a flow which are due to pressure changes to the forces due to viscous shear. Being a ratio of forces, this ratio has

---

* A similarly harmful effect is produced if we make the wing shape a too greatly cambered arch. Because of boundary layer separation problems, there is a limit to the amount of camber we can use in the design of wing section shapes.

no dimensions*, and it is called the Reynolds number. In testing an aeroplane model in a wind tunnel, for example, one should try to ensure that the Reynolds number for the test case is as near as possible to that calculated for the full-scale aeroplane in flight. Equality of the Reynolds number in the two cases then ensures that the test is truly representative of the full-scale situation. The Reynolds number, Re, turns out to be given by the following formula:

$$\text{Re} = \frac{\text{(fluid density)} \times \text{(flow speed)} \times \text{(representative length of body)}}{\text{(fluid viscosity coefficient)}} \tag{1}$$

My second comment concerns the ordered structure of the boundary layer itself. Usually, but not always, the boundary layer flow around the nose of a body is smooth and well-ordered, the layers of fluid moving over each other in a perfectly regular manner which is called laminar motion. However, small imperfections in surface shape, dirt in the fluid, irregularities in the approaching flow ahead of the body can cause the boundary layer to become a disordered turbulent mess. Boundary layers are particularly prone to this transition to turbulence in regions of rising pressure, regions which, as we have seen, occur inevitably toward the tails of bodies. Thus the aft parts of bodies are not only prone to separation but also to transition, a coincidence which has caused certain circles to confuse what are, in fact, two quite separate phenomena. However, the turbulent boundary layer, whilst producing higher skin friction and boundary layer pressure drag, is better able to withstand rising pressure situations, making it less prone to separation than the equivalent laminar boundary layer.

So far, I have avoided mentioning the effect of wing tips. Indeed, strictly speaking, I have been describing the flow about the distinctly hypothetical case of wings which have infinite span. For example, the flow field constructed in Fig. 3 represents the flow obtained around any section of such a wing: the circulation of the bound vortex is the same from section to section anywhere along this infinite span. But, in practice, we are interested in the flow about wings of finite span, or, to introduce a further technical term, wings of finite aspect ratio. The aspect ratio provides information about a wing's planform shape and is defined as the ratio of the wing span to the wing chord (or average chord, in the case of a tapered wing). Being the ratio of two lengths, the aspect ratio is a further dimensionless quantity. But having mentioned recently the constancy of the bound vortex circulation from section to section along our infinite wing, the question arises: what happens to this bound vortex, and its circulation, in the case of the real wing of finite aspect ratio which possesses tips? The answer I can give depends entirely on another of those theorems which make aerodynamics such a mathematical subject. This states, in simple terms, that a vortex cannot end and that its circulation must, in some way, be preserved. The smoke ring provides a good example of this: its vortex structure never ends because it exists as a closed toroidal loop having the same circulation all the way round. In the case of the real wing of finite span, the bound vortex becomes cast off toward the tips, the cast off pair of vortices turning themselves roughly through 90° so as to stream away downstream of the wing. Indeed, this vortex system also never ends. The so-called trailing vortices can be traced right back to the airfield where the loop is finally closed by the starting vortex (see Fig. 5) which is left behind at the airfield where the aeroplane originally generated sufficient lift to take off. The circulation of this vortex system is preserved all the way round the loop so that you notice that the starting vortex has a spin sense opposite to that of the wing's bound vortex. In fact, this starting vortex is generated by the sudden shedding of boundary layer vorticity at the instant the wing begins to move forward so as to generate its own bound

---

* The Mach number is another example of a dimensionless ratio, being the ratio of the body's speed through the fluid to the speed of sound in the fluid. For bodies moving with speeds very much less than the sound speed (the Mach number is then very much less than unity), it turns out that the air density is effectively constant. This was assumed to be the case in my statement of the mathematical model of the airflow given earlier. Thus this section is concerned with the subject of low speed aerodynamics.

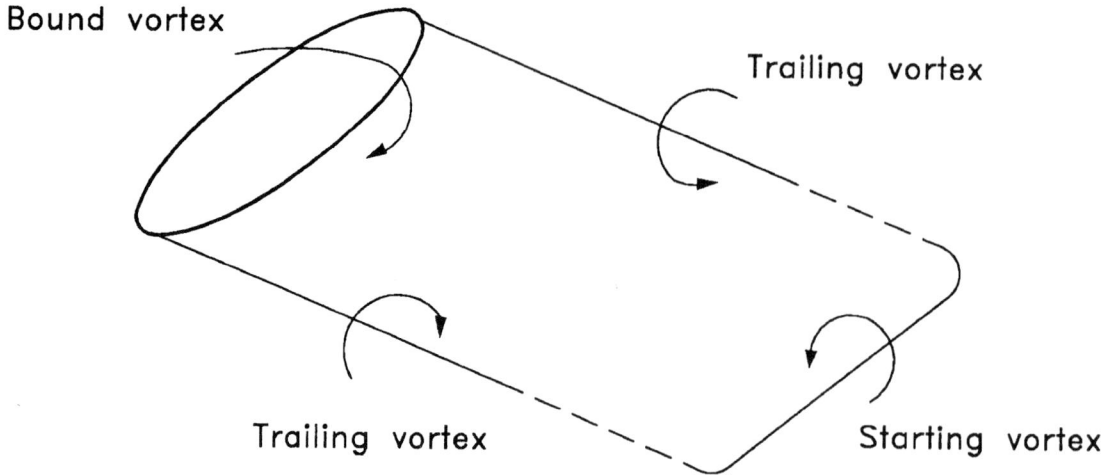

Fig. 5. The vortex system generated by a finite span wing

vortex, the boundary layer adjusting itself so as to insist on separating at the sharp trailing edge.

The trailing vortices behind high-flying aeroplanes are, of course, a familiar sight nowadays. If you look closely, you will see that the white vapour emanates from each engine exhaust, but these vapour trails quickly become swept into the trailing vortices so as to form the familiar pair of white trails which persist for miles.

The wing shown in Fig. 5 has a noticeable taper from root to tip. Consequently, the circulation at the root is larger than that outboard, the difference being made up by the circulation shed progressively from root to tip. This progressively shed circulation rapidly coalesces into the fairly well-defined trailing vortex pair. However, a consequence of this is that the lift generated by this wing of finite aspect ratio is not as large as that calculated for an equal spanwise segment of our hypothetical wing of infinite span. In simple terms, lift is reduced as wing aspect ratio is reduced.

A further consequence of this shedding of circulation is that wings of finite aspect ratio experience an additional drag force, over and above that directly attributable to the boundary layer and described earlier. This additional drag force is due to the fact that the moving wing inevitably creates an ever-lengthening pair of trailing vortices. The situation

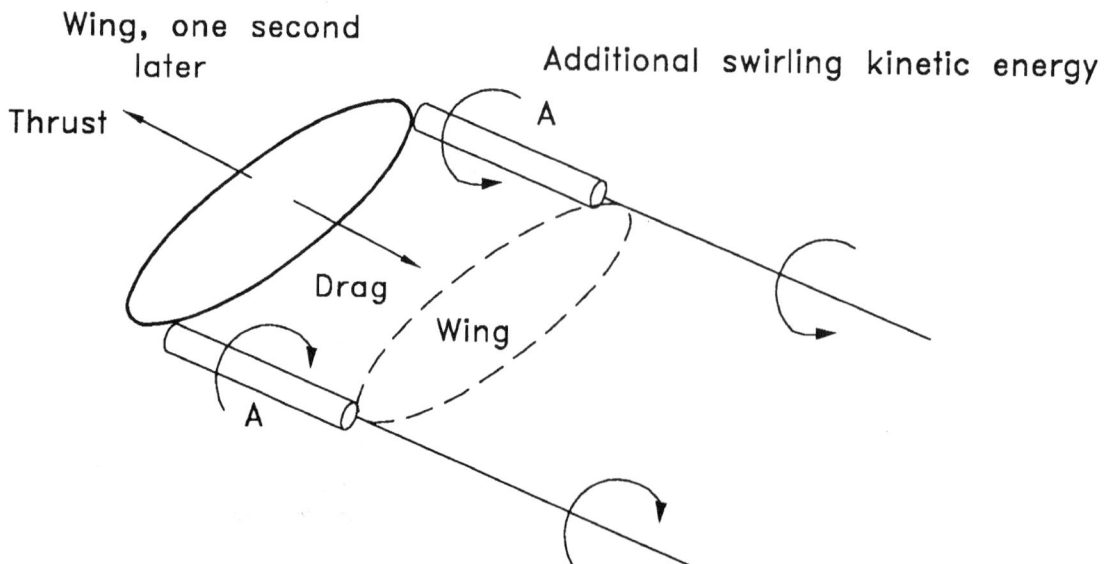

Fig. 6. Additional kinetic energy created by a wing's forward motion

is shown in Fig. 6, in which the wing is viewed at one instant and then, say, one second later. During this change with time, the wing leaves behind two further contributions to its trailing vortices (marked A in the figure). These additional chunks of swirling motion possess kinetic energy. Now there is a basic theorem in dynamics which requires that, in creating additional kinetic energy, work must be performed: a force must move through some distance. In the case of the wing, a propulsive thrust is required to push the wing forward so as to create this new swirling kinetic energy. But since the wing moves forward at constant speed, this propulsive force must be opposed by an equal drag force. This additional drag force is now called induced drag, the drag induced by the presence of the trailing vortices.

I will end this rather lengthy survey of modern aerodynamics with one observation. Much of this evidently involved and clearly difficult subject was unknown when Lanchester came to aeronautics in 1891. The miracle is that, working entirely on his own, in a little more than ten years he came to understand so much of it, as we shall see.

3 Lanchester's *Aerodynamics*

From what has gone before, you will appreciate that the basis of an understanding of aerodynamics rests on the two fundamental concepts of the boundary layer and the circulation theory of lift. Since Lanchester grasped, at least in part, both of these concepts, my task in the remainder of this chapter is to describe how far he progressed along the way. In doing so, I must attend carefully to what Lanchester himself wrote since it would be invidious to imply claims for discoveries which Lanchester did not make.

Sutton (1949) has provided an apt view of Lanchester's position in aeronautics, a view which I believe will be borne out here:

"A remarkable feature of Lanchester's work is its isolation. At no time did he form part of a team working at a large institution, nor was his work maintained by funds from external sources to any great extent. Throughout his life he remained an individualist, perhaps the last and possibly the greatest lone worker that aerodynamics will ever see."

These remarks preface an overview of the wing theory alone, whereas, in a later contribution, Sutton (1965) includes a description of Lanchester's boundary layer concept. However, in his desire to provide a modern view of the boundary layer, Sutton (1965) creates the impression that Lanchester's contributions went beyond what, in fact, he achieved, an example of that pitfall mentioned at the beginning of this section. In delivering the first Lanchester Lecture before the Royal Aeronautical Society, von Karman (1958) deals in broad terms with Lanchester's contributions to wing theory. In the rather more extensive review (Ackroyd, 1992) which provides the basis for this chapter, it may be felt that Lanchester has been painted rather more "warts" than "all", in which case I would ask that Sutton's comments be kept in mind.

Being the great individualist he was, Lanchester seems to have acknowledged few mentors. William Froude he mentions quite often, always with approval, and perhaps William John MacQuorn Rankine was held in similar regard. Yet it was Isaac Newton who had Lanchester's unreserved admiration. Indeed, in the discussion following one of his papers (Lanchester, 1937), Lanchester remarks that he is

"... always pleased to see words of recognition of the great example set to us by Sir Isaac Newton. It is my belief that even to-day a careful study of the *Principia* is an education in itself, and a great deal of Newton's work is better studied directly than through those who have attempted to improve upon his teaching by interpretation."

Having had the chastening experience of attempting to interpret Newton (Ackroyd, 1984),

I can only bow the head before such wisdom. As to Lanchester's study of Newton, Kingsford (1960) tells us that Lanchester spent much of his final year at the Normal School of Science (later, Royal College of Science; later still, Imperial College) in digesting Newton's ideas together with those of Rankine. As we shall see, this activity was to have a profound influence on Lanchester's approach to wing theory. However, since his boundary layer concept provides the justification for his wing theory, as, indeed, is the case in modern aerodynamics, I will deal first with his boundary layer concept.

3.1 *Lanchester's Boundary Layer Theory*

When Lanchester set himself the considerable task of investigating the behaviour of viscous fluids, little was known on the subject. As Lanchester discovered, our understanding of the subject begins with the hypothesis of Newton (1687) in which he postulates the existence of resistive forces between contiguous fluid layers which move at slightly different speeds. He applies this idea to the physical phenomenon of the vortex which is created by the steady rotation of a circular cylinder in a large expanse of fluid initially at rest. Without comment, Newton assumes that the no-slip condition must apply at the cylinder's surface. Whilst this is the first occasion on which this condition is applied correctly, you will realise that its application is absolutely necessary here, otherwise the fluid could never be set in motion. The existence of resistive, or shearing, forces is also necessary to an explanation of this phenomenon since it is through the action of such forces that the swirling motion near the cylinder's surface is communicated to the fluid far from the cylinder. A review of Newton's incorrect solution to this problem is given by Ackroyd (1992). Unfortunately, most of the correct and key physical ideas outlined above were ignored for nearly a century after Newton's death.

In some of the early subsequent analyses of fluid motion about bodies at rest it was recognised that a layer of slower moving fluid lies adjacent to their surfaces. For example, Daniel Bernoulli (1738) mentions this phenomenon in his statement of the continuity principle for pipe flows. Yet the fact that the no-slip condition must apply to fluid in contact with solid surfaces was not understood for a considerable length of time (see Goldstein (1938) for a detailed history of the debate).

The idea that shearing forces exist between fluid layers moving at different speeds resurfaces in the work of Navier (1827), in which the equations governing such laminar motions are set down. The definitive version of this is given by Stokes (1845) and here the idea is introduced that the shearing forces are proportional to the change with time of the shapes of fluid elements, in particular, the rates of strain experienced, as the fluid moves through space. The consequence of this is that the shearing forces within the fluid are proportional to the viscosity coefficient and to various velocity gradients existing in the flow. Even now, the resulting so-called Navier-Stokes equations governing the laminar motion of viscous fluids have no known general solution, although a few solutions have been obtained for certain very simple flow configurations. The greatest headway has been achieved by obtaining solutions to some appropriately chosen approximation to the full Navier-Stokes equations. Boundary layer theory is one such approximation and it is from this that our understanding of the physical behaviour of the boundary layer has developed. Nonetheless, whilst the main ideas concerning resistive forces internal to a fluid were being acquired during the first half of the nineteenth century, the key condition of no-slip at a solid surface had yet to be recognised. Thus, in the earliest analysis of laminar viscous motion in pipes, whilst Stokes (1845) obtains the correct form of the velocity variation across the pipe, he is unsure as to the application of the no-slip condition at the pipe wall. Later, however, this condition is accepted by Stokes (1851) in his analysis of the very slow motion of a sphere through a viscous fluid. This analysis is based on another form of approximation to the full Navier-Stokes equations (not the boundary layer form) and has the dual merit of not only predicting for the first time a drag force other than zero*, the value of which is now known to agree with experiment, but also a drag force which is directly proportional to the value of the viscosity coefficient. Thus, with hindsight, we can see that it was at this point in the mid-nineteenth century that the writing finally appeared on the wall so as to link drag to viscosity.

The explanation of the physical cause of the phenomenon we now call viscosity provides the culmination of the kinetic theory of gases given by Maxwell (1860), who shows that internal viscous forces are the consequence of the random motion of fluid molecules in a higher speed layer of fluid so that they collide with molecules in a layer of lower speed, or vice versa. And in reporting his measurements of the viscosities of gases, Maxwell (1866) remarks that, within the accuracy of his measurements, the no-slip condition appears to hold at the solid surfaces of his apparatus. Soon after the beginning of the present century, the combination of the no-slip condition and viscous shear provided the mathematical model proposed by Prandtl (1904) as a description of this slower moving fluid layer which surrounds solid surfaces, and which we now call the boundary layer. As already stated, Prandtl's mathematical model is an approximation to the full Navier-Stokes equations.

Independently of Prandtl, Lanchester (1907) also proposes a boundary layer concept, although it seems unlikely that a date can be given to its inception. All that seems to be known is that, in 1905, Lanchester was carrying out experiments with model gliders with the specific intention of estimating skin friction.

Throughout his first major publication on the subject of aerodynamics, Lanchester (1907) is careful to emphasise that, in all fluid motions, viscosity is of fundamental importance. At this point in history, it seems that Lanchester and Prandtl were virtually alone in adopting this crucial view. Lanchester (1907) then introduces the viscosity concept by way of the hypothesis of Newton (1687), mentioned above, and Maxwell's work. In particular, Maxwell's simple model of a viscous shear flow is used to demonstrate the essential principles involved. In this, a plate is moved with constant speed V parallel to a fixed solid surface, the gap of height h between the two surfaces being filled with a viscous fluid (Fig. 7). Here, Lanchester (1907) takes it as "well established" that the no-slip condition must be applied at both surfaces. As it happens, the solution of this flow problem is one of those few solutions of the full Navier-Stokes equations mentioned earlier, although the work of Navier and Stokes is not mentioned by Lanchester. The resistances experienced by both surfaces are shown to be proportional to the velocity gradient V/h which we see must be

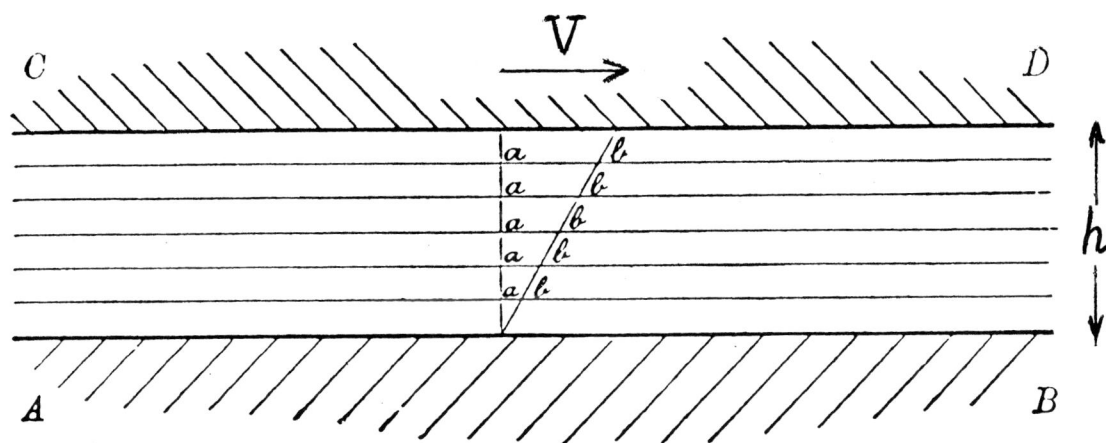

Fig. 7. *Maxwell's viscous shear flow (Lanchester, 1907). AB is the fixed plate, whereas the plate CD moves with constant speed V. The gap between the plates is h. In one second, say, a fluid element on each of the horizontal streamlines shown moves through the distance a–b. The velocity gradient is V/h*

* The zero drag result, mentioned in Section 2 in connection with the predictions of inviscid, irrotational flow theory (Fig. 3), bedevilled the development of fluid flow theory up until Stokes's era and beyond.

constant throughout this flow. Thus the surface resistances are proportional to V itself. Lanchester (1907) then turns to a problem which is intended to serve as a simple model for his future discussions of wing drag and in which the wing-like shape is reduced to its essentials. The problem is that of a flat plate set at zero incidence in a uniform stream of speed V. Lanchester (1907) argues that the simple proportionality for viscous resistance cited earlier for the Maxwell flow should no longer apply:

> "If, when the velocity V increases, we suppose that the layer of fluid affected to any given degree becomes thinner, and vice versa, it is clear that the viscous forces will rise in a greater ratio than directly as V, for the velocity gradient will be steeper, also the inertia forces will be less, for the mass of fluid to be set in motion will be less."

On the basis of this penetrating insight, Lanchester (1907) then begins his analysis. A detailed review of this analysis, together with its rather important omissions, is given by Ackroyd (1992). Suffice it to say here that, essentially, Lanchester (1907) balances the viscous force experienced within the plate's boundary layer with the fluid momentum loss rate in this layer (momentum, in this case, is the multiple of the mass of a fluid element and its velocity) so as to deduce correctly the result that the frictional drag, F, is given by

$$F \propto V^{3/2} \tag{2}$$

(Here $\propto$ means "is proportional to" and $V^{3/2}$ means that we take the uniform stream speed V, multiply it by itself three times – $V \times V \times V$, which is V^3 – and then calculate the square root of this triple multiple, which is indicated by the 1/2 index, ie $(V^3)^{1/2} = V^{3/2}$). In doing so, Lanchester produces a result which is tantamount to the statement that the thickness of the boundary layer is proportional to $1/V^{1/2}$. Although he does not comment on this, this result supports his original statement, quoted above, that as V increases the layer thickness decreases.

Lanchester (1907) recognises that the frictional drag, F, must also depend on other things in addition to V, such as fluid density, fluid viscosity and some characteristic dimension of the plate. So as to introduce these quantities into his statement of proportionality, relation (2), he turns to a mathematical technique known as dimensional analysis. First introduced by Fourier (1822), the essential idea here is that the dimensions of the terms on each side of an equation must balance. For example, in the relation (2) above, if quantities such as density, viscosity and length are to be introduced then they must combine with $V^{3/2}$ in such a way as to produce a term which has the dimension of force, thus balancing the dimension of force associated with F. In this way, Lanchester (1907) is able to give a correct proportionality between F and these other quantities. However, using these techniques alone, Lanchester (1907) is unable to obtain a value for the numerical constant (the "constant of proportionality") which changes his proportionality into a statement of equality, from which a numerical value of F could be calculated.

A rather inaccurate value of this crucial numerical constant had been quoted a few years earlier, however, in the first description of the boundary layer and its behaviour given by Prandtl (1904). In this astonishingly detailed paper, only eight pages long, Prandtl uses the term "boundary layer" once only, yet the paper contains many of the key ideas later elaborated by Prandtl and his associates at Göttingen University. Not only is the equation governing the boundary layer's motion deduced but the form of its solution is also obtained for the case of the flat plate boundary layer so as to give a rather more detailed version of Lanchester's later frictional drag result mentioned above. In order to achieve this detail, Prandtl (1904) makes use of the important deduction that the flat plate's boundary layer grows in thickness in proportion to the square root of the distance from the plate's leading edge. Lanchester (1907), in contrast, whilst recognising that boundary layer thickness probably increases from the leading edge, does not use this point. Indeed, it was not until some years later (Lanchester, 1915a) that he suggested a particular, though inaccurate, growth rate, and for the rather different case of a turbulent boundary layer. Meanwhile, the full and detailed solution of the laminar flow flat plate problem, including a far more accurate value for that important numerical constant mentioned above, had been given by one of Prandtl's earliest doctoral students at Göttingen (Blasius, 1908).

A wide acceptance of Prandtl's concept of the thin boundary layer and its mathematical theory was delayed by the onset of the First World War. The rather heated debate which ensued in Britain in the 1920s, for example, is described by Ackroyd (1989). A major question at that time concerned the nature of the mathematical approximation made by Prandtl in obtaining his boundary layer equations from the full Navier-Stokes equations. Indeed, this question did not become thoroughly settled until the 1950s, at which point a formal mathematical structure became developed so that Prandtl's approximation could be improved upon (see, for example, Van Dyke (1964)). Nonetheless, by the 1920s the usefulness of Prandtl's boundary layer ideas had gained some acceptance so that the validity of Lanchester's parallel thinking could at last be recognised.

To return to our main interest in this chapter, however, Lanchester (1907) then cites the now-famous experiments of Reynolds (1883)* to determine the onset of turbulence in pipe flows. From these experiments Reynolds had been able to deduce the crucial point that transition from laminar to turbulent flow depends entirely on the Reynolds number Re (see relation (1)). Thus the nature of a flow came to be seen as being fundamentally dependent on the value of the Reynolds number: at low values of Re the flow is laminar, at a higher value of Re transition to turbulence takes place so that at even higher values of Re the flow is fully turbulent. Lanchester (1907) uses this as an introduction to a discussion of the question of dynamic similarity in flows. Again using dimensional analysis, he demonstrates that

$$V = c\mu/(l\rho) \tag{3}$$

"... may be taken as the general equation connecting all similar systems of flow in viscous fluids."

Here again V is the flow speed whilst ρ is the fluid density, μ the viscosity and l is a characteristic length scale for the body. Lanchester's result (3) amounts to the statement that the Reynolds number, Re, must be the same for all situations of dynamic similarity (in effect, Re = c). Moreover, in effect Lanchester grasps the fundamental concept that it is not merely the nature of a flow (laminar or turbulent) which depends on Re but, more generally, the behaviour of a flow and the effects which that creates, such as the forces on a body, may also be dependent on Re and may vary continuously with variation of Re. However, Lanchester's subsequent arguments based on this idea are far less clearly expressed than the cogent, concise statement given by Lord Rayleigh (1910):

$$\frac{\text{Force}}{\rho V^2 l^2} \text{ depends only on Re.} \tag{4}$$

Since $\rho V^2 l^2$ has the same dimensions as Force, this quantity, Force/$(\rho V^2 l^2)$, is a further non-dimensional quantity. Rayleigh (1910) cites an earlier study (Rayleigh, 1899) in which he has used the idea of plotting experimental results for non-dimensional force against Re. Moreover, a paper by Zahm (1904) includes an appendix by Lord Rayleigh concerning the conditions for dynamic similarity. Not only is the form of the relation (4) given there but the appendix includes an even earlier reference (Rayleigh, 1892) in which similar ideas are proposed. Thus it seems a little invidious for Lanchester (see Kingsford (1960)) to claim

---

* Osborne Reynolds was appointed to the newly-established chair in Engineering in the University of Manchester in 1868. He established an undergraduate degree course in Engineering, one of the first to be offered in this country, which has developed into the degree programmes provided by the Manchester School of Engineering. Readers may also be interested to know that Reynolds's original apparatus is alive and well here in Manchester, and is frequently demonstrated to students and visitors.

sole discovery of the law of dynamic similarity for viscous flows. Nonetheless, Lanchester (1907) emerges as one of the earliest enunciators of it, although it is clear that he has not yet grasped fully the idea of expressing all forces in the form of non-dimensional quantities. As to the non-dimensional quantity $\rho Vl/\mu$ (see relation (1)), according to von Karman (1954) it was in 1908 that the German theoretical physicist Arnold Sommerfeld first named this quantity the Reynolds number.

Having dealt with laminar viscous motion and obtained the $F \propto V^{3/2}$ relation for skin friction (relation (2)), Lanchester (1907) then considers the effects of turbulence. For higher flow velocities (strictly, higher values of Re), he argues that the effects of turbulence should become dominant and that the skin friction should be greater than the laminar flow value. For turbulent flow, then, he suggests that the skin friction relation should take the form $F \propto V^m$ (m greater than 3/2, the laminar index, but no greater than 2). As evidence for this he cites the experimental results obtained in water by Beaufoy (1834) (m between 1.7 and 1.8) and Froude (1874) (mean value of m = 1.92). For practical application in the high Reynolds number problems of flight, however, Lanchester (1907) takes the view that a relation of the form

$$F \propto V^2 \tag{5}$$

might well be adequate. This he takes as representing what he terms "direct resistance", as distinct from the "aerodynamic resistance" which is the subject of the next section. On the basis of the V^2 relation (5) he defines the non-dimensional skin friction coefficient, ξ, as the ratio of skin friction to the resistance experienced by a plate held perpendicularly to a fluid stream. Lanchester (1907) recognises that the value of ξ decreases as the plate size and the flow velocity increase (in effect, ξ decreases with increasing Re). The values of ξ which Lanchester (1907) obtains from his small scale glider experiments are, however, rather inaccurate.

Lanchester returned to the subject of skin friction on a number of occasions (Lanchester, 1910, 1913a, 1915a, 1916, 1937), providing better estimates for ξ in the light of his own data and those of other investigators. Interestingly, Lanchester (1910) is rather wary of the empirical relation given by Zahm (1904) and obtained from wind tunnel experiments. Lanchester's criticism of the Zahm relation is, in effect, that the powers of l and V are not consistent with the general form of the relation (4). In his last major discussion of the topic, Lanchester (1937) shows a graph of the non-dimensional drag force plotted against Re in which he has included the various experimental data collected over the years. He notes that additional factors such as boundary layer transition and the various thickness-chord ratios of the bodies tested are the probable causes of the wide diversity shown in the data. Nonetheless, he is able to demonstrate that skin friction can now be estimated with reasonable accuracy.

In his discussions on the effects of viscosity, Lanchester (1907) reflects, no doubt unknowingly, a number of points central to the earlier seminal paper by Prandtl (1904). Thus, in one highly perceptive remark Lanchester (1907) suggests that, at the instant at which a body begins to move through a viscous fluid at rest, an irrotational flow field is generated. Thereafter, however, other flow regimes can develop due to viscous action. From his later comments it is clear that in this he includes what he calls the detachment of surfaces of discontinuity (as we would say, boundary layer separation), which thereby changes the whole character of the flow field. This Lanchester (1907) links to the resistance experienced:

"In all probability also the V^2 law, in cases involving other than pure skin-friction, is closely associated with the phenomenon of *discontinuity*. A system of flow of the discontinuous type is almost certainly accompanied by resistance following the V^2 law."

Lanchester (1907) is much concerned to explain the cause of discontinuous motion, and here again he demonstrates a remarkable feel for the problem, linking it to the behaviour of what he calls "the inert layer", or boundary layer (my explanatory additions in brackets):

"Now the surface of the body possesses regions of greater and regions of less pressure, and this inert layer will be steadily pushed along the surface from the regions of greater pressure to those of less. Therefore, taking the case of a normal plane (a flat plate held perpendicularly to the approaching stream), the surface current of fluid so formed will be available to "inflate" the surfaces of hydrodynamic flow (the inviscid, irrotational flow field exterior to the boundary layer) in the region of the edges, almost as if the edges of the plane were emitting fluid by volatilisation. This inflation of the surfaces of flow in regions of least pressure can be conceived to continue until the combined inflated region becomes one whole, the "dead water" (the slower moving wake), occupying the space in the rear of the plane. Similarly for other forms of body."

This might indicate that Lanchester (1907) is suggesting the not unreasonable criterion that boundary layer separation is located near pressure minima. However, in his earlier discussion of a photograph (Fig. 8) he has obtained by smoke visualisation of a circular cylinder flow, Lanchester (1907) remarks merely that the

"... discontinuity arises from a line some distance in front of the plane of maximum section."

Having emphasised the drag penalties of discontinuous motion, Lanchester (1907) then suggests the following perceptive definition of streamline form:

"A streamline body is one that in its motion through a fluid does not give rise to a surface of discontinuity."

In this context he investigates a variety of fish shapes, around which the flow field is, as he puts it, "comformable", or, as we would say, boundary layer separation occurs only at the tail of the body. Yet, in certain circumstances for such shapes, he recognises that conformability may not be maintained. Thus, in discussing the flow about the arched plate sections which form the basis of his wing theory, Lanchester (1907) shows three perceptive illustrations (Fig. 9) of non-conformability. In (a) the section is at slightly too great an incidence so that, although upper surface leading edge separation occurs, reattachment is envisaged aft of that edge. In (b) the incidence is too great for reattachment to be possible, whereas in (c) the section is at too low an incidence and undersurface leading edge separation is predicted.

3.2 *Lanchester's Wing Theory*

According to Kingsford (1960), Lanchester's active interest in flight began in 1891. Lanchester (1907) himself tells us that he

"... first formulated his theory in 1892, the basis being the study of the special case of an aeroplane of infinite lateral breadth."

Nomenclature being far from settled at that time, here Lanchester means a wing of infinite span.

Lanchester's public announcement of his interest in mechanical flight seems to have been the lecture which he gave before the Birmingham Natural History and Philosophical Society on 19th June 1894. From a summary of that lecture, Kingsford (1960) tells us that

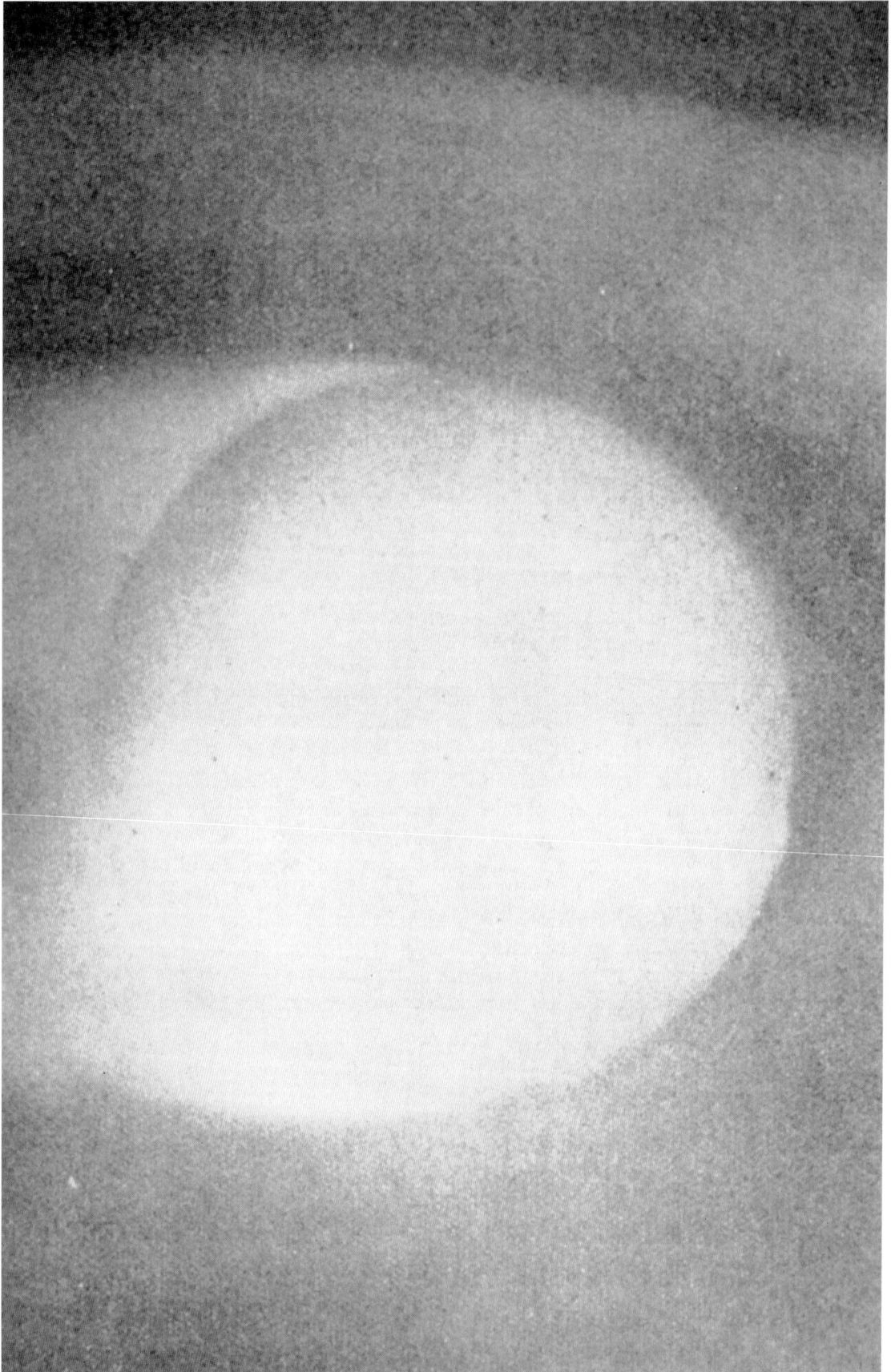

Fig. 8. Photograph of the flow about a circular cylinder (Lanchester, 1907)

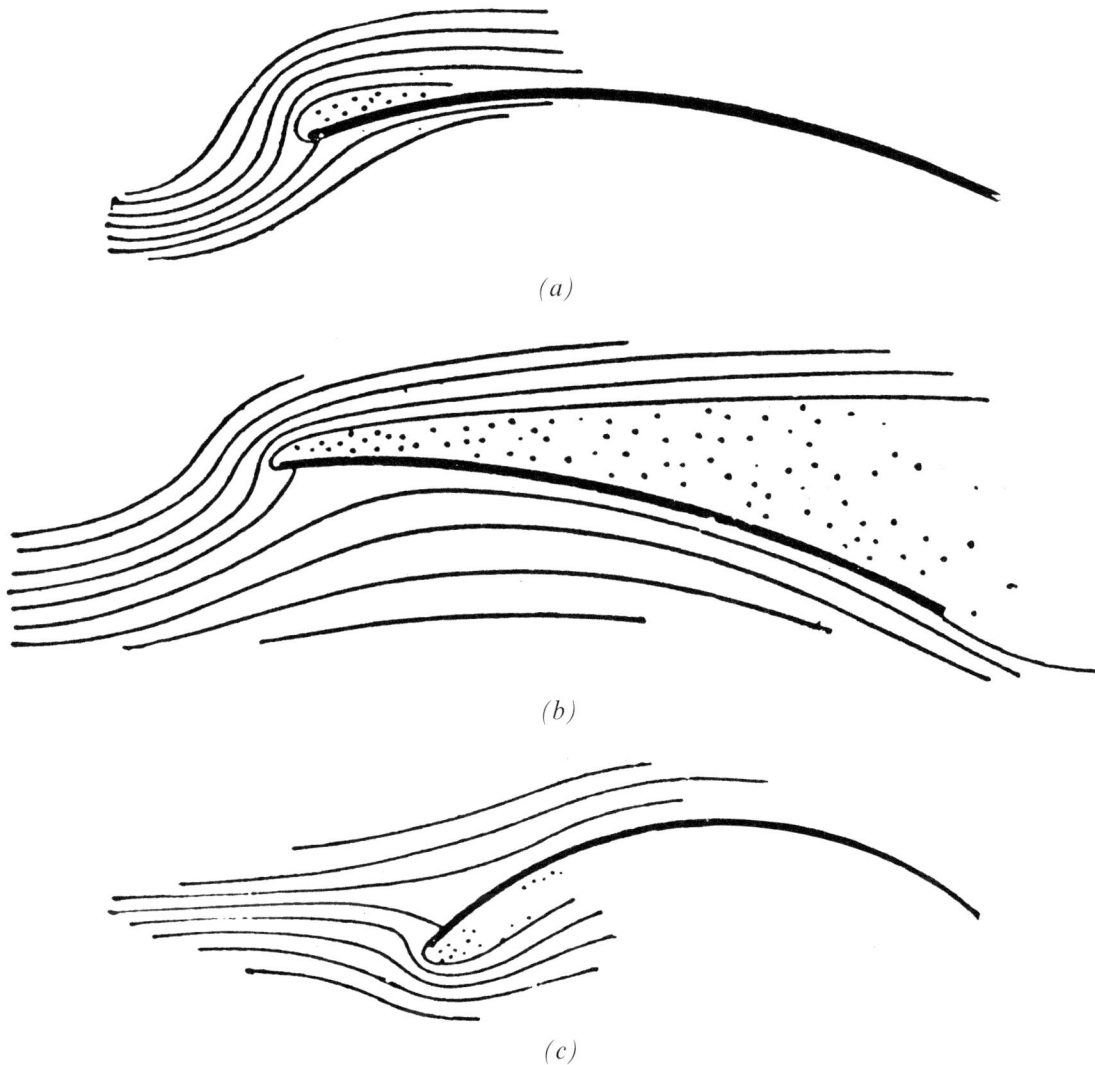

Fig. 9. Sketches of flows about an arched plate (Lanchester, 1907)

Lanchester covered three topics: his lift theory as it then stood (together with a description of the soaring of birds), a review of the aerodynamic experiments of Langley (1891), and a discussion of the power requirements for flight. Lanchester (1907) mentions that, in the years 1894–5, he had incorporated in his lift theory his ideas concerning the flow in the tip regions of wings of finite span. As to subsequent events, he explains that

"A more complete account of this work formed the subject matter of a paper offered to the Physical Society of London, but rejected (3rd September 1897)."

Lanchester (1907) then adds that, in the present work,

"... the main argument and demonstration are taken without substantial alteration from the rejected paper, the subsequent work being a revision of the theory on more orthodox hydrodynamic lines."

This revision, as we shall see, introduced a new idea which was of great significance.

Regrettably, the texts of both the Birmingham lecture and the rejected paper have been lost. Indeed, we have it on the authority of von Karman (1958), who obtained the information from Laurence Pritchard of the Royal Aeronautical Society, that, on moving house, Lanchester invariably made a bonfire of manuscripts and letters. Nonetheless, on the basis of the above quotations and information given by Lanchester (1926) regarding the figures used in the Birmingham lecture, it is possible to resurrect what I would suggest are the essential elements of Lanchester's argument for finite and infinite span wings as they stood in the period 1894–7.

The argument begins by asking us to consider a horizontal plate dropped vertically in still air (Fig. 10). The air must circulate around the plate edges so as to form what Lanchester calls the "vortex fringes" shown. Now superimpose on the plate a forward motion. The vortices are still present at the lateral edges of the plate, but we now obtain a situation in which the air incident at the plate's leading edge is in a state of upward motion relative to the plate, whereas, aft of the plate, the air is in downward motion. The lifting force on the plate is then not merely the reactive consequence of causing this downward motion of the air, but the reaction is further enhanced by the plate's destruction of the air's incident upward motion.

This correct, and recognisably modern, physical description of events thus appears to have emerged right at the beginning of Lanchester's investigation, and for the first time in history. As to the existence of the "vortex fringes", or trailing vortices, Lanchester (1907) recommends that this can be demonstrated experimentally

"... by the employment of an aerofoil under water and inverted ... the vortices being evidenced by the dimples in the surface of the water."

Thus Lanchester (1907) regards the flow field

"... as in the main consisting of two parallel cylindrical vortices ... which are being continually formed at the flank extremities, whose energy is continually dissipated in the wake of the advancing aerofoil."

Consequently, in the act of sustaining itself, the wing must inevitably create the swirling kinetic energy of these flanking vortices. Therefore, in creating continually this kinetic

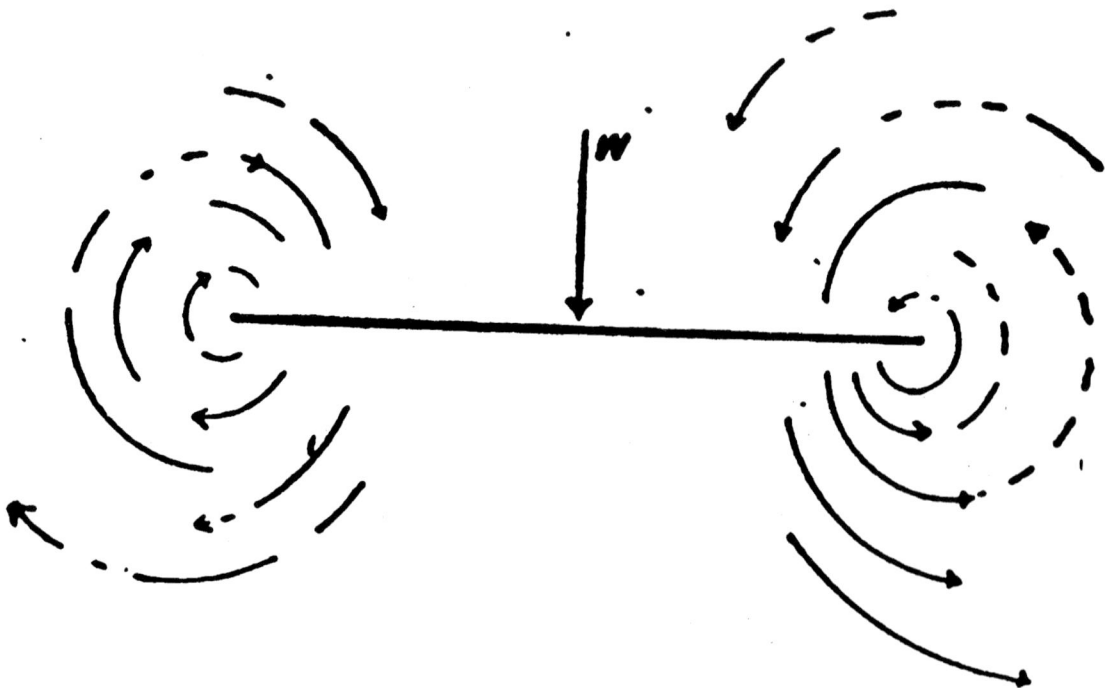

Fig. 10. Sketch of flow about a dropped plate (Lanchester, 1907)

energy, the wing must perform work. And to do work requires that motion must take place against a restraining force, as Lanchester (1907) puts it, "aerodynamic resistance", or, as we would say, induced drag.

As yet, however, our modern view of events is not quite complete. Because the kinetic energy of the flanking vortices is continually being dissipated by viscous action in the wake of the wing (Lanchester, 1907),

> "From another point of view, this loss of energy may be looked upon as a gradual spreading out and dissipation of the wave on the crest of which the aerofoil rides, and it becomes necessary that the aerofoil should constantly renew the diminished wave energy in order to maintain sufficient amplitude and support the given load."

Thus, at this stage in the development of the argument, the wing itself is not yet seen as creating its own lifting bound vortex, but is conceived as riding along on what Lanchester calls a "supporting wave". Lanchester (1897) repeats this conception and later (Lanchester, 1926) he confirms that this was his initial view. At this stage, then, the wing is seen to be rather like a surfboard riding the crest of a wave, yet a crest created and shaped by the shape of the board, or wing, itself.

As to wings of infinite span, Lanchester's argument is that the vortex fringes will be absent whilst the flow is subjected to a force field which is symmetric fore and aft. Thus energy is conserved and there will be no aerodynamic resistance. The wing, however, is again seen as riding on its supporting wave (Fig. 11), whilst, according to energy conservation arguments, flow streamlines must return to their initial levels and the air must return to its initial velocity. Because of this wave motion in the air, as to the best shape of wing section (Lanchester, 1907):

> "It would appear that any appropriate smoothly curved form, whose leading and trailing angles are conformable to the lines of flow, might be regarded as fulfilling the necessary conditions, the essential feature evidently being that neither edge shall give rise to a surface of discontinuity."

Hence, as Lanchester (1907) points out, the efficacy in lift generation of the "dipping front edge" sections devised and investigated by Phillips (1884, 1891) and also by Lilienthal

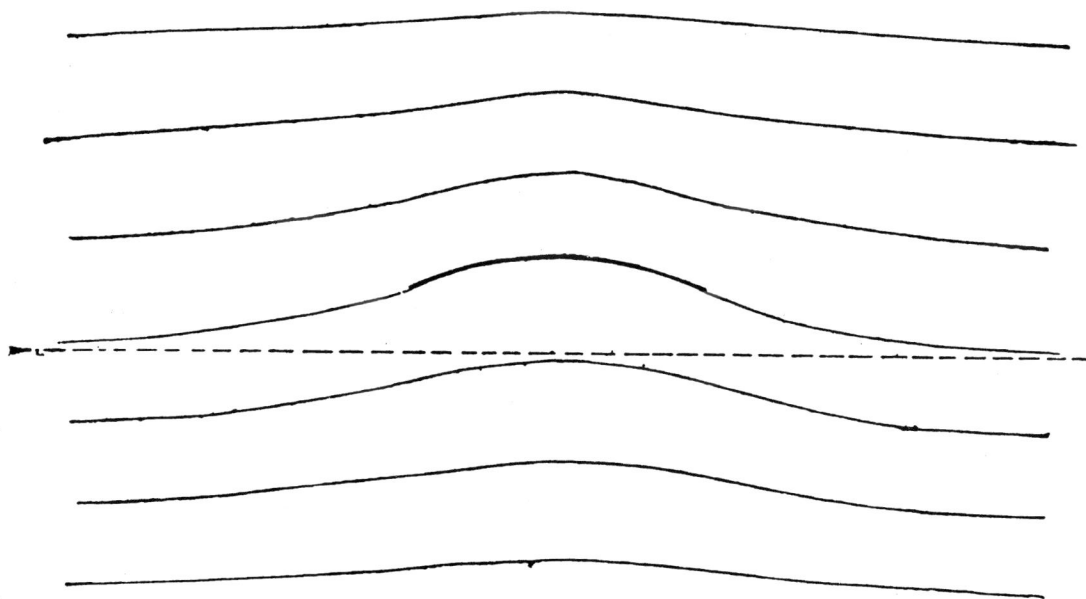

Fig. 11. Sketch of flow about a wing of infinite span at zero incidence (Lanchester, 1907)

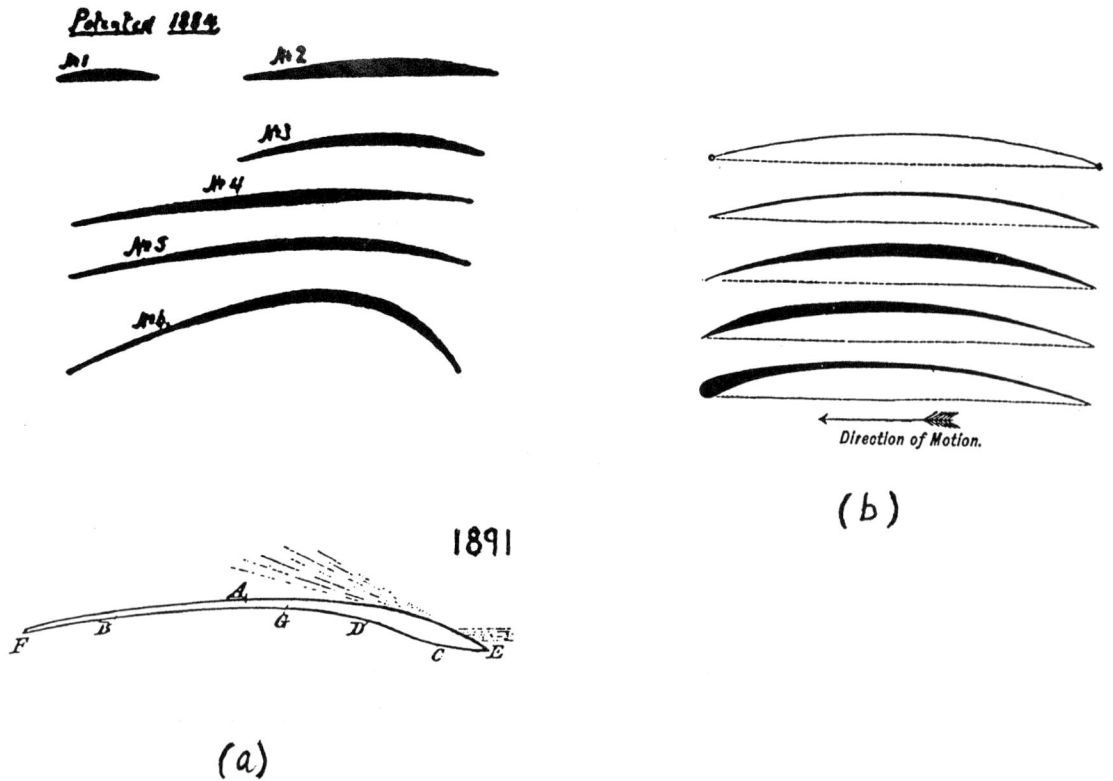

Fig. 12. Aerofoil sections investigated by (a) Phillips (1884, 1891) and (b) Lilienthal (1889)

(1889) (Fig. 12). However, as Lanchester (1907) correctly remarks, Phillips's explanation of the interaction between the dipping front edge and the incident air is entirely fallacious. Lanchester (1907) adds that, as early as 1894, using the arguments given above, he has constructed a wing of the section shown in Fig. 13. This wing, moreover, had an elliptic planform (Fig. 14) with the relatively high aspect ratio of 13. When tested later at the National Physical Laboratory (Lanchester, 1913b), a three-quarter size scale model of this wing produced the highest lift to drag ratio then known, a value of 17.1 at 4° incidence. Note, however, that whilst Lanchester uses a section shape of generally streamlined form, the leading edge has the sharp contour adopted since antiquity in the belief that this edge must cut the impinging air flow. As to his choice of planform geometry, Lanchester (1907) provides a rather speculative but perceptive argument for this on the basis that, from the

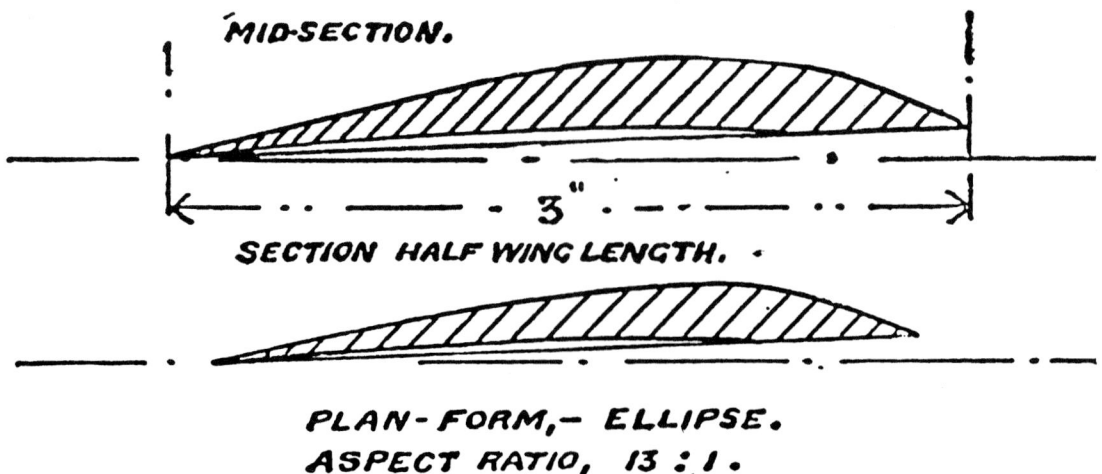

Fig. 13. Sections of Lanchester wing (1894) (Lanchester, 1907)

Fig. 14. Lanchester with rubber-powered model (1894)

point of view of reducing aerodynamic resistance, it is better not to have the kinetic energy of the trailing vortices concentrated at the tips but to spread this out along the span of the wing by choosing a wing planform which tapers from root to tip. He ends with the comment that bird wings have

"... ordinates that approximate more or less closely to those of an ellipse."

Presumably, the planform of the wing of 1894 was obtained on this basis.

As to the skeletal circular arc of Lanchester's section of infinite span depicted at zero incidence in Fig. 11, the flow field for this case was first analysed by Kutta (1902) using irrotational, inviscid flow theory along the lines depicted in Fig. 3. In his solution a vortex, bound to the section, is introduced so as to obtain smooth flow separation at the trailing edge, what we now call the Kutta-Zhukovskii *condition*, which determines the vortex circulation. Moreover, a version of what is now called the Kutta-Zhukovskii *theorem* is used so as to link the vortex circulation to the lift generated. A few years later, Zhukovskii (1906, 1907) independently obtained this same theorem. More details of the emergence of such analytical methods developed during the period of Lanchester's own work are given by Ackroyd (1992). It seems highly likely that Lanchester was unaware of these developments. And subsequent rapid developments in the theory of the wing of infinite span went considerably beyond what Lanchester had been able to achieve. Both Kutta (1910) and Zhukovskii's associate, Chaplygin (1911), derive the lift result for the circular arc wing when set at incidences other than zero. Chaplygin (1911) notes that, at such incidences, the flow velocity becomes infinite at the sharp leading edge and suggests the blunting of the leading edge as the panacea for this mathematical difficulty. In practice, as Lanchester (1907) understands, boundary layer separation at the leading edge is highly likely (see Fig. 9). A blunting of the leading edge cures this practical difficulty, although Lanchester never took this step. However, perhaps prompted by Chaplygin's remark, Zhukovskii (1910, 1912a) produced the theory of the thick wing section which has a blunt leading edge. Experimental confirmation of the lift predicted for these Zhukovskii sections came from Zhukovskii (1912b) himself and from Betz (1915) at Göttingen.

At some point, probably around the time of the rejection of the Physical Society paper and perhaps at the suggestion of that Society, Lanchester began a revision of his theory, as he says, "on more orthodox hydrodynamic lines". By this he means a revision which now incorporates some of the essential ideas developed in the mathematical theory of irrotational, inviscid fluid flows, a subject having a long and involved history.

As its name suggests, hydrodynamics had begun as the theoretical study of water flows. Daniel Bernoulli (1738), the Swiss mathematician who gave the subject its name, was much concerned with the flow of water in pipes. In this, his approach was based on the "vis viva" ("live force", or kinetic energy) principle of Newton's great German contemporary, Gottfried Wilhelm Leibniz. From this emerged some of the first steps toward the so-called Bernoulli equation linking pressure and fluid kinetic energy (or flow speed). Two further advances were introduced by Daniel's father, Johann Bernoulli (1743). The first was to envisage the fluid in a pipe flow as being divided into thin cross-sectional slices, an essential step toward the generalised idea of fluid elements. The second advance was to analyse the motion of an arbitrary slice using the dynamic principles of Newton (1687) which require that a change in motion must be caused by a force. In this case, the force was seen as being due to pressure imbalance across the slice. However, at this stage, the pressure concept was incompletely understood. A little later, the French mathematician Jean d'Alembert (1752) introduced the idea that a fluid flow can be considered as a field of continuous variations in speed and pressure. To assist in this, he also introduced the concept of flow streamlines. However, it was Daniel's friend and Johann's former pupil, Leonhard Euler, who brought together these various key ideas in hydrodynamics so as to obtain, for the first time, the correct and general equations of motion for a fluid acted on solely by internal pressure changes (Euler, 1752, 1755). From this, he went on to produce some of the simpler solutions to these equations, particularly for flows possessing no vorticity. Moreover, it was Euler who first produced the Bernoulli equation in its modern form. And whilst d'Alembert is often credited with the surprising yet correct result that a body exposed to a stream of inviscid, irrotational fluid will experience zero drag force (the so-called d'Alembert Paradox), it turns out that Euler had arrived at the same conclusion some years earlier (see Truesdell (1954) for a splendidly detailed account of this fascinating period in which this first great field theory of mathematical physics emerged). During the nineteenth century, hydrodynamics became much involved with the study of inviscid, irrotational flow fields about bodies of infinite span and constant sectional shape, solution methods advancing apace. For example, Stokes (1842) not only introduced more sophisticated methods but also extended the analysis so as to deal with bodies of circular cross-section. These methods were employed by Rankine (1864) in his theoretical study of irrotational flow about blunt oval shapes which he believed might serve in the design of ship hulls. Related ideas were used by Prandtl in his earliest work at Göttingen on airship envelope shapes, a particularly German preoccupation at that time. It was found by experiment (Fuhrmann, 1911) that, when the envelope shape was well-streamlined so as to ensure good boundary layer behaviour (or Lanchester's "conformability"), the measured pressure variations around the envelope agreed very closely with the predictions of hydrodynamic theory. Moreover, the measured drag force was small and could be conceived as being zero (the d'Alembert result of hydrodynamics) plus a small contribution accountable to the presence of the boundary layer. Apart from this contemporary Göttingen work, however, it was this formidably mathematical and extensive body of hydrodynamic theory which Lanchester set himself the task of absorbing.

A major consequence of this is that Chapter III of Lanchester's *Aerodynamics* (Lanchester, 1907) is largely devoted to a simple, generally lucid explanation of irrotational flow theory. There are one or two surprises, however. The Bernoulli equation is neither obtained nor stated, although a discussion of the link between pressure and kinetic energy serves in its place. More importantly, a number of other rather crucial mathematical results, although alluded to, are not sufficiently connected to his views on the crucial role of viscosity. These include the vortex theorems of Helmholtz (1858), one of which demands that a vortex cannot end, the result that circulation is preserved in an inviscid fluid (Kelvin, 1869), and

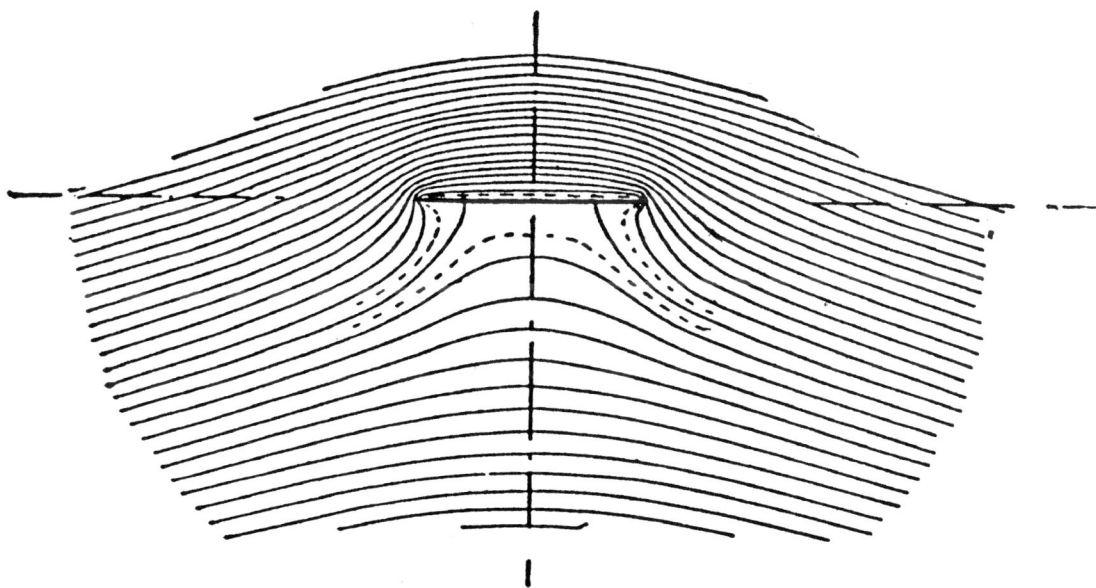

Fig. 15. Flow field about a flat plate, including vortex, calculated by Lanchester (1907)

the Thomson-Stokes theorem* which links circulation to the vorticity within the circulation circuit. Lanchester (1907) merely states that fluid rotation

"... is a quantity that in a perfect fluid can undergo no change. Conservation of rotation is an absolute law in an inviscid fluid."

Considering what follows, it may be that the fuller implications of the Helmholtz, Stokes and Kelvin theorems were not then a part of Lanchester's repertoire.

The main concern in the description of irrotational flows which Lanchester (1907) gives is to explain that lift is generated by the addition of a vortex, or "cyclic system", as he calls it, to such flows about a body. He does not obtain anything equivalent to the Kutta-Zhukovskii circulation/lift theorem and is able to calculate only certain flow fields which

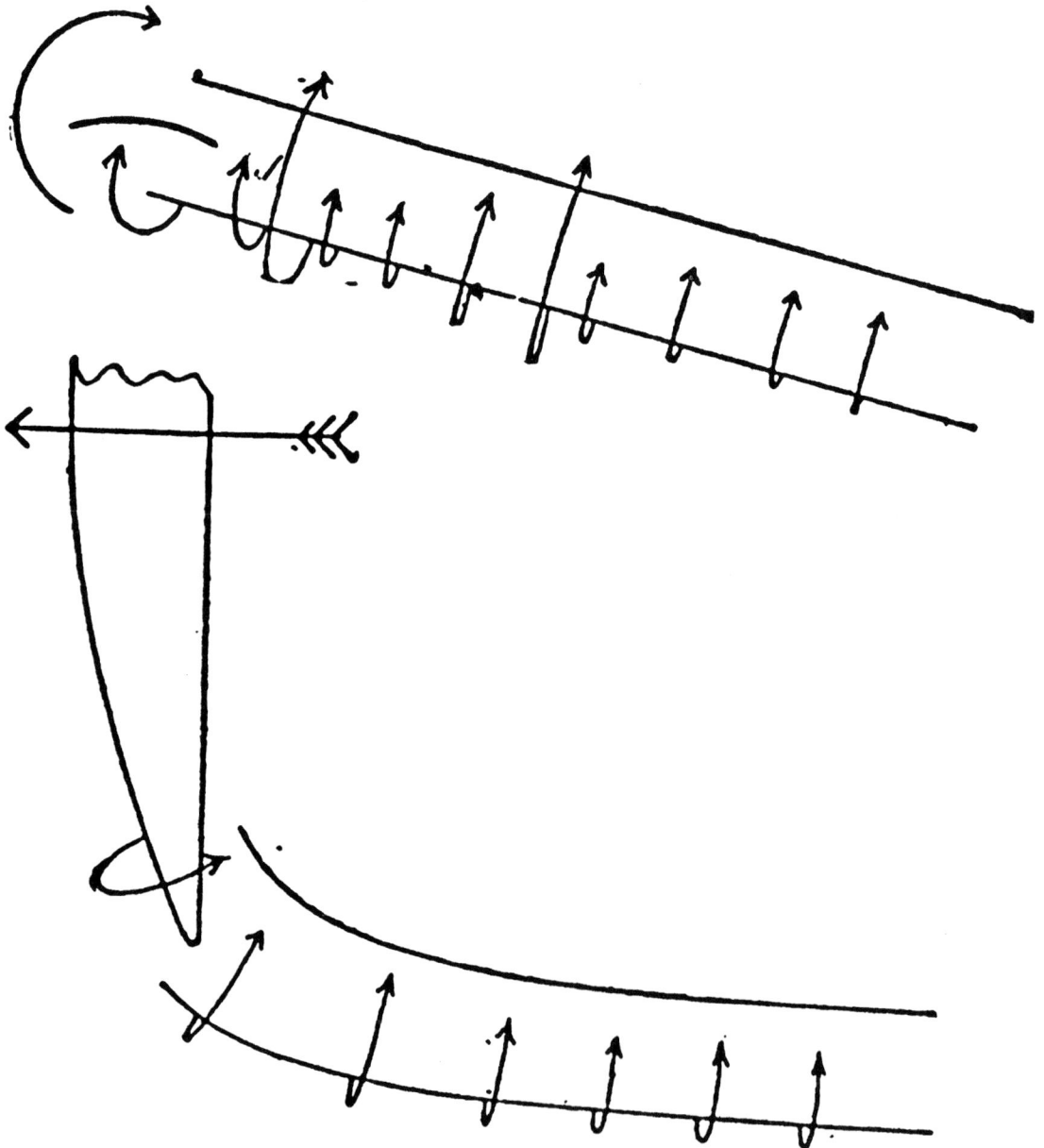

Fig. 16. Sketches of sideview (upper) and planview (lower) of wing flow (Lanchester, 1907)

* Apparently in 1850 William Thomson (later, Lord Kelvin) communicated the theorem to Stokes, who set the theorem for the Smith's Prize Examination for 1854 at Cambridge.

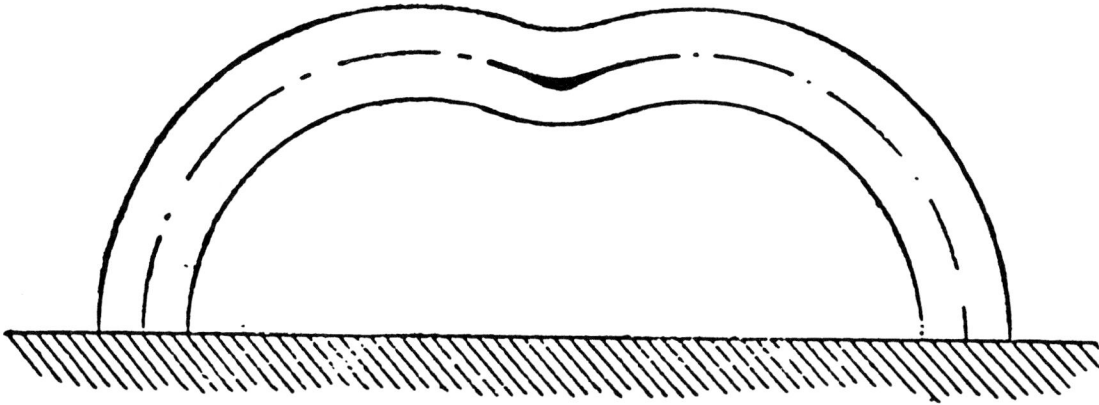

Fig. 17. Sketch of wing vortex system seen from in front (Lanchester, 1907)

include a vortex (Fig. 15). Clearly, these do not satisfy his requirement of "conformability". This he acknowledges, thereby revealing a feel for the Kutta-Zhukovskii condition of smooth flow at the trailing edge. Nonetheless, the upflow and downflow exhibited in these examples clearly fit his earlier picture of a "supporting wave". Thereafter in his *Aerodynamics* he replaces that picture with one in which he sees a wing as carrying with it a bound, lifting vortex (Fig. 16). This, then, is the major step prompted by his recent study of classical hydrodynamic theory. However, at this stage it is not entirely clear that he sees the strength, or circulation, of the bound vortex as being equal to the circulations of the trailing vortices. Perhaps his limited view of the Helmholtz and Kelvin theorems, mentioned earlier, deterred him from realising that link, although the following comment suggests that the link might have been made (Lanchester, 1907):

"... we would obtain, for an inviscid atmosphere, a system consisting primarily of a

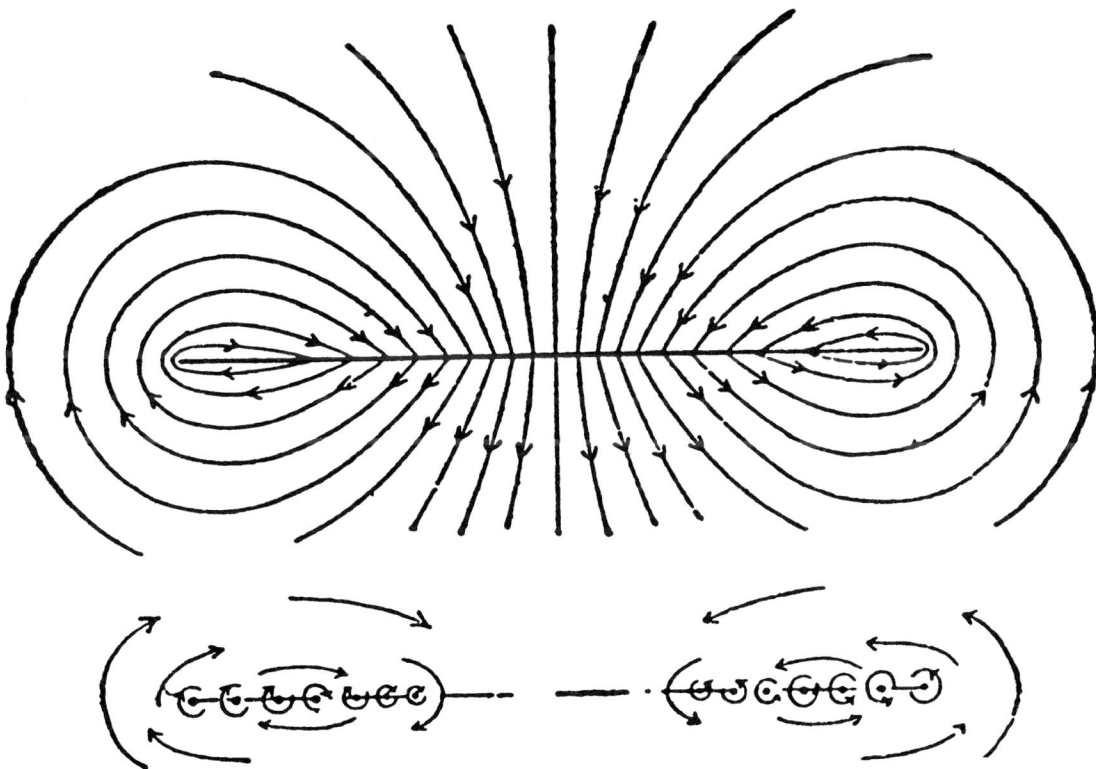

Fig. 18. Sketch of wing vortices, seen from rear (Lanchester, 1907)

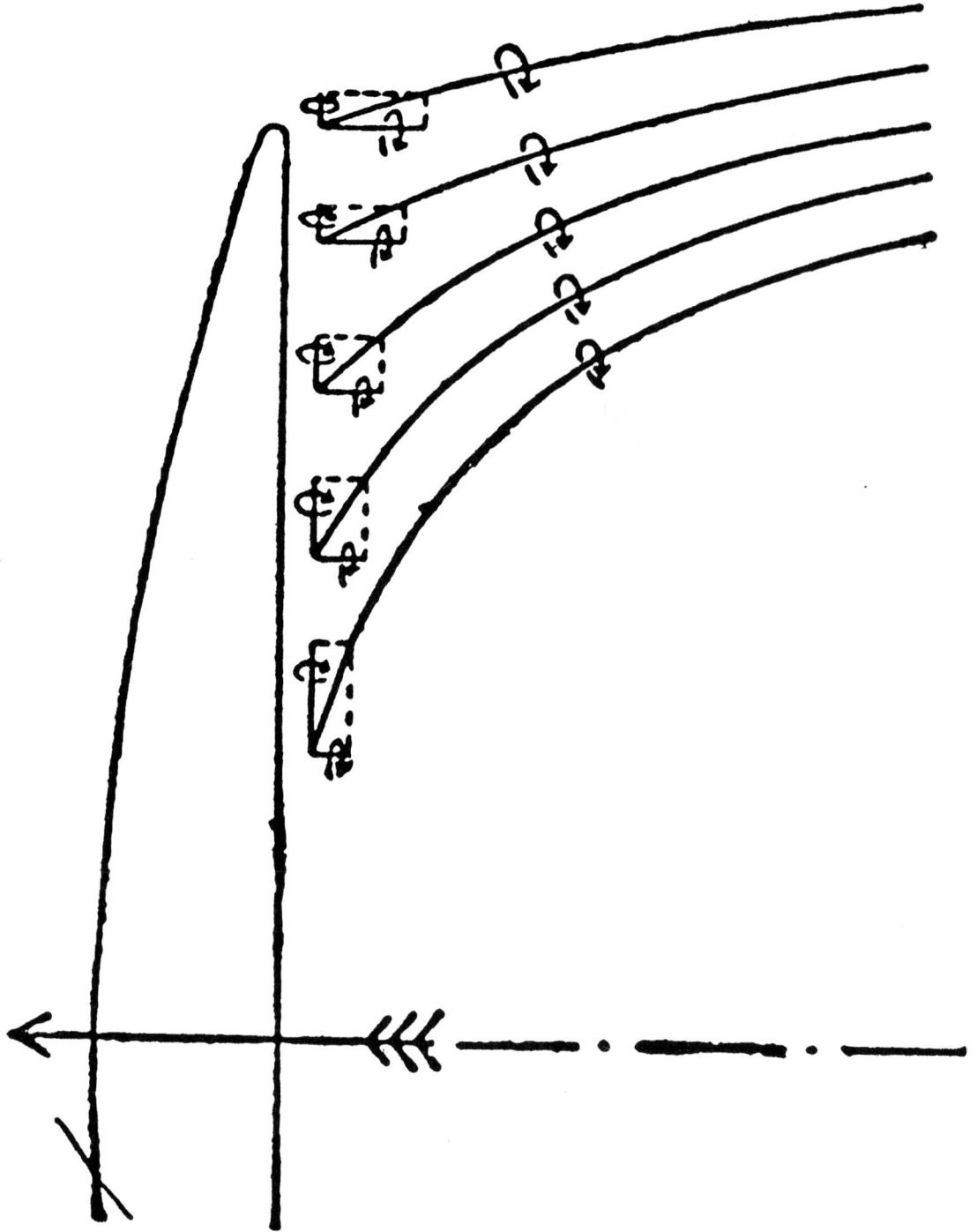

Fig. 19. Sketch of wing and trailing vortices (Lanchester, 1907)

vortex hoop or half ring, loaded in the centre by the aerofoil (Fig. 17), and whose energy will be perfectly conserved, the aerofoil and its supporting vortex lying in a plane at right angles to the direction of flight. Such a system in a fluid that is truly inviscid would be uncreatable and indestructible, just as in such a fluid a vortex ring is uncreatable and indestructible."

As implied here, Lanchester's concern in much of the discussion is to overcome objections raised by the theorem of Lagrange (1781) concerning the inability of an inviscid fluid to create vorticity. His method throughout is to speculate on the nature of flow fields which

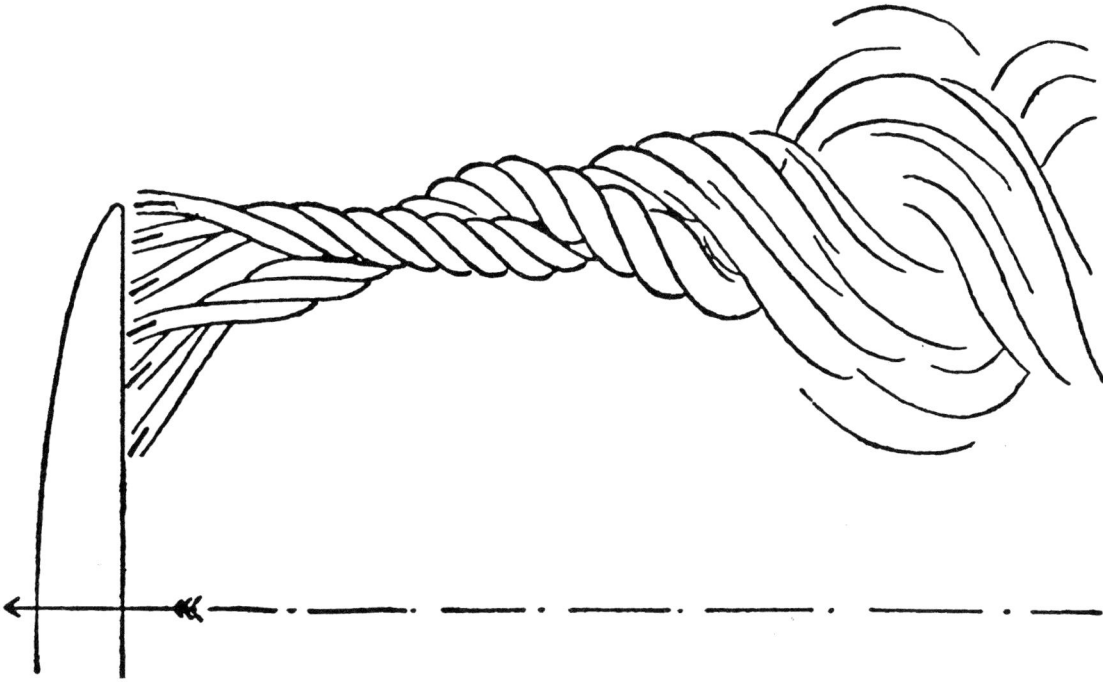

Fig. 20. Sketch of coiling of trailing vortices (Lanchester, 1907)

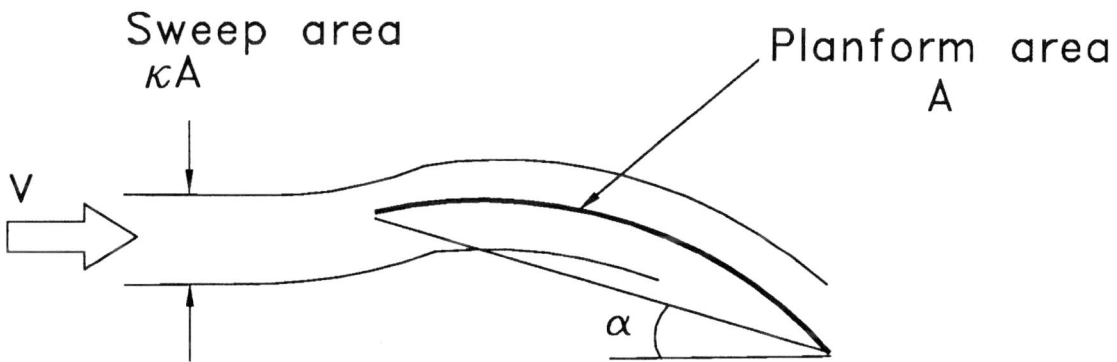

Fig. 21. Lanchester's conception of "sweep area"

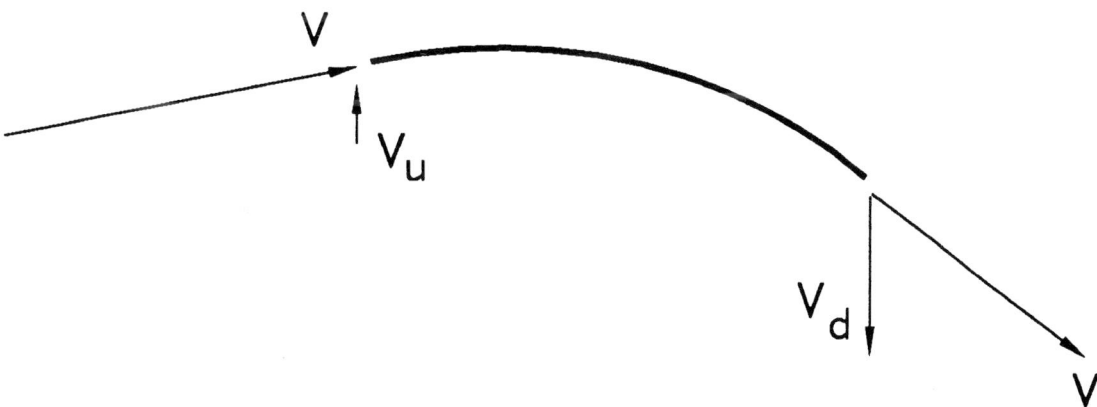

Fig. 22. Lanchester's conception of "conformable" flow about a wing section

are slightly viscous and then to argue, by extrapolation, to the case of an inviscid fluid. Here, once again, his approach parallels that of Prandtl (1904).

As to a more detailed picture of the flow field he is proposing, Lanchester (1907) suggests that, when viewed from the rear, a wing's streamline pattern will look like that shown in the upper diagram of Fig. 18. The flow at the trailing edge (lower diagram) is seen as a Helmholtz vorticity sheet. By the addition of flow speeds, this can be combined with the vortices of the wing itself so as to produce the plan view of the vortex filaments shown in Fig. 19. Thereafter, vortex filaments will interact, coiling around each other in the manner shown in the frequently-reproduced Fig. 20.

It may be that Lanchester's grasp of the concept of a bound vortex as the cause of wing lift generation pre-dates the use made by Kutta (1902) of this concept. Nonetheless, it must be emphasised that, in his calculation procedure for the prediction of wing performance, Lanchester (1907) makes virtually no use of the concept. Indeed, at two points only does Lanchester (1907) appeal, one crucially, the other peripherally, to the model of the flow field he has adopted. The crucial step is to apply elementary principles of dynamics to his realisation that, whilst downflow occurs aft of a wing, upflow is created ahead of it. Here his earlier "supporting wave" view would serve equally well. Indeed, it may well be the case that the calculation procedure given by Lanchester (1907) is, in its essentials, that used as early as 1894 for the design of his elliptic wing. The peripheral step is to elaborate an argument concerning the balance between the upward and downward momenta of the air passing over a wing and the corresponding momenta passing through the tip regions. Thus, whereas in the period 1894-1907 the model of the flow field has been developed, principally by the incorporation of a bound vortex, the calculation procedure may not have changed.

Lanchester (1907) begins his analysis by reviewing an older and well-known method developed from one of the two views of fluid flow adopted by Newton (1687): that of a "rare medium" composed of discrete particles which interact with a body solely by collision. Although physically unrealistic in this context, this approach has one major advantage over hydrodynamic theory. In contrast to d'Alembert's Paradox, it predicts a finite drag force. As to lift prediction, however, for small angles of incidence the Newtonian model results in the lift force being proportional to the wing's (incidence angle)^2. The debate as to the correctness of this conclusion rumbled on until the later years of the nineteenth century. By Lanchester's time it was largely accepted as an experimental fact that lift is not proportional to (incidence angle)^2 but rather to the incidence angle itself. The gradual emergence of this conclusion is described by Ackroyd (1992). To the non-mathematician it may seem surprising that the technical debate as to whether or not human flight was possible did, in fact, hinge on this crucial question concerning this apparently small difference in the two forms of the lift relation with incidence. Suffice it to say that, at small incidences, this difference is by no means small and the Newtonian result produces a serious underprediction of lift.

For wing flows, the device which Lanchester (1907) adopts so as to deal with the difficulties inherent in the Newtonian method is to assume that, whereas a definable streamtube of fluid is affected by the wing, this streamtube has a cross-sectional area which is independent of incidence (Fig. 21). This cross-sectional area Lanchester (1907) refers to as the "sweep area" of the wing. Lanchester (1907) notes that the experiments of Langley (1891) on superposed parallel plates set at various distances apart suggest that κ (see Fig. 21) might be near unity in value. Lanchester, himself, suggests that κ might lie between 1 and 2.

In order to model the interaction between this streamtube and a wing, Lanchester (1907) uses an adaptation of the method developed for propeller theory by Rankine (1865, 1867), W.Froude (1878) and his son R.E.Froude (1889). The method employs mass and momentum flux arguments which, in the latter respect, Lanchester (1907) tells us has been dubbed "the doctrine of the continuous communication of momentum". As to the interaction itself, Lanchester (1907) assumes that the air meets the wing leading edge "conformably" (see

Fig. 22). However, the rather crude assumption is made that the airflow meets the leading edge with its initial free stream velocity V, then subsequently leaves the trailing edge in like manner. Consequently, according to this mathematical model, the wing removes the upward component of V, V_u, and then imparts a downward component of V, V_d. Application of these assumptions results in an equation for the lift force, L. For our purposes, Lanchester's actual result can be re-written more conveniently as

$$L = \rho V^2 A a \alpha \tag{6}$$

Here again, ρ is the air density, whereas A is the wing planform area and a is the incidence angle of the wing. All of these dependencies given in equation (6) are correct; in particular, you notice that lift force increases as the square of the airspeed, V. As to the quantity a which incorporates κ above, Lanchester (1907) argues that this depends on the geometry of the wing's cambered arch and aspect ratio.

In order to obtain an expression for "aerodynamic resistance", or induced drag D_i, Lanchester (1907) uses the principle that the rate at which work is done by D_i must be equal to the rate of change of flow kinetic energy. From this, an equation for D_i is obtained directly. Again, rather than quote Lanchester's actual result, I will give the equivalent but rather more useful form,

$$D_i = \frac{1}{\rho A} b \frac{L^2}{V^2} \tag{7}$$

Like a in equation (6), Lanchester (1907) argues that the quantity b also depends on wing camber and aspect ratio. Equation (7) is similar in form to what became known as

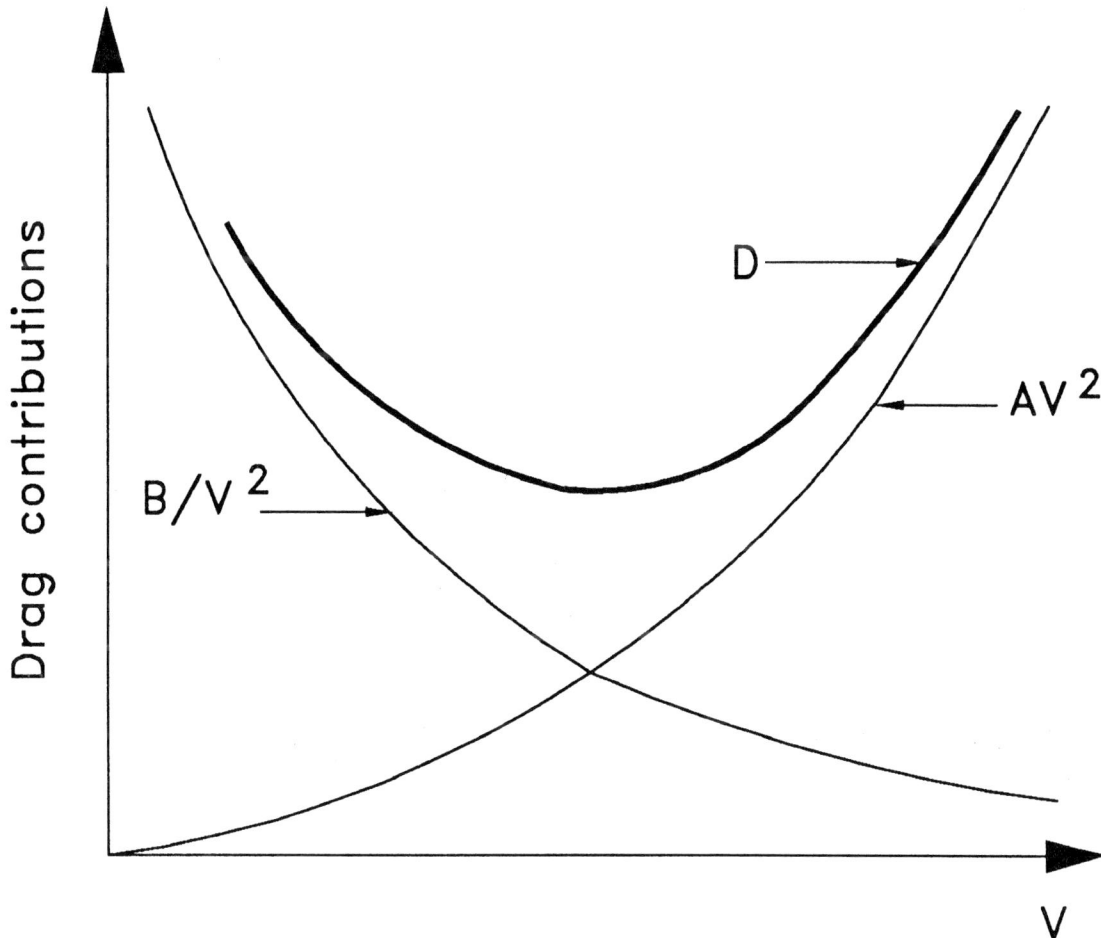

Fig. 23. Drag contributions plotted against an aeroplane's forward speed, V

"Langley's Law" (Langley, 1891), which states that, the faster one flies, the smaller is the total drag. Certainly, equation (7) states that, as V increases, D_i must be reduced but here we must remember that we are dealing only with induced drag, not the total drag. In fact, Lanchester (1907) takes issue with "Langley's Law" here, pointing out that Langley (1891) had concluded, erroneously, that skin friction is negligibly small. Lanchester's resolution of this we will come to presently.

The relations equivalent to equations (6) and (7) which Lanchester (1907) obtains for lift and induced drag have the correct forms. Their accuracy depends, of course, on Lanchester's ability to estimate correctly the quantities a and b, both of which, he recognises, depend on wing aspect ratio and camber. His arguments to obtain realistic values for these quantities, based, it must be emphasised, largely on flat plate data, are described in detail by Ackroyd (1992).

Equations (6) and (7) are the key results which can be obtained from the *Aerodynamics* of 1907. Comparisons between these and the results provided by the accurate lift theory of Prandtl and his Göttingen co-workers (Prandtl, 1918, 1919, 1921), developed predominantly during the period of the First World War, are given by Ackroyd (1992). Suffice it to say here that, whilst predicting the correct trend in the variation of lift with wing aspect ratio, Lanchester's results consistently underpredict lift by some 25%. The comparison is rather better when it comes to the prediction of induced drag, however, Lanchester's results being in almost precise agreement with the Göttingen work at a wing aspect ratio of 6, the difference widening to some 20% around an aspect ratio of 12.

In retrospect it is easy to criticise Lanchester's method: the over-simplistic mathematical model used and the similar status of the control volume employed in his momentum flux arguments, from which no great accuracy can be guaranteed; his enforced reliance on flat plate data; his use of an identity which he had no reason to suppose was valid (see Ackroyd (1992)), and so on. Nonetheless, despite all these deficiencies in his approach, Lanchester's results are seen to be remarkable for their time.

One final result from the *Aerodynamics* of 1907 must be mentioned, and this follows from Lanchester's realisation that Langley (1891) is in serious error in his neglect of skin friction. Using his result for "direct resistance", relation (5), in conjunction with equation (7) for induced drag, Lanchester (1907) states that the total drag, D, must be the sum of these two drag forces:

$$D = AV^2 + B/V^2 \tag{8}$$

Thus it is here that this basic relation for aeroplane drag makes its appearance for the first time in history. For an aeroplane in steady level flight at constant altitude, the quantities A and B both turn out to be constants. In this case, the effect of the summation given in equation (8) can be illustrated in simple graphical form. In Fig. 23 the two terms AV^2 and B/V^2 are shown plotted against the aeroplane's speed, V. Whereas AV^2 increases rapidly with increasing V, the reverse is the case for the term B/V^2 ("Langley's Law"). Summation of these two graphs produces the third curve marked D, representing the total drag. Clearly, this curve must possess a minimum value; in other words, an aeroplane must possess a flight speed at which its total drag has a minimum value. This Lanchester (1907) demonstrates by the use of calculus, showing that, at this minimum drag speed V_1, the skin friction and induced drags are equal to each other. At speeds significantly greater than V_1, however, skin friction drag dominates the effect of "Langley's Law". This illustrates the error made by Langley (1891) in neglecting skin friction. This was the point at issue, incidentally, in that argument between Lanchester and Wilbur Wright which I mentioned in my introduction; clearly, Lanchester was right and Wilbur, who supported Langley's assertion, was wrong. However, to return to the analysis of flight performance, Lanchester (1907) shows by similar methods that an aeroplane will also possess a further speed, V_2 ($= V_1/3^{1/4}$), at which its engine power is a minimum, the induced drag being now three times the skin friction drag. Going on to consider gliding flight in still air, he then obtains the now-accepted expression for minimum gliding angle. All this is astonishing when one takes account of the fact that, by that time, the powered aeroplane was generally perceived as having barely left the ground, and that the successful gliding pioneers, Lilienthal, Pilcher, Augustus Herring, the Wrights, could be counted almost on the fingers of one hand.

As far as Lanchester's aerodynamic theory is concerned, there the matter rested for a number of years. In 1908 Lanchester visited Göttingen, there to discuss with Carl Runge the details of a German translation of the *Aerodynamics*. There, too, Lanchester met Prandtl. Runge, whose mother was English and who spoke English fluently, acted as interpreter in their conversations. Prandtl (1927) describes his exposure to the *Aerodynamics* and his own contributions with careful honesty (my explanatory addition in brackets):

"The necessary ideas upon which to build up (the lift) theory ... had already occurred to me before I saw the book ... we in Germany were better able to understand Lanchester's book when it appeared than you in England. English scientific men, indeed, have been reproached for the fact that they paid no attention to the theories expounded by their own countryman, whereas the Germans studied them closely and derived considerable benefit there from. The truth of the matter, however, is that Lanchester's treatment is difficult to follow, since it makes a very great demand on the reader's intuitive perceptions, and only because we had been working on similar lines were we able to grasp Lanchester's meaning at once. At the same time, however, I wish it to be distinctly under-stood that in many particular respects Lanchester worked on different lines than we did, lines which were new to us, and that we were therefore able to draw many useful ideas from his book."

Certainly the British found some parts of the *Aerodynamics* difficult to swallow, as the press reviews cited by Kingsford (1960) attest. Indeed, one method adopted in dealing with Lanchester's wing theory was not to mention it at all. Otherwise, however, the reviews of *Aerodynamics* appear to have been welcoming of the work, constructive and reasonably fair. No doubt because of this, Lanchester became appointed to the British Advisory Committee for Aeronautics at its formation in 1909 under the presidency of Lord Rayleigh. Nonetheless, Prandtl (1927) is also correct in his statement that the British aeronautical establishment paid no attention to Lanchester's wing theory. As late as 1916, Lanchester (1937) recalls, a member of that Committee stated that "we do not believe in your theory". Having seen that theory as it stood in 1907, it is possible to appreciate that, at a purely technical level, people might well have been wary of it. But why not subject it to the test of experiment? Moreover, by, say, 1912, why not try to relate it to the more rigorous work emerging from Kutta, Zhukovskii and Prandtl himself? By then the Committee was publishing generally accurate and informative reviews of some of this work. Surely it must have been evident that a new, interesting, and perhaps seminal aerodynamic theory was emerging, a theory, moreover, which supported Lanchester's ideas? But this did not happen. Why this did not happen is far from clear. Perhaps part of the problem lay in the personalities involved. Lanchester, at times, could not restrain himself from being pointedly and publicly critical of some of the ideas held at government aeronautical establishments. At a more personal level, Lanchester had a deep distrust, to say the least, of Cambridge and London graduates, as the survey of Lanchester's correspondence given by Baxter (1966) reveals. And, again as Baxter's review suggests, this antipathy was perhaps mutual, and long-lived. Baxter (1966) cites an astonishing letter from Lanchester, dated as late as 1941, in which he quotes an unnamed correspondent who had written for advice but had then requested that this be kept quiet since

"... the action of asking your advice would be as fatal to my career as immorality or Communism."

An example of Lanchester's public criticism is contained in his next major paper on aerodynamics. Here Lanchester (1915b) complains quite justifiably about

"... the present-day neglect of theory ..., the lines of reasoning commonly followed ... are at fault, especially, for example, when it is sought to assign independent values to the upper and lower containing surfaces of the foil, or when, again, the question of aspect ratio is studied apart from that of camber."

In the discussion following the paper, this elicits a rather icy response from Leonard Bairstow, then in charge of aerodynamic testing at the National Physical Laboratory. However, it is in this paper that Lanchester introduces a significantly new approach to his aerodynamic theory. Here, at last, Lanchester (1915b) makes it clear that the circulation of a wing's bound vortex must be equal to that of the trailing vortices. Having illustrated his idea (Fig. 24), he concludes the discussion of this aspect of his revised theory with the remark that

"... the author regards the two trailed vortices as a definite proof of the existence of a cyclic component of equal strength in the motion surrounding the aerofoil itself."

Yet, whereas Prandtl had immediately, and much earlier, seen this as crucial to his mathematical modelling of the flow, even now Lanchester (1915b) makes little use of this aspect of the more precise and detailed physical model he has developed. Nonetheless, the new analytical approach which Lanchester (1915b) adopts has the advantage of removing his earlier reliance on flat plate experimental data.

His first step is to relate the "sweep area" of a wing to the effective area of the trailing vortices, what Lanchester (1915b) calls the "peripteral area". The approach here is hardly new, however, since Lanchester (1907) had introduced it in his discussion of propeller theory. Yet one rather crucial point in the argument given there is vague, whereas in

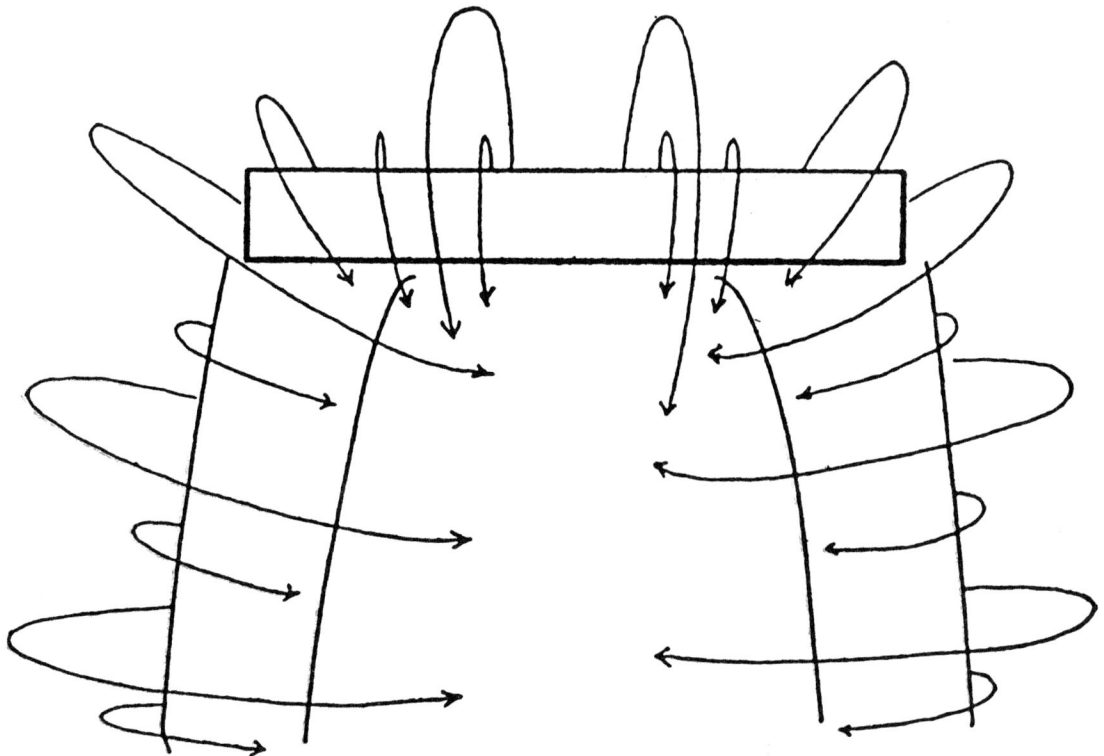

Fig. 24. Bound and trailing vortices (Lanchester, 1915b)

Lanchester (1915b) the complete argument is put with greater clarity. Even so, the physical principles involved are difficult to follow. Having written down general relations for the rates at which fluid momentum and kinetic energy flow through this peripteral area, Lanchester (1915b) sets these equal to the corresponding relations obtained from his earlier "sweep area" method. The flow through the sweep area is thus assumed to pass into the peripteral area. Lanchester's argument is then that the air, in passing over a wing, is pushed uniformly downward. This situation is analogous, he argues, to that in which a plate moves uniformly through still air. Here, you will note, Lanchester (1915b) has returned to the first step (Fig. 10) in his original explanation of wing flow. Appealing to the hydrodynamic theory given by Lamb (1879), Lanchester uses the result for the kinetic energy of the flow field which develops about this moving plate. Bringing all these steps together, Lanchester is able to produce a new version of his earlier expression for induced drag, equation (7), which is, as it turned out, identical to that produced by Prandtl (1918, 1919). Had he cared to do so, Lanchester (1915b) could have taken his lift expression a step further so as to produce a result in close agreement with Prandtl's work. Instead of this, however, Lanchester (1915b) continues to employ his earlier rather cumbersome method.

Ironically, by this time Prandtl had also arrived independently at this same relation for induced drag by considering the case of the lift distribution on a wing of elliptic planform shape. Later (see Prandtl (1919)) this shape emerged from Munk's analysis as that giving minimum induced drag.

Thereafter, Lanchester himself did not develop further his basic analysis. On two occasions he returned to review it (Lanchester, 1926, 1937), and on the last occasion gave a number of calculated examples demonstrating the accuracy of his theory in comparison with experimental results for wings. He was, however, solely concerned with drag, taking experimental results for lift so as to demonstrate by use of an equation similar to equation (7) that these, in combination with realistic values for skin friction, gave accurate results for total drag. By then, of course, much of Lanchester's thinking on aerodynamics had become superseded by the methods developed by Prandtl and his co-workers, which offered greater flexibility of application.

Sadly, Lanchester became rather touchy, I suspect, on the subject of Prandtl's successes. In an extraordinary exchange between himself and Gustav Victor Lachmann (a Göttingen doctoral graduate who then joined the Handley-Page company) during the discussion of this later review, Lanchester (1937) remarks that

"... he was still a little in the dark as to what was referred to when mention was made of the Prandtl theory."

Lachmann replies that

"He was referring to the mathematical edifice which had been built by Prandtl and his school on the conceptions which Dr. Lanchester had first put forward."

A little defensively, perhaps, Lanchester responds with the remark that

"His own results had been obtained by simpler means."

Lachmann then presses his point:

"The expressions used in practice were based on the mathematical formulae developed by Prandtl."

Lanchester then closes the exchange by remarking that

"He had drawn a blank once more."

It is astonishing that Lanchester should display such a lack of awareness of Prandtl's theory. Being the great individualist he was, perhaps Lanchester could not bring himself to study the matter.

We do not complain that Elgar is not Beethoven. Equally, I believe, we should not complain that Lanchester is not Prandtl. Life, it seems to me, teaches us that genius should be revered wherever we may find it. Lanchester's position is now secure as the originator of much of the vortex theory of wing lift and induced drag, although it has to be admitted that, as a mathematical modeller, Lanchester does to rank in the Prandtl class. Doubtless this can be attributed, at least in part, to deficiencies in the training of British engineers, deficiencies not present in the Continental system in which sophisticated mathematical modelling and advanced techniques of mathematical analysis had long been taught. Over our particular period of interest, however, other than Prandtl, Kutta and Zhukovskii I can find no other persons displaying Lanchester's insights in aerodynamics. Nonetheless, it is evident that during that period, and for long afterward, "official science" in Britain failed to recognise the essential merits of Lanchester's contributions to aerodynamics. Equally it has to be admitted that foreign theoretical work in aerodynamics made little impression in Britain. During the crucial early years of practical aeronautics, 1900 to 1912, German work, in particular, began to take the lead in aeronautical science. Reviews of some of this work were appended to the official reports of the British Advisory Committee for Aeronautics, yet, accurate and informative though these summaries were, they stimulated little in the activities undertaken on behalf of that Committee. Justification for Lanchester's early ideas in aerodynamics was then left to Prandtl and his colleagues at Göttingen. Eventually, this German work received due recognition in Britain, primarily as a result of a visit made by the Farnborough scientist, Hermann Glauert, to Göttingen in 1919 and the subsequent publicity he gave to that work through his extensions of it. And from that, together with additional pressure from people such as A.R.Low through the Royal Aeronautical Society, Lanchester's earlier work came to be recognised. Nonetheless, it is a matter of deep regret and considerable shame that Britain's full recognition of Lanchester's extraordinary achievements in aerodynamics took so long.

In reviewing his contributions to wing theory, Lanchester (1926) expresses the view that the allocation of credit to the various contributors to that theory is a task for the technical historians of the future, and that he is content to leave the matter in their hands. Lanchester's words have been very much in my mind, and on my conscience, in attempting this survey. Unfortunately, despite its length, this survey can give an assessment in only certain areas of Lanchester's multifarious contributions to aerodynamics. Vast as the corpus of Lanchester's writings is, the technical historian is faced with a daunting task in assessing its many facets. Unreservedly, my hope is that, in this small way, Lanchester might feel that I have done him justice.

References

Ackroyd, J.A.D. (1984): *The Science of aviation – a history* (video series), Manchester University Television Productions

Ackroyd, J.A.D. (1989): Sydney Goldstein, F.R.S., Hon.F.R.Ae.S. – An Appreciation and Bibliography, *Aerospace*, **16,** No. 6, 26–30

Ackroyd, J.A.D. (1992): Lanchester, the Man (The 31st Lanchester Lecture), *Aeron. Jour*, **96,** 119–40

Baxter, E.G. (1966): *Catalogue of the private papers of F.W.Lanchester in the Library of Lanchester College of Technology, Coventry*, Thesis submitted for Fellowship of the Library Association

Beaufoy, M. (1834): *Nautical and hydraulic experiments, with numerous scientific miscellanies*, London

Bernoulli, D. (1738): *Hydrodynamica, sive de viribus et motibus fluidorum commentari*, Strasbourg

Bernoulli, J. (1743): Hydraulica nunc primum detecta ac demonstrata directe ex fundamentis pure mechanicis; Anno 1732, *Opera omnia IV*, 387–488, Lausanne-Geneva

Betz, A. (1915): Untersuchung einer Joukowskyschen Tragfläche, *Zeit. für Flugtechnik und Motorluftschiffahrt*, **6,** 173–9

Blasius, P.R.H. (1908): Grenzschichten in Flüssigkeiten mit kleiner Reibung, *Zeit. für Mathematik und Physik*, **56**, 1–37

Chaplygin, S.A. (1911): On the pressure exercised by a parallel two-dimensional current on a submerged body, (In Russian) *Mathematical Collections of Moscow*, **28**, 120–66

D'Alembert, J. le R. (1752): *Essai d'une nouvelle théorie de la résistance des fluides*, Paris

Euler, L. (1752): Principles of the motion of fluids, (see Truesdell, C.A. (1954))

Euler, L. (1755): Sequel to the researches on the motions of fluids, (see Truesdell, C.A. (1954))

Fourier, J.B.J. (1822): *Théorie analytique de la chaleur*, Paris

Froude, R.E. (1889): On the part played in propulsion by differences of fluid pressure, *Trans. Naval Architects*, **30**, 390–405

Froude, W. (1874): Report to the Lords Commissioners of the Admiralty on Experiments for the Determination of the Frictional Resistance of Water on a Surface, under Various Conditions, Performed at Chelston Cross, under the Authority of Their Lordships, 44th Report of the British Association for the Advancement of Science, 249–55

Froude, W. (1878): On the elementary relation between pitch, slip, and propulsive efficiency, *Trans. Naval Architects*, **19**, 47–57

Fuhrmann, G. (1911): Widerstands und Druckmessungen Ballonmodellen, *Zeit. für Flugtechnik und Motorluftschiffahrt*, **2**, 165–6

Gibbs-Smith, C.H. (1970): *Aviation: An historical survey*, H.M.S.O., London

Goldstein, S. (Ed) (1938): *Modern developments in fluid dynamics*, Vol. II, Oxford Univ. Press

Helmholtz, H.L.F. von (1858): Über Integrale der hydrodynamischen Gleichungen, welche den Wirbelbewegung entsprechen, *J. für die Reine und Angew. Math.*, **55**, 25–55. [On the integrals of the hydrodynamic equations, which express vortex motion, (Trans. Tait, P.G.), *Phil. Mag. Ser 4*, **33**, 485–510.]

Karman, T. von (1954): *Aerodynamics*, Cornell Univ. Press.

Karman, T. von (1958): Lanchester's contributions to the theory of flight and operational research, *J.R.Ae. S.*, **62**, 80–93

Kelvin, Lord (Thomson, W.) (1869): On vortex motion, *Trans. Roy. Soc. Edin.*, **25**, 217–60; Papers IV, 13–66

Kingsford, P.W. (1960): *F.W.Lanchester: A life of an engineer*, Edward Arnold, London

Kutta, M.W. (1902): Auftriebskräfte in strömenden Flüssigkeiten, *Illus. Aeronautische Mitteilungen*, **6**, 133–5

Kutta, M.W. (1910): Über eine mit den Grundlagen des Flugproblems in Beziehung stehende zweidimensionale Strömung, *Sitzungsberichte der königlich Bayerischen Akademie der Wissenschaften*, **40**, 1–58

Lagrange, J.-L. de (1781): Mémoire sur la théorie du mouvement des fluides, *Nouv. Mém. De l'Acad. de Sci. de Berlin*, **12**, 151–88

Lamb, H. (1879): *Hydrodynamics*, Cambridge Univ. Press

Lanchester, F.W. (1897): *Improvements in and relating to aerial machines*, Patent Specification No. 3608, H.M.S.O., London

Lanchester, F.W. (1907): *Aerodynamics*, Constable & Co. Ltd., London

Lanchester, F.W. (1910): *Notes on the resistance of planes in normal and tangential presentation and on the resistance of ichthyoid bodies*, Advisory Committee for Aeronautics, R. & M. No. 15, (Part 1)

Lanchester, F.W. (1913a): *Surface cooling and skin friction*, Advisory Committee for Aeronautics, R. & M. No. 94

Lanchester, F.W. (1913b): *Aerofoils of high aspect ratio*, Advisory Committee for Aeronautics, R. & M. No. 109

Lanchester, F.W. (1915a): *A Note on the subject of skin friction*, Advisory Committee for Aeronautics, R. & M. No. 149

Lanchester, F.W. (1915b): The Flying machine: the aerofoil in the light of theory and experiment, *Proc. Inst. Auto. Eng.*, **9**, 171–259

Lanchester, F.W. (1916): *The Flying machine from an engineering standpoint*, Constable & Co. Ltd., London

Lanchester, F.W. (1926): Sustentation in flight, *J.R.Ae.S.*, **30**, 587–606

Lanchester, F.W. (1937): The Part played by skin-friction in aeronautics, *J.R.Ae.S.*, **41**, 68–131, 322–3

Langley, S.P. (1891): *Experiments in aerodynamics*, Smithsonian Institution, Washington

Lilienthal, O. (1889): *Der Vogelflug als Grundlage der Fliegerkunst*, Berlin [*Birdflight as the basis of aviation*, (Trans. Isenthal, A.W.), Longmans Green & Co., London (1911).]

Maxwell, J.C. (1860): Illustrations of the dynamical theory of gases. Part 1. On the motions and collisions of perfectly elastic spheres, *Phil. Mag.*, **19**, 19–32

Maxwell, J.C. (1866): On the viscosity or internal friction of air and other gases, *Phil. Trans. Roy. Soc.*, **156**, 249–68

Navier, L.M.H. (1827): Mémoire sur les lois du mouvement des fluides, *Mémoires de l'Académie Royale des Sciences*, **6**, 389–416

Newton, I. (1687): *Philosophiae naturalis Principia Mathematica*, London [*Mathematical principles of natural philosphy*, Trans. Third Latin Edit. by Motte, A., London (1729).]

Phillips, H.F. (1884): *Blades for deflecting air*, Patent Specification No. 13768, H.M.S.O., London

Phillips, H.F. (1891): *Flying machines*, Patent Specification No. 13311, H.M.S.O., London

Prandtl, L. (1904): Über Flüssigkeitsbewegung bei sehr kleiner Reibung, *Verhandlungen des dritten internationalen Mathematiker-Kongresses, Heidelberg*, 484–91, Leipzig

Prandtl, L. (1918): Tragflügeltheorie, I. Mitteilungen, *Nach. der kgl. Gesellschaft der Wiss. zu Göttingen, Math.-Phys. Klasse*, 451–77

Prandtl, L. (1919): Tragflügeltheorie, II. Mitteilungen, *Nach. der kgl. Gesellschaft der Wiss. zu Göttingen, Math.-Phys. Klasse*, 107–37

Prandtl, L. (1921): *Applications of modern hydrodynamics to aeronautics*, N.A.C.A. Rep. No. 116

Prandtl, L. (1927): The generation of vortices in fluids of small viscosity, *J.R.Ae.S.*, **31**, 718–41

Rankine, W.J.M. (1864): On plane water lines in two dimensions, *Phil. Trans. Roy.Soc.*, **154**, 369–91

Rankine, W.J.M. (1865): On the mechanical principles of the action of propellers, *Trans. Inst. Naval Arch.*, **6**, 13–39

Rankine, W.J.M. (1867): On the theoretical limit of the efficiency of propellers, *The Engineer*, **23**, 25

Rayleigh, Lord (Strutt, J.W.) (1892): On the question of the stability of the flow in fluids, *Phil. Mag.* Ser. 6, **34**, 59–70

Rayleigh, Lord (Strutt, J.W.) (1899): Investigations in capillarity: The size of drops. The liberation of gas from supersaturated solutions. Colliding jets. The tension of contaminated water-surfaces, *Phil. Mag. Ser. 6*, **48**, 321–37

Rayleigh, Lord (Strutt, J.W.) (1910): *Note as to the application of the principle of dynamic similarity*, Advisory Committee for Aeronautics, R. & M. No. 15 (Part 2).

Reynolds, O. (1883): An Experimental investigation of the circumstances which determine whether the motion of water shall be direct or sinuous and of the law of resistance in parallel channels, *Phil. Trans. Roy. Soc. A*, **174**, 933–82

Stokes, G.G. (1842): On the steady motion of incompressible fluids, *Trans. Cam. Phil. Soc*, **7**, 439–55; *Math. and Physical Papers*, **1**, 1–16

Stokes, G.G. (1845): On the theories of the internal friction of fluids in motion, and of the equilibrium and motion of elastic solids, *Trans. Cam. Phil. Soc.*, **8**, 287–305; *Math. and Physical Papers*, **1**, 75–129

Stokes, G.G. (1851): On the effect of the internal friction of fluids on the motion of pendulums, *Trans. Cam. Phil. Soc.*, **9**, 8–106; *Math. and Physical Papers*, **3**, 1–141

Sutton, O.G. (1949): *The Science of flight*, Penguin Books Ltd., Harmondsworth

Sutton, O.G. (1965): *Mastery of the air*, Basic Books Inc., New York

Truesdell, C.A. (1954): *Rational fluid mechanics, 1687–1765, Leonhardi Euleri Opera Omnia, Series II Opera Mechanica*, **12**, Swiss National Science Society, Lausanne

Van Dyke, M. (1964): *Perturbation methods in fluid mechanics*, Academic Press, New York

Zahm, A.F. (1904): Atmospheric friction on even surfaces, *Phil. Mag. Ser. 6*, **8**, 58–66

Zhukovskii, N.E. (1906): De la chute dans l'air de corps légers de forme allongée, animés d'un mouvement rotatoire, *Bulletin de l'Institute Aerodynamique de Koutchino*, **1**, 51–65

Zhukovskii, N.E. (1907): On adjoint vortices, (In Russian), *Trans. Physical Section of the Imperial Society of the Friends of Natural Science, Moscow*, **13**, 12–25

Zhukovskii, N.E. (1910): Über die Konturen der Tragflächen der Drachenflieger, *Zeit. für Flugtechnik und Motorluftschiffahrt*, **1**, 281–4

Zhukovskii, N.E. (1912a): Geometrische Untersuchungen über die Kutta'sche Strömung, *Trans. Physical Section of the Imperial Society of the Friends of Natural Science, Moscow*, **15**, 36–47

Zhukovskii, N.E. (1912b): *The Theoretical bases of aeronautics*, (In Russian), Moscow

Chapter Six

LANCHESTER'S *AERODONETICS*

by Eur. Ing. J.A.D. Ackroyd

Lanchester's *Aerodynamics* (Lanchester, 1907), reviewed in the previous chapter, constitutes the first of his two volume treatise entitled *Aerial flight*. The second volume, *Aerodonetics* (Lanchester, 1908), appeared scarcely before the ink was dry on the *Aerodynamics*. Whereas the *Aerodynamics* had set down Lanchester's conceptions of the boundary layer and the vortex theory of lift, so that the drag and lift forces on a wing could be estimated, the *Aerodonetics* sought to address the next major problem in aeronautics: how to ensure that an aeroplane is stable in flight. Like the *Aerodynamics*, *Aerodonetics* is coloured by Lanchester's highly individualistic and often unusual approach so that his ideas and the arguments which he developed were at the time, and are even now, often difficult to understand. A major difficulty for his contemporaries was that the *Aerodynamics* proposed ideas which were not only new but often highly controversial. Nonetheless, they paralleled, even foreshadowed, near-contemporary advances elsewhere, notably the boundary layer theory of Prandtl (1904), the lift theories of Kutta (1902, 1910) and Zhukovskii (1906, 1910) and the wing theory of Prandtl (1918, 1919). The *Aerodynamics* thereby later gained credence from such more sophisticated advances. The *Aerodonetics*, in contrast, proved to be rather less seminal in its effect. Indeed, apart from one notable contribution which will be the main subject of this section, much of the work presented in the *Aerodonetics* turned out to be inappropriate to the needs of a developing aeronautical science.

As it happened, an alternative approach to the problem of aeroplane stability had emerged a few years earlier in the work of Bryan and Williams (1904) and this was later to be taken to a higher degree of sophistication by Bryan (1911). This approach was, in fact, a fairly straightforward extension of methods already developed to analyse the stability of dynamic systems. Over the early years of the present century, similar methods appeared in other countries and it is from this common approach to aeroplane stability that the present-day method has emerged. Put at its simplest, the method is as follows. The aeroplane is taken initially to be in steady level flight at constant altitude. In these circumstances, the aerodynamic force of lift precisely opposes the aeroplane's weight whilst the aerodynamic force of drag precisely opposes the propulsive thrust of the engine. There is then no net force acting on the aeroplane. Provided that one further condition is fulfilled, the aeroplane continues in its straight line motion at constant speed. This further condition, vitally important, is that there is no net moment of these aerodynamic and thrust* forces acting about the aeroplane's centre of gravity so that there is no tendency for the aeroplane to rotate about its centre of gravity. This condition of zero net moment is referred to as the trim condition for the aeroplane. It is then supposed that the aeroplane experiences a small disturbance caused, for example, by encountering a slight gusting of the wind or by the pilot nudging the control column. The consequence, however, is that the aerodynamic forces of lift and drag no longer precisely oppose the aeroplane's weight and thrust force. More importantly, the moments of these aerodynamic forces are no longer in balance about the

---

* For simplicity in what follows, it will be assumed that the thrust force acts through the centre of gravity and therefore contributes no moment about that point. A similar assumption will also be made about the drag force. The dominant moment considered will then be due to the lift force alone.

aeroplane's centre of gravity and the aeroplane loses its trimmed condition. Under the action of these out of balance forces and moments, the motion of the aeroplane continues to be disturbed. If this disturbed motion grows in amplitude as time progresses then, clearly, this is an unstable situation. Usually this is also deemed an unacceptable situation requiring appropriate re-design of the aeroplane, although certain types of instability are accepted provided that they grow sufficiently slowly for the pilot to have time to control them. However, if the disturbed motion decays in amplitude then this is a stable, and acceptable, situation. The mathematical analysis of these disturbed motions often reveals that the disturbances take the form of sinusoidal oscillations either growing in amplitude (unstable) or decaying in amplitude (stable) as time progresses. Alternatively, the disturbance might grow continually with time (a divergent instability) or it might decay continually (stable); in such cases, no oscillation is involved.

Such disturbed motions often result in changes in the aeroplane's wing incidence angle relative to the local airstream. Clearly, in order to carry out the analysis outlined above, it is necessary to know how the aerodynamic forces and moments vary with these wing incidence changes. Such forces and moments are due predominantly to the distribution of the air pressure around a wing's surface. It is the pressure distribution's variation with incidence angle which causes the variations of aerodynamic forces and moments. Thus it is a deepening partial vacuum on a wing's upper surface which is the main cause of the lift force increasing steadily with increase of incidence, at least in the small incidence range (Fig. 1a). As to the aerodynamic moment, this is a little more complicated and here it is convenient to introduce the concept of the centre of pressure (Fig. 1b), the point on the wing at which the lift force can be conceived as acting. It is now know that, for the reasonably well-streamlined, slightly arched (cambered) wing section shapes coming into

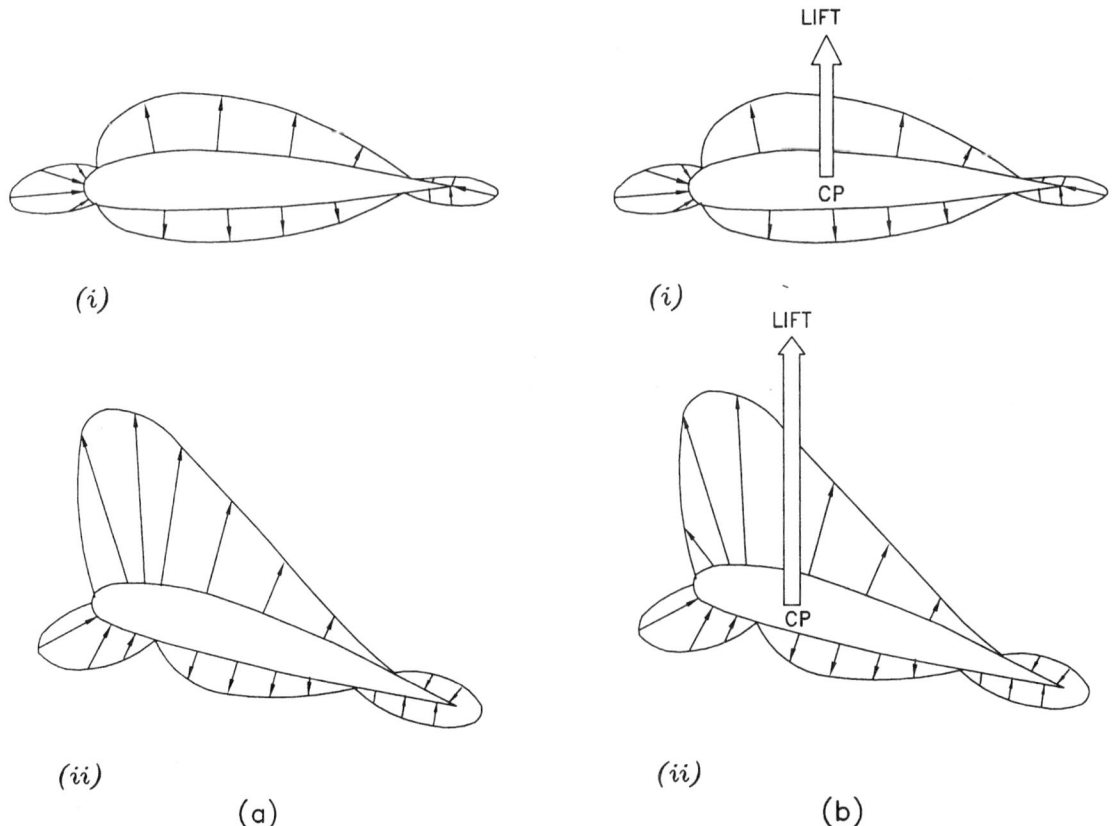

Fig. 1a. The pressure distribution around a wing, (i) at small incidence, (ii) at larger incidence. Inwardly directed arrows indicate overpressure (pressure above ambient pressure), outwardly directed arrows indicate underpressure, or suction, partial vacuum (pressure below ambient pressure).
Fig. 1b. The centre of pressure (CP) on a wing. Note that the centre of pressure moves forward with increasing incidence.

service in Lanchester's era, the centre of pressure position moves forward as wing incidence is increased. It is this behaviour which has a dominant influence on the disposition of an aeroplane's aerodynamic surfaces in order to achieve stability, as we shall see presently. As far as I am aware, this forward movement of the centre of pressure with incidence increase was first established as an experimental fact by the Wright brothers (McFarland, 1953) who, thereafter, kept this to themselves. This behaviour was not known to Bryan and Williams (1904) or to Bryan (1911) in their early developments of aeroplane stability theory. This deficiency detracts somewhat from their findings. This centre of pressure behaviour became recognised as a result of the rapid expansion of the aerodynamic testing of wing section shapes begun in the early stages of the First World War. Moreover, as the lift theories of Kutta and Zhukovskii came to be digested, these were seen to confirm this behaviour. I should add, however, that this could not be predicted from the relatively simple mathematical models adopted by Lanchester in his two major contributions to lift prediction (Lanchester, 1907, 1915).

I have veered into this historical consideration for one reason: the confusion surrounding this issue during our period of interest. What was widely known at the time was that, for flat plates, the centre of pressure moves in the opposite direction, rearward, with incidence increase. It is now known that this is due to the poorly streamlined shape of the flat plate. As explained in the earlier chapter on the *Aerodynamics*, at very small incidences the boundary layer separates from the sharp leading edge, leaving the plate in a permanently stalled condition with a pressure distribution significantly different to that obtained with the better streamlined wing. The flat plate's rearward centre of pressure movement had been established as an experimental fact by, amongst others, Langley (1891). Yet the flat plate had rapidly become displaced as a suitable lifting surface by the slightly cambered surfaces investigated by, for example, Lilienthal (1889). Indeed, it was Lilienthal's work, in particular, which became the major influence in producing a wider recognition of the fact that cambered, rather better streamlined surfaces produce greater lift and less drag (more importantly, a higher lift to drag ratio) than flat plates at the same incidence. However, the crucial point is that it was widely assumed at this time that, for these new cambered surfaces, the centre of pressure movement with incidence change would have the same direction as that established for the flat plate. This erroneous assumption became the source of much subsequent confusion.

To clarify the situation with regard to aeroplane stability, let me now describe a particular example. Fig. 2a shows an aeroplane in steady level flight. The wing lift L_W acts through the centre of pressure (CP) which is located at a distance x_W ahead of the aeroplane's centre of gravity (CG). Although this disposition is not typical of an aeroplane at small incidence, it makes the subsequent argument easier to understand and does not change the main conclusions. The tailplane lift L_T is located at a distance x_T behind CG. There is no net moment acting about CG because the two aerodynamic moments, the clockwise $M_W = L_W x_W$ and the counterclockwise $M_T = L_T x_T$, precisely oppose each other. Now suppose that the aeroplane becomes slightly disturbed, pitching nose up about CG as shown in Fig. 2b. As a result, the position of CP moves forward so that x_W increases, and so does the lift L_W. Thus the clockwise M_W (still equal to $L_W x_W$) increases, thereby attempting to increase the nose-up rotation. This effect is therefore de-stabilising. However, the incidence of the tail has also increased, producing a greater L_T and therefore a greater counterclockwise M_T (still equal to $L_T x_T$). This attempts to rotate the nose downward again, apparently tending to stabilise the aeroplane. The question then arises: which moment dominates? If the wing moment M_W dominates then this will clearly produce an increasingly nose-up rotation, a divergent motion without oscillation which can be unacceptably rapid. On the other hand, if the tail moment M_T dominates then the nose will pitch down again. However, in this the nose may well descend below that position shown for the zero net moment condition of Fig. 2a. Thereafter, the aeroplane could experience a period of rotational oscillation which might increase in amplitude (an unstable situation) or a decaying oscillation

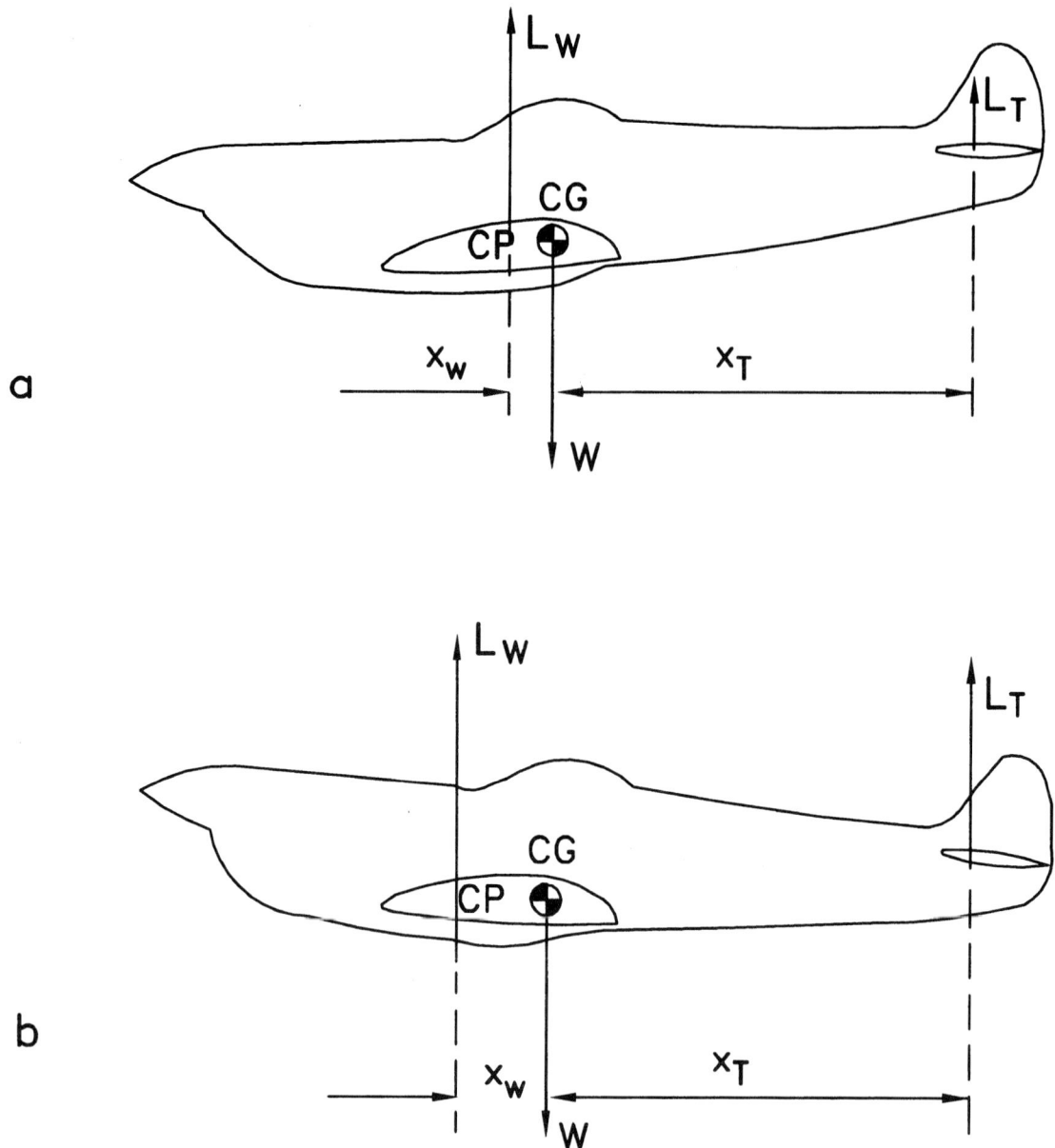

Fig. 2. Aeroplane in steady level flight, (a) undisturbed, (b) after pitching nose-up.

(a stable situation). One immediate conclusion to be drawn from this is that an otherwise conventional straight-winged aeroplane with camber, but without a tailplane*, can never be stable. Admittedly, such an aeroplane can be flown so that its wing's centre of pressure coincides with its centre of gravity. There is then, at this condition, zero net moment about the centre of gravity and the aeroplane is at its trimmed condition. However, the slightest nose-up disturbance will move the centre of pressure forward so that M_W will always be such as to increase the disturbance. The provision of a tailplane, attached at a sufficiently large distance x_T, is then essential to the production of a dominant M_T which gives some hope of stability.

---

* Here I exclude the "flying wing" type of tailless aeroplane which has swept back wings, the wings being slightly twisted from root to tip. In this case, the rearward tips, turned up at their trailing edges, perform the stabilising function of the tailplane. The overall CP location for the complete wing then moves rearward with increasing incidence in a manner similar to that of the flat plate, although such wings do not possess the poor boundary layer behaviour of the plate.

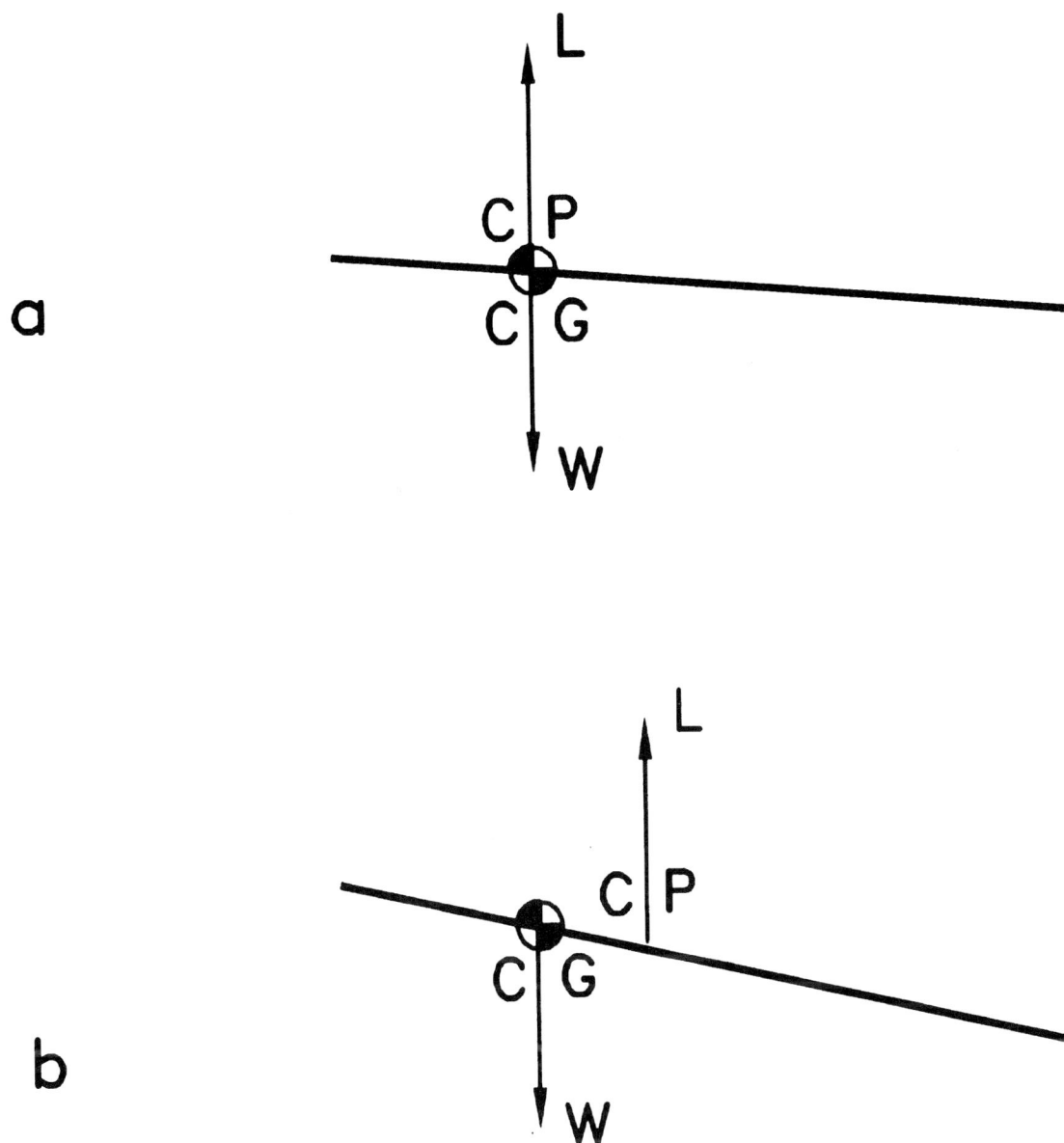

Fig. 3. The flat plate in steady level flight, (a) undisturbed, (b) after pitching nose-up.

The situation for a flat plate at a small incidence angle, however, is significantly different. The plate is shown in Fig. 3a, loaded at the nose so that its centre of gravity (CG) coincides with its centre of pressure (CP). There is then no net moment about CG. When disturbed (Fig. 3b) so as to pitch nose-up, the plate's CP location moves aft, causing a nose-down restoring moment. In fact, the plate alone, whilst being a poor lifting surface, can execute stable motion and no tailplane is required.

To take the aeroplane argument a stage further, imagine that the aeroplane's CG location is moved aft; lead weights, for example, might be placed at the tail (see Fig. 4). Comparing this situation with that of Fig. 2, we see that the tailplane's moment arm x_T has been reduced whereas the wing's moment arm x_W has been increased. Thus the tailplane's ability to produce a restoring moment has been reduced. Indeed, it is possible to identify one particular point, called the neutral point, beyond which any further rearward location of CG will always result in instability. In such instabilities, the aeroplane's incidence continues to increase in an unstable divergent manner. It is a simple matter to calculate the location of the neutral point, which depends solely on the aeroplane's aerodynamic characteristics

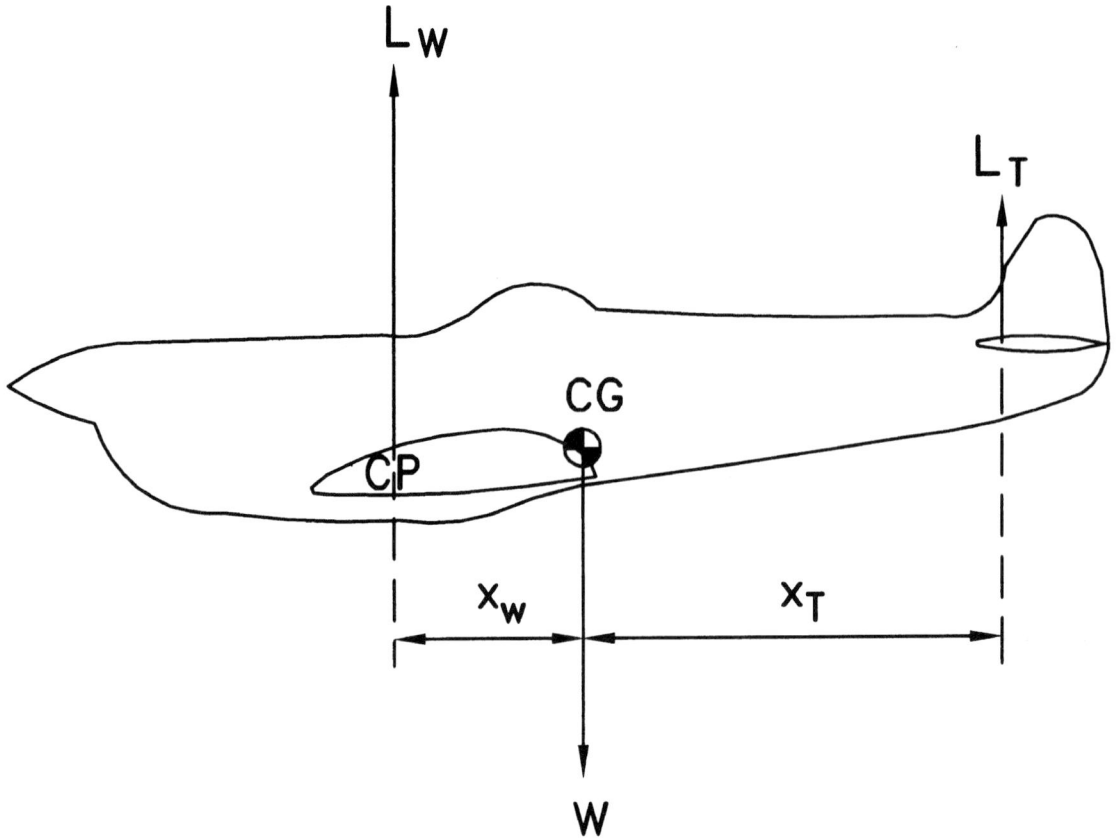

Fig. 4. Aeroplane in steady level flight, but with rearward centre of gravity.

(specifically, the rates of change of wing and tail lift with incidence) and the geometry of the aeroplane (wing and tailplane planform areas and the dispositions of these surfaces). However, when CG is located ahead of the neutral point there is then some hope that the aeroplane will be stable. Indeed, provided that CG is located significantly ahead of the neutral point, stability theory reveals that the aeroplane's disturbed motion divides itself into two distinct phases. In the first of these, there is an oscillation in which, predominantly, the wing incidence changes. This oscillation is usually of high frequency and, if stable, is rapidly damped. Crudely speaking, the aeroplane then finds itself at the correct incidence again but at a slightly different speed or, more to the point, a slightly different kinetic energy (which is proportional to the square of the speed). The second phase is then one in which the aeroplane rises and falls in the air, exchanging kinetic energy for potential energy (proportional to altitude) in an oscillation which is lightly damped. During this phase the aeroplane maintains approximately constant incidence to the local airstream. The rising and falling oscillation usually occurs quite slowly, taking as much as one minute or more to execute one complete oscillation. Thus, if desired, the pilot has time to correct this behaviour.

On the basis of the above preamble, we can now examine the contents of *Aerodonetics*.

Lanchester (1908) begins his discussion of aeroplane stability by describing his own demonstrations performed with the flat plate glider, a thin sheet of mica, shown in Fig. 5. This is loaded at the nose with split lead shot so as to have the forward centre of gravity position indicated. He repeats an argument given in the *Aerodynamics* to show that this tailless flat plate glider can be made to fly in a straight line at constant speed and at some shallow angle to the horizontal. To achieve this, however, he recognises that two conditions must be fulfilled. Firstly, the glider must have an incidence such that the centre of pressure coincides with the centre of gravity, thereby achieving the zero net moment condition (our

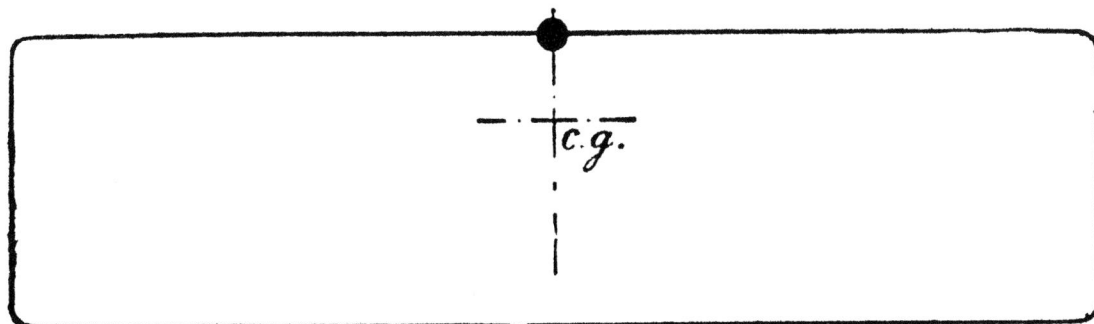

Fig. 5. Flat plate glider (Lanchester, 1908).

condition of trim). Secondly, the glider must move at a speed such that its lift force (proportional to the square of the speed) is sufficient to sustain the glider's weight (strictly, this should be the glider's weight component perpendicular to the glider's line of motion). Should the glider's incidence become disturbed so that trim is lost, Lanchester (1908) then gives essentially the argument associated with Fig. 3 so as to show that the centre of pressure movement will be such as to restore stable motion. On the other hand, if (Lanchester, 1908) (my change underlined)

"... the velocity is deficient, so that the weight is insufficiently sustained, the gliding angle and the component of gravity in the line of flight automatically increase and the <u>glider</u> undergoes acceleration. Conversely, if the velocity is excessive the gliding angle (and so the propulsive component) diminishes, and the velocity is thereby reduced."

In such cases of incidence disturbance or glider projection at speeds other than what Lanchester (1908) calls this "natural velocity", then the glider (Lanchester, 1908) (my change underlined)

"... will perform a wave-like trajectory, oscillating about the glide path of natural velocity, the oscillation gradually diminishing in amplitude, and the path of the <u>glider</u> approximating more and more closely to a uniform glide (Fig. 6)."

For this rising and falling oscillation he invents the term "phugoid", derived from the Greek and meaning "flight-like". Later he acknowledges this as not entirely felicitous since this Greek term for "flight" here has the sense of "fleeing". Nonetheless, the term has stuck and it is to Lanchester that we owe the identification of this phugoid oscillation having the long time period described earlier.

Lanchester (1908) then describes his earliest practical investigations of aeroplane stability. For these he used his glider of 1894, the planform of which is shown in Fig. 7; to give scale to the drawing, the wing span was 1.58 m. A rubber powered, twin propeller version of this, also used to investigate stability and also from this year, is shown as Fig. 14 of the

Fig. 6. The wave-like trajectory of a glider (Lanchester, 1908).

chapter on the *Aerodynamics* (p. 83). I should point out immediately here that this glider, and the powered version, were likely to be highly stable. For the glider, my estimate of its neutral point places this at approximately three wing root chords aft of the wing leading edge, whereas the centre of gravity location given by Lanchester (1908) lay one third of the chord aft of the leading edge. In other words, the glider's centre of gravity was located significantly ahead of its neutral point. A major contribution to this was the generous distance between the wing and the tailplane.

Lanchester's experiments of 1894 were conducted from the back first floor window of "Fairview", St. Bernard's Road, Olton, the flights being generally in a westerly direction so that the glider often overflew the nearby Kineton Green Road. Having given these aeroplanes the name "aerodone"*, Lanchester (1908) describes their flights as follows:

"When the weather was calm the aerodones carved their way through the air as if running on invisible metals, without the smallest visible fluctuation or quiver. When on the other hand the flight was made in a gale of wind, the flight path took the form of a bold sweeping sinuous curve, without a momentary suggestion of loss of equilibrium, but rather with the appearance of some set and intelligent purpose."

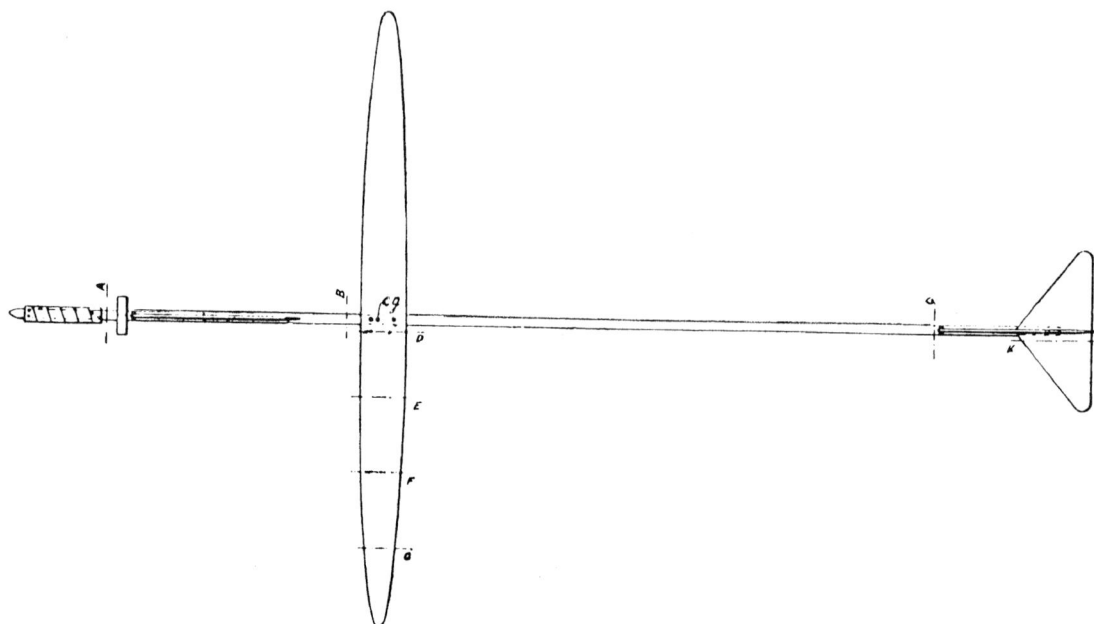

Fig. 7. The glider of 1894 (Lanchester, 1908).

* Another Greek derivation, meaning "tossed in mid air; soaring". Hence the title of the book.

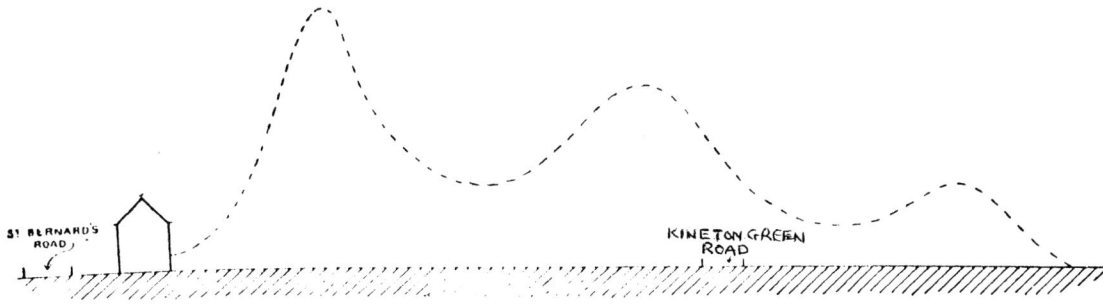

Fig. 8. Flight No. 2, 24th June 1894 (Lanchester, 1908)

One such "bold sweeping sinuous curve" is shown in Fig. 8, which depicts a phugoid of huge amplitude. The glider had a high "natural velocity", requiring catapult projection, and this, in combination with projection into a "high wind with powerful gusts", no doubt caused the huge disturbance to its motion. The consequence of such observed disturbances, however, was that Lanchester's interests in stability then became concentrated on the mathematical description of the flight paths of such huge phugoids. Indeed, five of the ten chapters of the *Aerodonetics* are devoted to this. For the progress of aeronautical science, this can be seen now as something of an unwarranted diversion since, in the case of the piloted and controlled aeroplane, such huge long period oscillations would be rapidly corrected. Nonetheless, certain aspects of Lanchester's investigation merit attention.

Lanchester (1908) begins his analysis of phugoid motion by considering an aeroplane restricted so as to move only in the vertical plane. Further assumptions are imposed, most notably that the sum of the kinetic and potential energies is conserved during the motion (or that a thrust force is supplied which, at all times, precisely opposes the drag) and that the aeroplane's mass is concentrated at its centre of gravity (the effect of the aeroplane's moment of inertia is neglected). Additionally, the lift force is assumed to act at all times through the centre of gravity. Consequently, no knowledge of centre of pressure movement is required for the analysis. For the curved flight path of the phugoid, in effect Lanchester (1908) sets down the correct requirement that the multiple of the aeroplane's mass and its centripetal acceleration must be equal to the difference between the aerodynamic lift force and the aeroplane's weight component perpendicular to the instantaneous line of flight. The lift force is taken to be proportional to the square of the aeroplane's instantaneous speed, as set down correctly in the *Aerodynamics*. From the mathematical expression of these physical principles, he obtains two results, one an equation for the aeroplane's instantaneous flight direction relative to the horizontal, the other an equation for the instantaneous radius of curvature of the flight path. He then devises an ingenious graphical method by which the shape of the phugoid flight path can be constructed from these two equations. A selection of his resulting phugoid paths is shown in Fig. 9. In this, curve 1 (a straight horizontal line) indicates the undisturbed condition, the curves carrying ascending integers indicating phugoids of increasing amplitude. Thus, for example, curves 2 and 3 are small amplitude motions realisable in practice. Curve 7, however, consists of a sequence of cusped curves, each of which turns out to be a semi-circle. Lanchester (1908) recognises that such a sequence is unlikely to be achieved in reality since the aeroplane, having zero velocity at the point of each cusp*, is assumed to turn itself instantaneously through 180° before descending. We recognise now that, should the cusp be reached in practice, a "tail slide" is the likely outcome, after which considerable aerodynamic complexities ensue. Similarly, curves 8 to 12, showing looped flight paths, are likely to involve related

---

* The topmost straight line in Fig. 9 is the point from which it is supposed the aeroplane has been dropped vertically, solely under gravity, so as to achieve the constant speed in the straight line motion of Curve 1. Thus the peak of each cusp of Curve 7 lies on this topmost line.

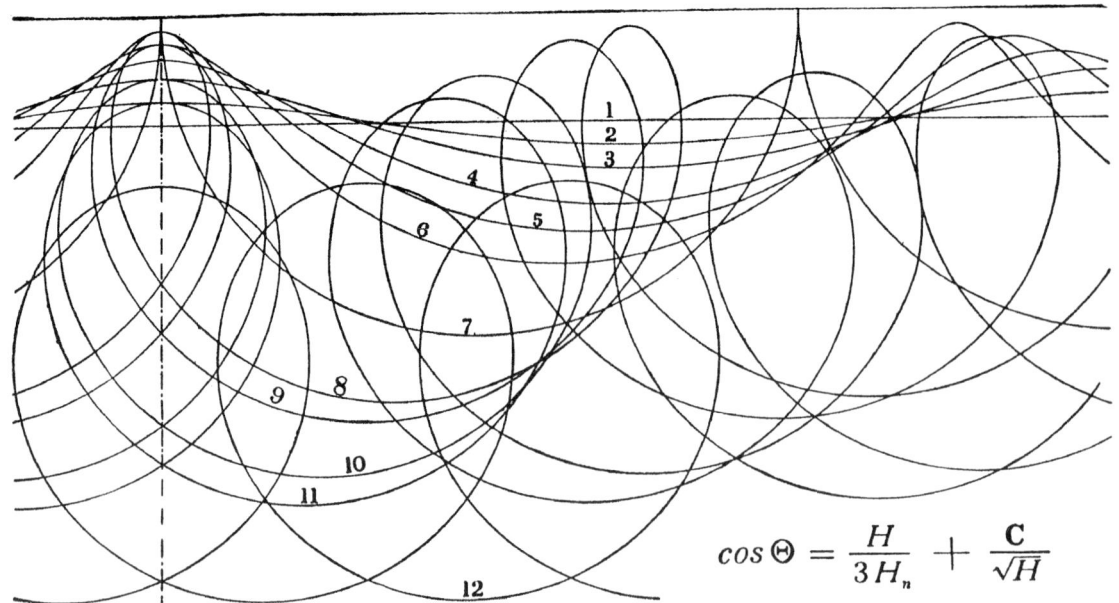

$$cos\,\Theta = \frac{H}{3H_n} + \frac{C}{\sqrt{H}}$$

Fig. 9. Phugoid curves (Lanchester, 1908).

complexities, not least the stalling of the airflow over the wings. Nonetheless, Lanchester (1908) expends some effort in investigating the mathematical properties of these curves, showing, for example, that in certain cases they are related to a curve called the trochoid.

The important result to spring from Lanchester's analysis concerns the small amplitude phugoid. In this case it is now known that Lanchester's initial assumptions are valid. For this Lanchester (1908) derives the correct results that the phugoid path is a sine curve and that the period for one complete oscillation, P, is given by

$$P = \sqrt{2}\,\pi\;U/g. \tag{1}$$

Here U is the aeroplane's undisturbed speed and g is the gravitational acceleration. Thus, for example, an aeroplane flying at 100 m/s would, when slightly disturbed, experience a phugoid oscillation of period 45 s.

It is also apparent that Lanchester (1908) recognises the presence of the short period, high frequency oscillation, described earlier, which prefaces the phugoid oscillation. Commenting on observations by Mouillard (1881), Lanchester (1908) remarks that (my addition in brackets)

"... it is quite possible that the oscillations recorded by Mouillard were merely vibrations about the attitude of equilibrium, that is to say, oscillations of much quicker pitch than those of the flight path, due to some initial disturbance of the balance between the *attitude* (incidence) of the plane and the position of its centre of pressure."

Later, Lanchester (1908) adds, correctly,

"The author has noticed quick pitch vibrations occasionally when experimenting with mica models, which are in no way related to the phugoid oscillation."

Earlier, Lanchester (1908) turns to the question of whether or not such phugoids would be damped in practice. He concludes, correctly, that the effects of viscous friction will always be such as to dampen the disturbed motion. He then argues, incorrectly as it turns out, that the distribution of the aeroplane's mass about its centre of gravity (the effect of the aeroplane's moment of inertia) is such as to increase the disturbance in the small amplitude phugoid. in fact, modern stability analysis shows that the moment of inertia influences the short period oscillation, but not the long period phugoid. In contrast, Lanchester (1908) arrives at a phugoid stability criterion dependent on which of these two effects, friction or moment of inertia, dominates.

Seeking to provide verification for his finding, Lanchester (1908) describes a number of experiments he has conducted with small gliders. Thus, for example, observation of the time period of small amplitude phugoids is found to agree with equation (1). Agreement for the time of the motion is achieved even in the case of the almost semi-circular path (Curve 7 of Fig. 9) obtained when the glider is released from rest, nose pointing vertically downward. In this case, however, as Lanchester (1908) adds in a footnote,

"Owing to the damping factor the flight path does not form a *cusp*, but rather a crest, the curve becoming inflected."

As to verification of his stability criterion, for this Lanchester (1908) describes tests conducted with five glider configurations designed so as to bracket this criterion. This they duly do, their flights being recorded as "just stable" or "scarcely stable" in accordance with his criterion. Yet, in this, the agreement with his stability criterion is rather misleading. One such glider is shown in planform in Fig. 10, its wing span being 644 mm. You notice the short tail arm. An estimate of its neutral point position places this almost coincident with the centre of gravity location which, according to Lanchester's comments, probably lay at 0.4 of the wing chord aft of the leading edge. As Lanchester (1908) records, this glider was "just stable". However, it achieved this not on the basis of Lanchester's stability criterion but because its centre of gravity was very close indeed to its neutral point. It now becomes possible to see what Lanchester had done. He had selected glider configurations, their tail arms in particular, so that their moments of inertia would have values which provided conditions neatly bracketing his stability criterion. Unknowingly, in doing this he had arrived at configurations possessing neutral points very close indeed to, and either side of, their centres of gravity. In all five cases, their recorded behaviour fits the modern argument based on neutral point placement relative to the centre of gravity. However, I should add that, for cases in which the centre of gravity is slightly aft of the neutral point, the instability problem becomes rather complicated and it is probable that Lanchester's observed instabilities were of a divergent nature.

Lanchester (1908) then goes on to apply his stability criterion to certain species of birds and, more interestingly, to the hang glider type in which Otto Lilienthal met his death in 1896. Lanchester (1908) provides a drawing of this glider (Fig. 11) in which you will note not only the short tail arm but also the relatively small tailplane area. According to Lanchester's stability criterion this glider is far from stable. More to the point, a rough estimate of the glider's neutral point places this at one third chord aft of the leading edge, almost coincident with the probable centre of gravity location. This, of course, is a potentially dangerous situation since the glider would have been sensitive to the least

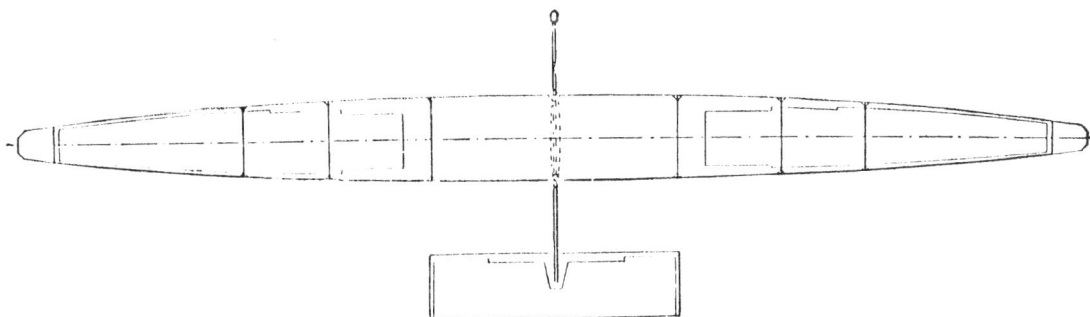

Fig. 10. Model glider, mica, 1905–7 (Lanchester, 1908).

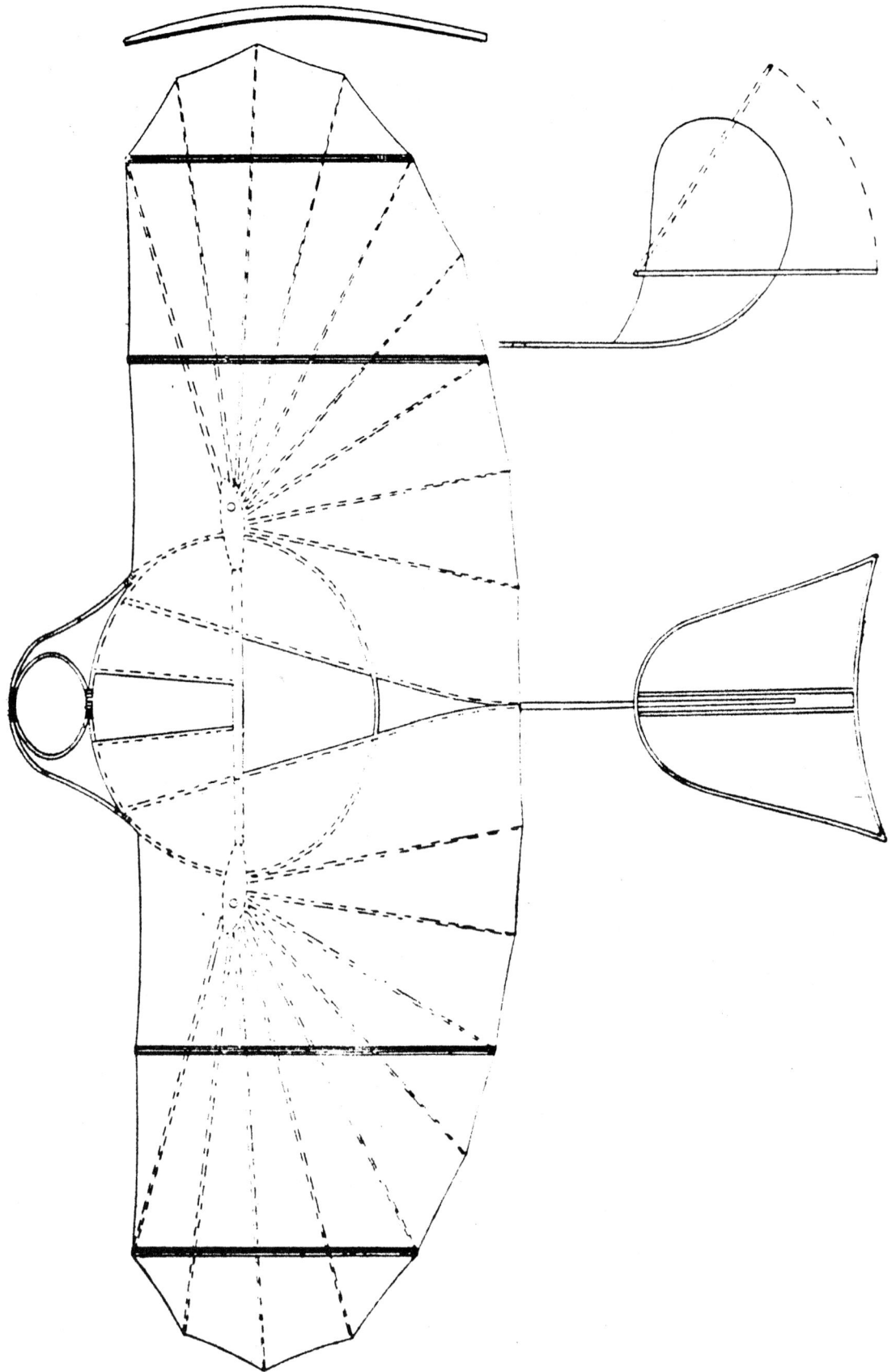

Fig. 11. Lilienthal glider No. 11 (Lanchester, 1908).

disturbance. However, this also made the glider more sensitive to control, an important requirement when the only means of control employed was to produce small changes in centre of gravity location by the pilot swinging his legs in the required direction. However, what exacerbated the instability situation was the upwardly moving tailplane, freely hinged at its leading edge as shown in Fig. 11. Without going into Lilienthal's reasons for adopting this dangerous feature, I will point out that the glider was designed so that, under normal gliding situations, its wing's centre of pressure lay slightly aft of the forward centre of gravity. To achieve trim under these circumstances, the tailplane was intended to, and indeed did, produce a slight download (a negative tailplane lift). This tailplane load pushed the tailplane downward so that it remained lying against stops and therefore acted as a stabilising surface. However, at large incidences near the stall the wing's centre of pressure moved forward, slightly ahead of the centre of gravity (as in Fig. 2), so that the tailplane began to experience an upload (positive lift). This caused the tailplane to rotate about its forward hinge so as to behave merely like a wind vane, no longer a stabilising surface. Thus the glider became effectively tailless. This decidedly unstable situation must have been very difficult to control by leg movements alone. It is known that it was a particularly vicious stall which caused Lilienthal's death. I should add that, throughout his articles on gliding, Lilienthal persisted in the belief that the centre of pressure movement had the same direction as that for a flat plate. Moreover, throughout the *Aerodonetics* I can find no indication that Lanchester believed otherwise. It may be that, because of this, at this stage he did not quite understand in their entirety all of the vital functions required of the tailplane, a difficulty exacerbated by his being sidetracked by his specious argument concerning phugoid stability.

Throughout the *Aerodonetics*, Lanchester (1908) takes the view that an aeroplane's motion in the vertical plane, its longitudinal motion, can be considered separately from any other form of motion. This is correct and it is to Lanchester's credit that he thus isolates and explores the related problems of longitudinal stability, most notably the phugoid. Moreover, Lanchester (1908) also recognises that an aeroplane can become disturbed so as to roll about the fore and aft axis, yaw about a vertical axis and move laterally in sideslipping motion. In dealing with such disturbances, however, it has to be said that Lanchester (1908) only begins to scratch the surface of what is, in fact, a highly complex subject. And although he tries to explain such disturbances separately, a not unreasonable stance in this first attempt at the subject, he nonetheless admits, correctly, that usually such disturbed motions are coupled together. Thus, for example, Lanchester (1908) must be credited with the first observation of the so-called Dutch roll in which the aeroplane exhibits a rolling, wallowing gait comprising coupled disturbances in roll, yaw and sideslip. Spiral instability is also observed in which the aeroplane, disturbed slightly in roll, begins a long spiralling dive, an instability nowadays acceptable provided that it develops sufficiently slowly. Yet in describing the dispositions of the aerodynamic surfaces required to ameliorate such disturbances, the arguments which Lanchester (1908) advances become something of a curate's egg. Thus, for example, in introducing the concept of wing dihedral Lanchester (1908) begins with a misconceived explanation for the correction of the sideslipping motion of his mica glider, or, as he calls it, the "ballasted aeroplane". When disturbed in roll (Fig. 12) the glider experiences a side force to the right, causing a sideslipping motion in the direction of the arrow. This is correct, but Lanchester (1908) then states that

"... the resulting motion causes the centre of pressure to move laterally, so that the necessary restoring couple arises and the plane returns in due course to its horizontal position."

Quite how this lateral movement of the centre of pressure is achieved in the case of the flat plate wing, he does not say. I shall return to this point presently. He then adds:

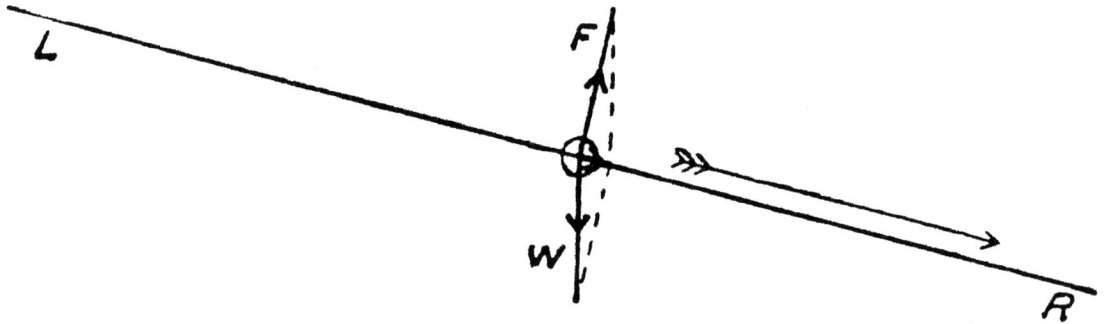

Fig. 12. Flat plate glider disturbed in roll (Lanchester, 1908).

"Similarly, in the author's flight models ..., the vertical fins perform the same function as is performed in the ballasted aeroplane by the change in the position of the centre of pressure; when the aerodone acquires a lateral motion the pressure that arises on the upwardly projecting fins acting above the mass centre of the aerodone supplies the righting moment."

Here he is correct and examples of the fins mentioned above can be seen in the photograph of his rubber-powered model (Fig. 14 in the chapter on the *Aerodynamics*, p. 83). The rear fin Lanchester (1908) clearly sees as also providing directional, or weathercock, stability in which any tendency to yaw about the vertical axis is automatically corrected. The forward fin is present, in part, so as to aid in this correction of the roll disturbance described above. Immediately, then, Lanchester (1908) introduces the dihedral concept:

"There is another modification of the design by which the same effect can be produced. If an upward inclination is given to the two wings of the aerofoil, Fig. 13a, or to the extremities only, Fig. 13b, a restoring couple arises as an immediate result of any attempt on the part of the aerodone to capsize."

Considering the arguments advanced earlier, the implication here is that Lanchester (1908) sees the interaction of sideslip and dihedral as causing the wing's centre of pressure movement necessary to provide the restoring moment. As such, this marks the beginning of the correct explanation for the function of dihedral, dihedral having been first described by Cayley (1809, 1810) although there the explanation had been incorrect since it did not

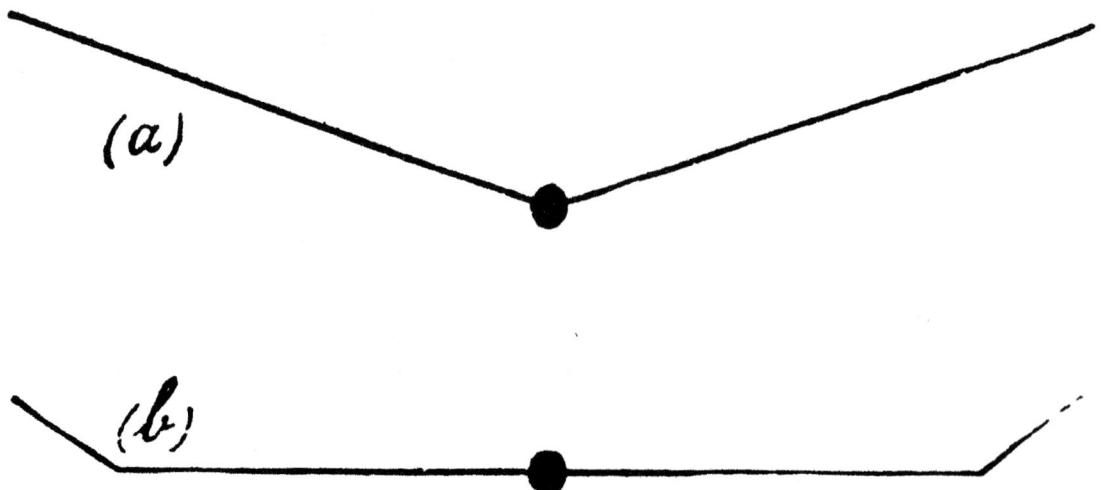

Fig. 13. Dihedral examples (Lanchester, 1908).

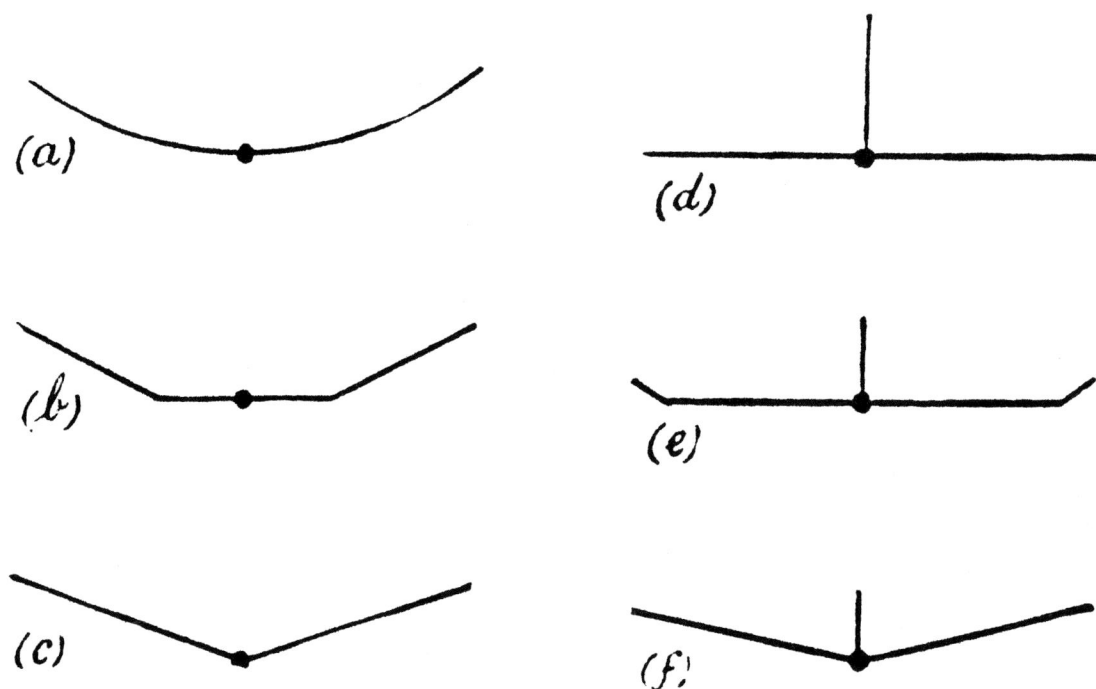

Fig. 14. Dihedral and vertical fin combinations (Lanchester, 1908).

include the crucial element of sideslip. The correct explanation is that the interaction of sideslip, forward motion and dihedral puts the wing on the sideslip side at a slightly higher incidence, and hence higher lift, than the other wing. This difference in the lifts of the two wings, or lateral movement of the centre of pressure of the combined wings as Lanchester (1908) puts it, produces the restoring moment. But one is then at liberty to ask: how is this difference in lifts achieved in Lanchester's original description of the behaviour of the flat plate mica glider possessing no dihedral? Lanchester (1908) then shows a variety of dihedral and fin combinations (Fig. 14) for the correction of roll disturbances, example (d) being the one he usually employs. Using example (a), however, he then produces a rather suspect argument in which the aeroplane is likened to a pendulum, oscillating along a circular path coincident with the curved wings. The purpose here, by analogy with his earlier specious argument for phugoid stability, is to discuss the effects of viscous friction and moment of inertia on roll damping.

A considerable part of Lanchester's discussion of lateral and directional stability is devoted to what Lanchester (1908) describes as "direction maintenance". In the case of the piloted, controlled aeroplane this function is performed primarily by the movable rudder so that Lanchester's concern here is rather inappropriate. However, Lanchester (1908) begins this topic by suggesting the use of what he calls an "abutment fin" placed vertically at, or near to, the aeroplane's centre of gravity. The intention is that this forward fin makes the aeroplane insensitive to side gusts, as well as assisting in the correction of roll disturbances (in combination with, or as an alternative to, dihedral). If I have understood Lanchester's subsequent argument correctly, he then attempts to achieve a lateral balance to side gusts by arguing that the forward fin should be placed ahead of the centre of gravity so as to compensate for the side gust effect on the rear fin. Hence the forward placement of the front fin in Fig. 14 of the chapter on the *Aerodynamics* (p. 83). However, the problem here is that, whilst a rear fin corrects yaw disturbances, a forward fin exacerbates such disturbances. It is probable, then, that Lanchester's forward fin detracted from the stabilising action of the rear fin in yaw. I am not certain that Lanchester (1908) has understood this point.

The penultimate chapter of the *Aerodonetics* is lengthy and is devoted to soaring flight.

Here the arguments expand on a basic dictum proposed by Rayleigh (1883) which Lanchester (1908) expresses as follows:

"In order that a bird should remain continuously in flight without performing work and without loss of altitude, that is to say, in order that a bird should soar, either –
(A) The wind being uniform possesses an upward velocity component; or,
(B) The wind is not uniform."

The obviousness of this is more apparent now than it was in Lanchester's era, I suspect, since Lanchester (1908) goes to some lengths to explain the upward motion of the air as it approaches hills, cliff faces and other obstacles as well as the "thermals" caused by the temperature differentials between the sea, the land and the air.

The final chapter of the *Aerodonetics* describes the construction of Lanchester's gliders and the experimental techniques employed in, for example, the determination of moments of inertia. *Aerodonetics* closes with a lengthy sequence of appendices, many of which amplify arguments given in the main body of the text. In addition, however, there are interesting articles on gyroscopes, the boomerang and the rifled projectile.

From the above description, it will be seen that the *Aerodonetics* does not achieve the stature of the *Aerodynamics*. Nonetheless, the achievements of the *Aerodonetics* are remarkable in comparison with what had gone before, particularly in the identification of the phugoid, the marking out of areas for concern in lateral stability and the attempts to develop mathematical models for disturbed motions so as to promote the view that aeroplane stability could be expected to be quantifiable. All this represented solid advances at the time, particularly when we recall that the powered aeroplane had barely left the ground.

A more rigorously developed view of aeroplane stability began to emerge in the work of Bryan (1911). However, Bryan's work has its limitations and these, interestingly, have a bearing on the results presented in the *Aerodonetics*. Unaware as yet of the aerodynamic behaviour of cambered wings, Bryan (1911) takes the wings and tailplane of his aeroplane to be flat plates. Thus the aerodynamic terms required for his comprehensive small disturbance equations are obtained from the, by then, well-known and largely empirical formulae for flat plates. Moreover, in an attempt to reduce the proliferation of terms in his analysis, Bryan (1911) makes the oversimplifying assumption that the centre of pressure movement with incidence change, albeit for the flat plate, can be ignored. One consequence of all this, of course, is that the potentially destabilising nature of the cambered wing is entirely absent from the argument. Thus nothing akin to the simple and highly useful idea of the neutral point emerges from his analysis. Nonetheless, Bryan (1911) is able to show that longitudinal disturbances divide into the two cases described earlier, the short and the long period oscillations. However, the interesting point is that, as a further consequence of his rather unrealistic assumptions outlined above, Bryan (1911) arrives at a stability condition for the short period oscillation which happens to be identical to the stability criterion given erroneously by Lanchester (1908) as governing the phugoid. This coincidence no doubt compounded the confusion at the time and it was to be some years before the correct analysis of the short period stability problem emerged, introducing significant changes to the Bryan/Lanchester criterion. Meanwhile, in a further publication on the longitudinal stability problem, Lanchester (1914) takes heart from what he sees as confirmation for his stability criterion in Bryan's work whilst apparently failing to notice that Bryan's result applies to the short period oscillation, not the phugoid. In this paper, Lanchester (1914) also expands on earlier arguments in the *Aerodonetics* in an attempt to account correctly for the effect on the tailplane of the air's downward motion caused by the wing's vortex system. He also incorporates a modification to his stability criterion suggested by Bryan (1911) (attributed to Bryan's colleague, Mr. E.H.Harper) to account for the effect of gliding angle on his glider observations used to test this criterion.

References

Bryan, G.H. and Williams, W.E. (1904): The longitudinal stability of aerial gliders, *Proc. Roy. Soc, A*, **73**, 100–16

Bryan, G.H. (1911): *Stability in aviation*, Macmillan Co., London

Cayley, G. (1809, 1810): On aerial navigation, *Nicholson's Journal of Natural Philosophy, Chemistry and the Arts*, **24**, 164–74 and **25**, 81–7, 161–9. [For corrected version, see Gibbs-Smith, C.H. (1962), *Sir George Cayley's Aeronautics, 1796–1855*, H.M.S.O., London.]

Kutta, M.W. (1902): Auftriebskräfte in strömenden Flussigkeiten, *Illus, Aeronautische Mitteilungen*, **6**, 133–5

Kutta, M.W. (1910): Über eine mit den Grundlagen des Flugproblems in Beziehung stehende zweidimensionale Strömung, *Sitzungsberichte der königlich Bayerischen Akademie der Wissenshaften*, **40**, 1–58

Lanchester, F.W. (1907): *Aerodynamics*, Constable & Co. Ltd., London

Lanchester, F.W. (1908): *Aerodonetics*, Constable & Co. Ltd., London

Lanchester, F.W. (1914): *Note on the stability of the flying machine as affected by considerations relating to propulsion*, Advisory Committee for Aeronautics, R. & M. No. 115

Lanchester, F.W. (1915): The Flying machine: the aerofoil in the light of theory and experiment, *Proc. Inst. Auto. Eng.*, **9**, 171–259

Langley, S.P. (1891): *Experiments in aerodynamics*, Smithsonian Institution, Washington

Lilienthal, O. (1889): *Der Vogelflug als Grundlage der Fliegerkunst*, Berlin. [*Birdflight as the basis of aviation*, (Trans. Isenthal, A.W.), Longmans Green & Co., London (1911).]

McFarland, M.W. (Editor) (1953): *The Papers of Wilbur and Orville Wright*, McGraw-Hill, New York

Mouillard, L-P. (1881): *L'Empire de l'air; essai d'ornithologie appliquée à l'aviation*, Paris

Prandtl, L. (1904): Über Flüssigkeitsbewegung bei sehr kleiner Reibung, *Verhandlungen des dritten internationalen Mathematiker-Kongresses, Heidelberg*, 484–91, Leipzig

Prandtl, L. (1918): Tragflügeltheorie, I Mitteilungen, *Nach. der kgl. Gesellschaft der Wiss. zu Göttingen, Math.-Phys. Klasse*, 451–77

Prandtl, L. (1919): Tragflügeltheorie, II Mitteilungen, *Nach. der kgl. Gesellschaft der Wiss. zu Göttingen, Math.-Phys. Klasse*, 107–37

Rayleigh, Lord (Strutt, J.W.) (1883): The soaring of birds, *Nature*, **27**, 534–5

Zhukovskii, N.E. (1906): De la chute dans l'air de corps légers de forme allongée, animés d'un mouvement rotatoire, *Bulletin de l'Institute Aerodynamique de Koutchino*, **1**, 51–65

Zhukovskii, N.E. (1910): Über die Konturen der Tragflächen der Drachenflieger, *Zeit. für Flugtechnik und Motorluftschiffahrt*, **1**, 281–4

Chapter Seven

LANCHESTER'S REVIEW OF AEROPLANE DESIGN AND CONSTRUCTION

by Eur. Ing. J.A.D. Ackroyd

With the First World War already two years old, Lanchester deemed it appropriate to produce in book form a collection of his more recent papers on the essential technical aspects of aeronautics. As Lanchester (1916a) remarks in the Preface to this,

> "Since the outbreak of war, from considerations of national secrecy, very little, indeed, of a technical character has been added to the stock of public information, and thus the position existing immediately prior to the war has become a matter of more permanent interest than the author anticipated at the time his Lecture was prepared."

The Lecture mentioned here was his "James Forrest" Lecture (Lanchester, 1914) delivered before the Institution of Civil Engineers three months prior to the outbreak of war, and much of the book is devoted to a slightly revised version of this. Although most of the technology described is that achieved by 1914, some of the other articles in the book, presented as appendices, provide an indication of more recent advances. However, it must be emphasised that all of the material presented is based on Lanchester's own work, mainly that described in his two volume treatise *Aerial flight* (Lanchester, 1907, 1908) and reviewed elsewhere in this volume, or on test results provided by the Royal Aircraft Factory, the National Physical Laboratory and laboratories elsewhere (Eiffel in Paris, Prandtl at Göttingen). As to such test results, I should also emphasise that these are seen through the discerning eye of not only the author of *Aerial flight* but also an eminent engineer who has been a member of the Advisory Committee for Aeronautics since its foundation in 1909.

Both the book and the Lecture are aimed at the practising engineer who has, as yet, little knowledge of this now rapidly expanding technology. In both publications the exposition is devoid of mathematical development, the emphasis being on physical principle and practical experience. Lanchester (1914, 1916a) begins by pointing out that the appeal of the current aeroplane lies significantly less on economy than on expediency. As proportions of their weights, air resistance figures for the Wright and Voisin aeroplanes are quoted as being a rather uneconomic 13%, whereas the corresponding traction resistance figure for a train is 1%. The aeroplane's advantage lies, of course, in its speed, despite the limitations imposed by the velocity-squared air resistance law. One such limitation, compounded by the further limitation of fuel capacity, is the aeroplane's range although here Lanchester (1914, 1916a) presents a reasoned argument to suggest that, with current technology, an Atlantic crossing has

> "... a possible chance of successful achievement."

As to utilisation of the aeroplane, he remarks that the main limitation here lies in the inadequate provision of airfields. However, the military authorities have already admitted the usefulness of the aeroplane for reconnaissance purposes and he envisages more belligerent employment as being inevitable. Recall here that such passages were written originally before the outbreak of the war.

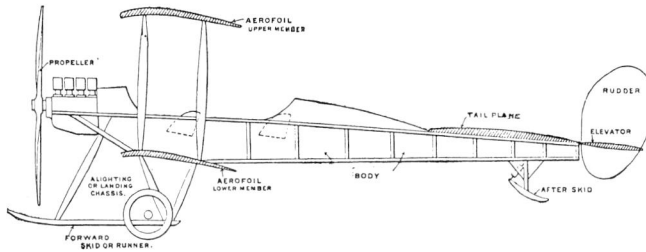

Fig. 1. The Royal Aircraft Factory B.E.2 Aeroplane (Lanchester, 1914, 1916a).

As to the aeroplane itself, based on a drawing of the Royal Aircraft Factory B.E.2 (Fig. 1) Lanchester (1914, 1916a) outlines some of the main technical considerations: weight, resistance, flight velocity, gradient of climb, engine power and propeller performance, structural stresses and undercarriage loads. He then turns to a description of possible flight paths and the question of stability. Most of this is taken from his phugoid theory given in Lanchester (1908) (here, and in what follows, it is useful to refer to this volume's article on the *Aerodonetics*, Chapter 6). The problem of instability leads him to a discussion of what he calls "Catastrophic Instability", the subject of two earlier publications (Lanchester, 1913a, 1913b) in which he had attempted to explain the fatal accident of Major G.C.Merrick at the Central Flying School, Upavon, on 3rd October 1913. A modern interpretation of this accident given by Bruce (1982) suggests that Major Merrick, unrestrained by the straps provided whilst in a steep descent to land, slipped forward onto the controls. The aeroplane, the Short S.62 (R.F.C. No. 446), then steepened its dive so quickly that Merrick was thrown out. The aeroplane became inverted, righted itself, but then turned over on crashing. If this is correct then the accident can be laid at the door of "pilot error", but the consequence was that Lanchester then began the misguided quest for an explanation of how an aeroplane, of its own volition, might suddenly become inverted. His argument is based on his experience with his earlier flat plate mica gliders. Because these gliders were symmetrical top to bottom they had no "upside down" and, if inadvertently projected at a slightly negative incidence, would zoom downward in a near semi-circular path so as to fly away, inverted, in a direction opposite to that of the original projection. The explanation which he gives of these events is correct but he then argues, mistakenly, that a disturbance of sufficient magnitude and duration could have had the same effect on Merrick's aeroplane. However, the consequence of all this is that it leads Lanchester (1914, 1916a) into the description of a wing's aerodynamic pitching moment variation caused by the movement of the centre of pressure with wing incidence change. He concludes correctly that, in order to avoid instability,

"... it is necessary to bring the centre of gravity appreciably in front of the most forward position of the centre of pressure of the aerofoil, so that the tail-plane will under all conditions carry a slight negative load."

This, in fact, is in line with the modern dual requirement that the aeroplane should be longitudinally stable and trimmable. The result is a highly stable aeroplane and here, no doubt, Lanchester is reflecting the thinking behind the development programme associated with later series of the B.E.2 which produced just such an aeroplane. But, in taking this discussion further, Lanchester (1914, 1916a) gives recent experimental results (Fig. 2) which at last clearly show the centre of pressure on a cambered wing moving forward as the incidence is increased (the curve a-a in Fig. 2). Moreover, he adds in a footnote (my addition in brackets),

"... the position of the point defined by the line gg (in Fig. 2, marking an unnecessarily adjusted furthest forward position of the centre of pressure), is an important landmark;

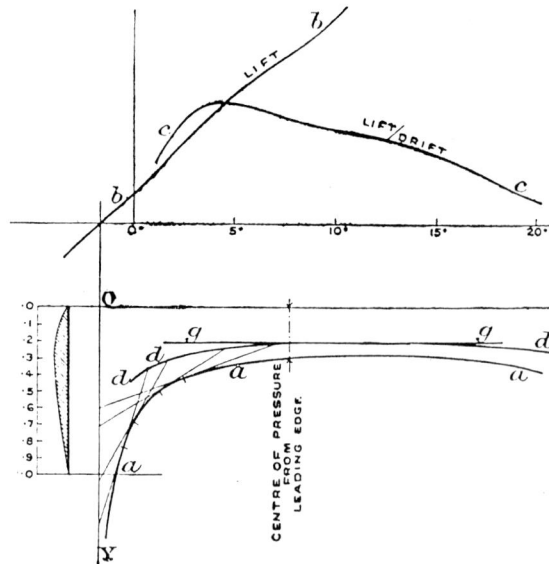

Fig. 2. Lift, lift to drag ratio, and centre of pressure position for a cambered wing (Lanchester, 1914, 1916a).

it should be ascertained for every individual aerofoil and should form the datum in relation to which the position of the centre of gravity is specified."

The wing theory of Kutta and Zhukovskii predicts that this point lies at, or very close to, the quarter chord point aft of the leading edge, i.e. very close to the tangent to Lanchester's curve a-a. As to aeroplane stability, the modern theory of the neutral point does indeed take the quarter chord point as such a datum for the centre of gravity. To this is then added a further aftward distance, contributed by the stabilising action of the tailplane, through which the centre of gravity can be moved before the aeroplane becomes longitudinally unstable. Thus Lanchester's dictum is not only very much on the safe side but is also, itself, a landmark in the gradual acquisition of these more advanced ideas.

Lanchester (1914, 1916a) then turns to the subject of air resistance. Here the review is based largely on his earlier ideas presented in Lanchester (1907) (for these, see this volume's Chapter Six on the *Aerodynamics*), although there is some updating of his estimates for skin friction provided by recent test results. The form of the induced drag relation is merely stated but its combination with skin friction is shown to lead to the total drag curve for a wing (see Fig. 23 in the chapter on the *Aerodynamics*, p. 91) which exhibits a minimum. The practical consequences of these ideas are explored in some detail but in simple terms, particular emphasis being placed on the minimum gliding angle achievable. One important addition to the story here, however, is provided by some recent lift results for flat plates and cambered wings given by Eiffel which illustrate what we now call wing stall. As wing incidence is increased, the lift increases steadily until stall is reached, at which incidence the lift peaks and then falls away for any further incidence increase. As Lanchester (1914, 1916a) rightly comments (my change in brackets),

"The breakdown of the (lift relation) at these limiting angles puts a very definite limit to flying at low speed."

The additional drag contributions from the fuselage, undercarriage and other exposed parts are then discussed, particular emphasis being placed on the drag advantages obtained from streamlined shapes. Again, recent test results are quoted, and in terms of the equivalent areas of flat plates held perpendicularly to the airstream so as to provide ease of understanding for the practical engineer. The sum total of all these drag contributions has then to be equal to the engine thrust so that the aeroplane achieves steady level flight. However, as Lanchester (1914, 1916a) then emphasises, there must be a sufficiency of excess thrust so as to produce a worthwhile rate of climb. Indeed, he goes further, remarking correctly that

"... any machine with an insufficient rate of ascent is intrinsically dangerous."

He then quotes rate of climb specifications for recent aeroplanes.

This leads him to the subject of propulsion, particularly the problem of propeller design. Here he draws a clear distinction between the characteristics of the marine propeller and those of the airscrew. As he rightly says, the marine propeller is limited in its design by the need to avoid the phenomenon of cavitation and this results in broad blades of relatively small aspect ratio. No such phenomenon is present in the case of the airscrew although, as Lanchester (1914, 1916a) comments with some prescience, something akin to this may occur if the airscrew blade moves close to the speed of sound. He then presents a brief sketch of the main considerations in airscrew design based on the propeller theory developed in Lanchester (1907). Although not mentioned in this volume's earlier chapter on the *Aerodynamics*, this theory is a fairly natural development of the wing theory given in the *Aerodynamics*. The propeller theory received some updating (Lanchester 1915a) in line with his more advanced ideas on wing theory (the paper listed as Lanchester (1915b) in this volume's chapter on the *Aerodynamics*). Lanchester (1914, 1916a) remarks that

"The author's method of propeller-design has been adopted and employed for some years by the Superintendent and staff of the Royal Aircraft Factory, with satisfactory results ..."

However, in a footnote he adds, no doubt with some irritation (my additions in brackets),

"... the method in question (as cited in the relevant Advisory Committee Report) is described as due to Drzewiecki (1909), but both the terminology and method are those given by me in "Aerial Flight", vol. i., in 1907 (ie. Lanchester (1907))."

As to airscrew efficiency, Lanchester (1914, 1916a) concludes that,

"Everything considered, the author is disposed to put the limit of efficiency ... at somewhat less than 85 per cent ..."

This is a fair conclusion, but he later adds (my addition in brackets),

"... at present there is but little available information on the question of efficiency, owing to the fact that the arrangements up till now (1914) at the disposal of the Royal Aircraft Factory do not permit of the effective testing of full-sized propellers."

There then follows an authoritative assessment of the various advantages and disadvantages of the engine types (rotary, fan assisted air-cooled, water-cooled) available at the time. The importance of a low weight to power ratio is emphasised, of course, but also the importance of achieving compatibility with propeller characteristics. The Wright brothers receive due praise in this last respect, particularly with regard to their design of a propeller of high efficiency* and that their twin propeller arrangements were correctly geared down using a chain drive from the engine.

---

* The Wrights independently developed their own propeller theory in 1903. As such, this is one of the first, if not the first, soundly established airscrew theories and is based on their own experimental tests on a wide variety of wing shapes. Sadly, it was never published in their lifetimes (see McFarland (1953)).

Wing layout is then discussed, descriptions being given of monoplane and biplane forms; in the latter case, the effects of wing gap and stagger are described. Lanchester (1914, 1916a) then turns to the methods of structural analysis currently employed for wing mainspars and interplane struts. He points out that, whilst the well-known methods used in bridge design (treating the struts as pin-jointed members) are still in use, more advanced and accurate methods are now coming to the fore. He cites, in particular, the work of Bairstow and MacLachlan (1913) in which the analysis employs the "theorem of three moments" (Clapeyron, 1857) to calculate the bending moments and stresses in braced monoplane and biplane configurations. Two short sections follow in which Lanchester (1914, 1916a) describes the known resistance properties of struts and bracing wires, and the need for sufficient fin area to ensure lateral stability. In this last respect, he comments that account must also be taken of the effect of the aeroplane's other vertical surfaces (struts, the fuselage itself and the wheel disks) as well as the effect of the propeller disk.

In view of the prevailing accident record of the aeroplane, the next section was no doubt of cogent interest at the time. Here Lanchester (1914, 1916a) discusses the dynamic load factor and factor of safety, giving arguments to show that, in the pull-out from a vertical dive, an aeroplane might be expected to experience loads as much as ten times its own weight. Even in a banked turn of 70° angle, the load is three times the aeroplane's weight. Such extra loads must be taken into account in the design of the aeroplane's structure. Moreover, a correct distribution of such loadings must be ascertained in the case of the wing's front and rear mainspars. In determining this distribution of the mainspar loads, account must be taken of the variation of the centre of pressure position with wing incidence change.

The landing gear is then described, both for land-based and marine aeroplanes. For the land-based aeroplane the advantage of a rubber suspension is emphasised since this is capable of absorbing a far greater impact energy than steel. Moreover (Lanchester, 1914, 1916a),

"... the signs of fatigue in rubber are evident to the most casual observer and the material is cheaply and easily replaced."

As to the structure of the landing gear itself, this must be of sufficient strength whilst avoiding excessive air resistance. The layout of the landing gear of the Bleriot 11 aeroplane is then described in some detail. As to marine aeroplanes, what little is currently known about the use and behaviour of floats and stepped hulls is briefly discussed.

This is followed by a short section dealing with what Lanchester (1914, 1916a) calls "acentric types of machine". By this he means aeroplanes for which the line of the propeller thrust or the line of action of the drag do not pass through the aeroplane's centre of gravity. This topic had been investigated in the *Aerodonetics* (Lanchester, 1908) and here it is discussed in similar terms employing phugoid theory.

In the final section, Lanchester (1914, 1916a) deals briefly with the stability and control of the aeroplane. Essentially, he lists, and gives his opinions of, current aeroplane stability theories. He favours his own work in the *Aerodonetics* (Lanchester, 1908), of course, but also the recently available work of Bryan (1911). As to contemporary theories produced on the Continent, he does not rate them highly in comparison with that of Bryan.

Lanchester (1916a) then continues with four Appendices. In the first of these he gives his current thinking on the subject of skin friction in which he attempts crudely to account for the variation of airspeed around a wing section. Appendix II is a reproduction of an earlier paper on aeroplane stability which has been mentioned in this volume's Chapter Six on the *Aerodonetics* and is there listed as Lanchester (1914). Appendix III reproduces the report of the testing of a three-quarter scale model of Lanchester's wing of 1894; this is dealt with in this volume's Chapter Five on the *Aerodynamics*. The final Appendix describes an eleopneumatic system for an undercarriage designed by Lanchester and built by the Daimler

Co. Ltd. for Messrs. White and Thompson, of Middleton, Sussex, in 1909–10. As is usual, this shows every evidence of Lanchester's careful attention to detail in the design, and represents one of the earliest of such undercarriage systems. Unfortunately, as Lanchester (1916a) remarks,

"... no opportunity has occurred of giving it a trial."

Lanchester (1916a) closes with the republication of an earlier paper (Lanchester, 1915b) which had suffered at the hands of the committee in charge of the 1915 International Engineering Congress held in San Francisco. As Lanchester (1916a) explains in the Preface,

"This paper was so badly edited by the Committee, and, as issued, contained so many errors and mistakes, that many of its passages were rendered meaningless."

The paper itself is a non-mathematical review devoted almost entirely to Lanchester's earlier work. Thus the subject of skin friction is discussed on the basis provided by the report listed as Lanchester (1915a) in this volume's chapter on the *Aerodynamics* (Chapter Five). Lanchester's more recent advances on the vortex theory of lift (listed as Lanchester (1915b) in the *Aerodynamics* chapter) are summarised in like manner. The paper closes with a brief mention of Lanchester's current thinking on propeller theory (Lanchester, 1915a).

During the period of the First World War, Lanchester made two further significant contributions to the development of aeronautical technology. It therefore seems appropriate to include these in this article. The first of these (Lanchester, 1915c) concerns flight at high altitude. Specifically, Lanchester (1915c) sets out to ascertain the drag and engine thrust characteristics of an aeroplane capable of sustaining steady level flight at an altitude of 18,000 ft. As to the drag, he considers the aeroplane flying at its minimum drag speed so that, using the results obtained in Lanchester (1907), he is able to compare minimum drag conditions at 18,000 ft. with those at sea level. As to engine performance, he argues that, although engine power is approximately proportional to the density of the combustible mixture, because of the effect of temperature on carburation it is more realistic to assume that engine power is directly proportional to the ambient pressure which, itself, decreases with altitude. He then presents a reasoned and correct argument to show that an aeroplane, possessing the drag and engine characteristics outlined above so that it (Lanchester, 1915c)

"... will just maintain itself in horizontal flight at 18,000 ft. height, will have twice the torque and thrust necessary for horizontal flight when at about sea level."

Briefly he considers engine supercharging but rejects this on the grounds of extra complexity and, more importantly, because (Lanchester, 1915c) (my addition in brackets)

"... no such provision is really necessary ... we have seen that the machine just able to fly horizontally at this height will have twice the bare thrust necessary at sea level. But this is just what is being called for in the modern military machine (in order to achieve current specifications on rate of climb)."

As he later explains, that current aeroplanes possessing these sea level rates of climb cannot, as yet, ascend to 18,000 ft. is a circumstance which can be attributed to such factors as carburettor icing and the need to run the engine at significantly higher and damaging speeds. All such problems must be addressed if high altitude flight is to be achieved. However, in his view a major problem remains, which is that current engines are too heavy for the power produced. Lanchester (1915c) then provides a detailed survey of engine mechanical design in the search for lower engine weight. He argues that a multi-cylinder engine possessing small individual cylinders is preferable. He then turns to engine cooling, pointing out that current air-cooled engines have less than half the cooling system weight per horsepower of water-cooled units. He anticipates a major weight advantage in the use of an efficient air-cooling system and therefore produces a lengthy and detailed analysis of the heat transfer characteristics of cooling fins. His analysis is based on his earlier work (listed in this volume's chapter on the *Aerodynamics* as Lanchester (1913a)) in which he had produced a relationship between heat transfer and skin friction similar to that proposed by Reynolds (1874). However, arguing that copper is the best material for the cooling fins, he shows that a very significant weight saving is, indeed, the likely outcome of a carefully designed air-cooled engine. Thus, many of the features described by Lanchester (1915c) point the way to the future development of the aeroplane engine, and not only for the provision of high altitude flight.

My final item in this article shows Lanchester tackling, and tackling highly successfully, a newly discovered problem area in aeronautics. In the *Aerodonetics*, Lanchester (1908) mentions briefly the possibility of significant distortion of the aeroplane caused by the elasticity of its structure. The problem which Lanchester (1916b) describes is one in which structural elasticity and aerodynamic forces conspire together to produce the potentially disastrous event called aeroelastic flutter. According to Garrick and Reed (1981), this was the first occasion on which the problem was identified and solved. The aeroplane concerned was the Handley Page 0/100 biplane bomber, which incorporated a biplane tailplane carrying four elevator surfaces, two (a right one and a left one) to each tailplane. The right and left elevators on each tailplane were essentially independent, being connected to the pilot's control column by separate, flexible, cable runs. Lanchester (1916b) describes the resulting problem as follows:

"I have been consulted by Squadron Commander Babbington with regard to a difficulty which has been experienced with a large machine. Arising out of an interview with him, I visited the sheds and made examination of the machine in question.
(1) The difficulty experienced is that at certain critical speeds of flight a tail wobble is set up, involving heavy torsional stresses on the fuselage, the type of vibration being an angular oscillation approximately about the axis of the fuselage; I am informed that the angular magnitude of this oscillation amounts at times to something approaching 15°, and is undoubtedly extremely dangerous to the structure of the machine. I gather that the experience of the pilots when this vibration is at its worst is terrifying."

Lanchester (1916b) goes on to argue that the aeroplane's structural strength is more than adequate for normal loadings and that the oscillations observed are of too low a frequency to be attributed to engine vibration or to propeller wash. He concludes, in effect, that the oscillation must be self-exciting and that he is (Lanchester, 1916b)

"... therefore disposed to think that it is in the dynamics of the tail that the trouble must be sought.... I also think that the backlash and spring which exists between the right and left pairs of elevator flaps is a contributory source of trouble; in other words, we may suspect this elastic system of entering into the total dynamic effect."

To interpose a little more explanation, what was probably happening here was that the flexible control cables allowed the elevators to oscillate about their hinges under aerodynamic load. At certain speeds, the coupling between the aerodynamic forces and the cable flexibility allowed the right and left elevators to achieve antisymmetric oscillations which then excited the torsional oscillation of the whole rear fuselage. Lanchester (1916b) suggests remedies for this, the most significant of which being

"That the two upper flaps on the right and left should be made as one continuous flap, and be stiffened by a torsional member, consisting of, say, weldless tube of $2\frac{1}{2}$ inches or

3 inches diameter, and about 14 or 16 s.w.g. Also that the upper and lower flaps should be coupled by a strut link instead of by the system of wires at present used."

The contribution which Lanchester (1916b) makes to the solution of this problem is followed by a mathematical analysis, probably the first such analysis, of aeroelastic flutter provided by Bairstow and Fage (1916). In this, they set down the equations governing the motion of the tail assembly and then use the small disturbance techniques employed by Bryan (1911) so as to predict the onset of flutter. Their findings largely support Lanchester's conclusions. Lanchester's suggested remedy, quoted above, was adopted and thus cured the problem. According to Garrick and Reed (1981), an epidemic of tail flutter problems arose one year later with the de Havilland DH-9 aeroplane. Here again, Lanchester's remedy was adopted and, ever since, the torsionally stiff connection between the right and left elevators has remained an important design feature. This seems an entirely appropriate point at which to conclude my survey of the extraordinary insight which Lanchester brought to aeronautical research and development.

References

Bairstow, L. & Fage, A. (1916): *Torsional vibrations of the tail of an aeroplane*, Advisory Committee for Aeronautics, R. & M. No. 276 (Part (ii))

Bairstow, L. & MacLachlan, L.A. (1913): *A preliminary note on methods of calculation which may be employed in the determination of the stresses in the spars of aeroplane wings*, Advisory Committee for Aeronautics R. & M. No. 83

Bruce, J.M. (1982): *The Aeroplanes of the Royal Flying Corps (Military Wing)*, Putnam, London

Bryan, G.H. (1911): *Stability in aviation*, Macmillan Co., London

Clapeyron, B.P.E. (1857): Calcul d'une poutre élastique reposant librement sur des appuis inégalement espacés, *Comptes Rendus,* **45**, 1076–80

Drzewiecki, S. (1909): *Des hélices aériennes; Théorie générale des propulseurs hélicoidaux, et méthode de calcul de ces propulseurs pour l'air*, Paris

Garrick, I.E. & Reed, W.H. (1981): Historical development of aircraft flutter, *J. of Aircraft*, **18**, 897–912

Lanchester, F.W. (1907): *Aerodynamics*, Constable & Co. Ltd., London

Lanchester, F.W. (1908): *Aerodonetics*, Constable & Co. Ltd., London

Lanchester, F.W. (1913a): *Catastrophic instability*, Advisory Committee for Aeronautics, R. & M. No. 114

Lanchester, F.W. (1913b): Report on Catastrophic instability in aeroplanes, *Engineering*, 574–5

Lanchester, F.W. (1914): The flying machine from an engineering standpoint, *Proc. Inst. Civil Eng.*, **98**, 3–96

Lanchester, F.W. (1915a): The screw propeller, *Proc. Inst. Auto. Eng.*, **9**, 263–354

Lanchester, F.W. (1915b): A discussion concerning the theory of sustentation and expenditure of power in flight, Paper contributed to the International Engineering Congress, San Francisco, Cal.

Lanchester, F.W. (1915c): *Report on high altitude flying and the development and improvement of the aeronautical motor*, Advisory Committee for Aeronautics, R. & M. No. 220

Lanchester, F.W. (1916a): *The flying machine from an engineering standpoint*, Constable & Co. Ltd., London

Lanchester, F.W. (1916b): *Torsional vibrations of the tail of an aeroplane*, Advisory Committee for Aeronautics, R. & M. No. 276 (Part (i))

McFarland, M.W. (Editor) (1953): *The papers of Wilbur and Orville Wright*, McGraw-Hill, New York

Reynolds, O. (1874): On the extent and action of the heating surface for steam boilers, *Proc. Manchester Lit. Phil. Soc.*, **14**, 7–12

Chapter Eight

AIRCRAFT AT WAR:
STRATEGY, TACTICS AND RELATED TOPICS

by Dr. Barry D. Hobson

Amongst the earliest references to Lanchester (1894) there is one which relates to an address given to the Birmingham Natural History and Philosophical Society, concerned with the soaring flight of birds and the possibilities of mechanical flight. By way of introduction, I think it worthy of note that, at this time, some nine years before the first recognised powered flight by the Wright brothers, Lanchester is quoted then as showing,

"... that at the present time scientific investigation as to the phenomena of flying has to a great extent taken the place of oversanguine and imaginative speculation: and although we cannot say definitely whether man is any nearer flying than he ever was, it is certain that we are beginning to have an insight into the underlying principles."

Within the same paper Lanchester deals with the maintenance of equilibrium and its problems and he continued with these interests through the publication of his books *Aerodynamics* (1907), *Aerodonetics* (1908), comparison of flying machine performance (1909a), and the delivery of another paper on the flight of birds (1909b), to the same society as in 1894.

The importance of the performance and stability of flying machines was thus addressed in their development and in fact the attributes of different aircraft stability were eventually related to their operational use for strategic purpose where the benefits of a stable flying platform and reduced pilot demands were advantageous. Lanchester was also quite concerned about the safety aspects arising due to unskilled piloting and flying machine* reliability with regard to mechanical flight (1909c).

The issue of practical flying experience compared to scientific investigation was always a controversial subject, and quite often involved Lanchester in strong discussion throughout the earlier days of aeronautical progress. In fact he bemoans the lack of research and laboratory exploration, particularly into aerodynamics within the U.K. universities (in 1909) but supports the systematic work initiated at the National Physical Laboratories (1909d). It is interesting that this latter reference mentions the early use of wind tunnels, for example, by Prandtl, Stanton and the Voisin Frères.

Now at this stage, apart from the aeroplane Lanchester considers the dirigible such as the Zeppelin, as having its advantages for example with regard to payload raised, duration of flight and rate of climb, but gives the aeroplane the edge in terms of flight velocity, compactness and potential for development. I quote his words (1909c) with regard to both aircraft as "... each being employed in certain definite directions, to which it will be confined by its limitations." These are prophetic words in the light of consequent hostilities in 1914, and in fact, he was soon involved with engine development and improvement with regard to high altitude flight (1915) which he saw as an advantage for combat purposes.

The preceding comments form a brief background to Lanchester's work, from the point

---

* The use of the terms "flying machine" and "dirigible" was succeeded by the name "aircraft", which was interpreted to cover both aeroplane and dirigible.

of view of his aeronautical interests leading up to the articles on aircraft in warfare in 1914. The latter constituted the basis for his book (1916a) and I have concerned myself, in the main, with this book entitled *Aircraft in warfare: the dawn of the fourth arm*. However, further references are included herein which relate both directly and indirectly to the use of aircraft and air warfare as seen by Lanchester. Of course, these references can only comprise a selection of the additional addresses, lectures, papers and letters attributed to this widely referenced man, but the information obtained provides a fascinating insight into the current philosophy and future predictions of that time, in the light of progress to the present day.

Lanchester was never backward at coming forwards, and apart from his technical contributions, he kept in touch with aeronautical matters in general. At times, his comments could be adjudged as an over-reaction to a situation or sweeping in context, but he attempted to inform and keep people aware of the facts within contentious situations, often misrepresented by the Press and also by persons with vested interests, particularly in aeronautical matters.

In *Aircraft in warfare*, Lanchester presents an appraisal of the issues involved in the use of aircraft for military and naval purposes, both currently and for the future. However, to add the continuation of the title, i.e., "The Dawn of the Fourth Arm" is effectively to summarise this treatise in the sense of the Air Force formation and to put it in context with regard to the events which followed. It was indicative of the way ahead that Lanchester quotes in his author's note that "... it cannot today be disputed that the immediate future of the flying machine is guaranteed by its employment by the Army and Navy." He goes on to say that typical duties would include reconnaissance work and bomb, torpedo or gun-carrying. There is no doubt that the onset of the 1914–18 war concentrated the mind with regard to the use of airpower and in particular illustrated, as might be expected, the initial lack of public interest in, awareness of, and enlightenment on such matters. Of course it should be noted that aviation as such, was still in its infancy and that the war started only just over a decade after the Wright brothers' first recognised flight in a powered aircraft in the U.S.A. With the consequent interest and flying in Europe, the public became quite enthralled with the growing band of intrepid aviators, and hardly cognisant of the potentially more sinister aspects of aircraft usage. Indeed, recognition of the aircraft as having a place within the theatre of war would take time.

Hence it is worth establishing at the outset that Lanchester's reference to the "Fourth Arm" is a definition which follows on from the first three, as constituted by the Infantry, Cavalry and Artillery. The reader needs to be well aware of the era in the sense of manpower, bayonets, horses and guns as traditionally used within strategic and tactical manoeuvres. Viewed in this light, the development of a fully-fledged Air Service would prove to be a demanding task, and Lanchester saw the primary function of the aircraft service as one of relating to the other three branches of the service. As such it would co-operate with the Army for example, providing support through its mobility and observation rôles, particularly with regard to reconnaissance/scouting: in this latter respect it was much better than the cavalry. In addition it could provide an attack or defence potential relative to hostile aircraft, and this was defined as a secondary function.

As aircraft development improved then advantages were to be gained technically and also from more efficient strategy resulting from these improvements. Comparison between the two types of aircraft, i.e., aeroplane and dirigible, showed natural divisions in per-formance and duties. The dirigible formed a good base for observation purposes but was slower than the aeroplane and as well as requiring large storage sheds, presented a larger target to the enemy. Of course, at that time the dirigible, as exemplified by the Zeppelin, had reached a near maximum size and inherently had the capacity for large flight duration times. Lanchester saw much more room for aeroplane size development but early on tended to advocate the smaller aeroplane, looking towards larger numbers of such types rather than building more unwieldy machines with a greater fire-power.

Again, Lanchester's views, at that time and throughout his book, showed that he saw the aeroplane range subject to definite limits, whilst the dirigible allowed greater potential for development. The superiority of the aeroplane at low level was demonstrated by its ability to carry out scouting duties over large areas even with cloud problems, and for directing gun-fire, whilst the dirigible could remain at high altitude and stay in visual contact with the enemy for long periods. Certainly Lanchester felt that the dirigible was of little use for bomb-dropping on land and could not be effectively armoured against the enemy. In fact, he advocates tactics of diving steeply from above on non-rigid airships and attacking with a "... hundred yards or so of barbed wire trailed beneath the aeroplane." Such tactics would tear the envelope and hence cause great problems. Furthermore, when considering dirigibles of the rigid type, i.e., Zeppelins, he feels that,

"... the structure would not stand up under ... blow from ... steel bar ... of 70 lbs. ... dropped from height of 200 ft. or 300 ft."

On a personal note, I find this combat situation fascinating in the sense that Lanchester considers a steep dive as a dangerous feat of airmanship but then goes on to say,

"... It is an open question whether airmen will be found ready to step forward at the critical moment to go to certain death, and so the general feasibility of ramming tactics must, for the time being, remain in doubt."

The aeroplane emerges as the aircraft which would be developed significantly in the future and in looking towards its strategic and tactical uses, there is an inbuilt comparison with the rôle of the cavalry at that time.

Hence the strategic scout aeroplane would be used for reconnaissance to locate and monitor enemy forces, it being used as an informer rather than a fighter. Such a machine requires high speed for evasion purposes, long range for large area sweeps but no heavy armour since it would not be acting in a defensive rôle. I would thus think the pilot not too unhappy in operating at high altitude where possible, otherwise there would be a risk of shrapnel damage at low level. Rifle or machine gun fire from the ground would be relatively ineffective above a maximum of, say 5,000–7,000 ft., but artillery gun fire potentially more dangerous with a vertical range of approximately 12,000 ft. Generally, it would be difficult anyway to hit aeroplanes with the then state of the art gunnery, hence attack from ground-fire at high altitude would be impracticable. Closer to home of course, Lanchester states that,

"... in our climate at least, not more than one day in four is sufficiently clear to render high-altitude shooting possible."

Thus the strategic scout aeroplane would be found very useful for scouting as a prelude to cavalry reconnaissance, and hence the flight or squadron may be attached to the independent cavalry.

Turning to the tactical rôle, such an aeroplane would be employed for local reconnaissance, locating or directing gun fire and engaging enemy aeroplanes. This would involve flight at low altitudes, often under fire and in that respect the aeroplane may have "... almost as much to fear from its friends as its foes." When monitoring and directing gun fire at enemy positions, then several methods of signalling may be used, including a sharp turn over the objective, release of a smoke bomb, or lights if of sufficient strength, employed in daylight.

In the matter of protection and defence from gun fire, Lanchester considers two lines of approach: the first is to use armour plate and the second, a transparent construction. The latter design allows bullets to pass through the aeroplane components without causing significant damage, but some protection is still needed for the pilot! These remarks apply to operational flight at, say, less than 5,000 ft. when attack by, typically, machine gun and rifle fire becomes important. To put the design problems in perspective, Lanchester suggests the need for 6 mm steel plate at 1,000 ft. and 3 mm thickness at 2,000 ft., below the latter is considered low altitude flying. The weight problem becomes prohibitive for 6 mm plate if designing an aeroplane for ordinary tactical duties, but he goes on to consider very low altitudes of 100/200 ft. where the element of surprise would be large, but where 9 mm plate would be necessary and hence the area of armour cover would have to be reduced to a minimum, with a fair degree of risk accepted. Ultimately, when these weight problems are overcome for low altitude flying, then the aeroplane tactics would have a significant effect on the man and horse cavalry operations.

Lanchester looks towards the utilisation of a few squadrons of aeroplanes, each with one or two machine guns and/or bombs and hand grenades, to be used at a critical phase of the battle e.g. in an enemy retreat situation. This is, for example, where the Lewis machine gun used at very low altitude could create havoc amongst the enemy troops. Such tactics lead to the requirement for cavalry supported by its own aeroplane auxiliary to counter enemy aeroplanes. Lanchester foresees problems in effecting this combination but states that the Air Arm would emerge as a "... fighting force of no mean importance." He goes on to evaluate the significance of the n-square law, which may be quoted here for reference as,

Fighting strength of a force is proportional to (Numerical strength)^2 × fighting value of the individual units (see Chapter Nine, below)

and is one of the main reasons for Lanchester's recommendation for large numbers of smaller aeroplanes rather than a small number of large machines. He also feels that for example, the aeroplane with three guns is not as efficient as three smaller aircraft with one gun each. The question of size is discussed further (1916b, c, d, f) with respect to its cost, wing weight, operation, armament etc., it is interesting to see that the manufacturing cost, including the engine, was 15 shillings per lb. at that time: I leave the reader to convert this figure to present-day values! At this stage it is interesting to note that Lanchester's views are contested in the columns of *Engineering* (1916e), by an anonymous "aeroplane designer-constructor", amongst others.

In discussion of armament, I consider it important to note that during this war period, the aeroplane designer was hard pressed to obtain a significant range for his machine, never mind payload demands from gun, armour, shell weight and bomb requirements. Lanchester discusses such matters in detail when considered in the light of aeroplane tactics of attack and defence, i.e. the secondary function of the Air Arm. Points of interest arise, in that maximum movement of fire was required in a gun installation within structural limitations, initially to fire downward (for example on Zeppelins as well as cavalry) with consequent use of armour protection for gun fire from below. Of course, progress indicated that such armour would be needed also for protection from above, and additional design thoughts included the use of fixed shields on moveable gun installations to protect the gunner. He emphasises the advantage of height in attacking the enemy, the aeroplane having better protection from below and also the advantage of height for positioning and manoeuvring for air combat. Events in the Second World War illustrated the efficacy of this doctrine and the advice "beware the hun in the sun" became very relevant. I think it is pertinent to quote the man here, with regard to securing a height advantage, in that it,

"... will probably prove to be, and will remain, the key or pivot in which every scheme of aeronautical tactics will, in some way or other be found to hang."

On this latter point, he considers the possibility of the use of direct-lift machines to climb vertically (1917j), and feels that,

"... the day when the propeller thrust can be caused at will to exceed the total weight of the machine – will be the day on which, ... the conquest of the air will be complete."

He envisages the defending scouts to be 'sitting on their tails', and ready to lift off at short notice. Landing would be accomplished by an appropriate tail-slide – the pilot "... alighting backwards on to his tail ..." Furthermore, and most importantly he realises,

"... that it would be dangerous to rise vertically from the ground if the machine were incapable of control in the interval during which it is acquiring velocity."

He goes on to specify the requirements for slipstream effects from the propeller, relatively large tail assemblies and more, as unknowingly he predicts in principle, the basis for the Ryan vertijet and similar designs, with variants such as the Flying Bedstead and the "Harrier" type of aircraft, all built in later years.

Continuing his account of weapon usage, it is of interest to note that humanitarian issues came under discussion and the words "anti-personnel" are used with regard to hand-grenades and steel darts. In fact, the additional weapons considered include bombs, rockets, explosive/expanding bullets, shells and an air-borne torpedo design. The latter, which was basically a high-velocity gliding bomb utilising an impact fuse, was proposed by Lanchester in 1897, but not developed since it was specifically for airship destruction (as were rocket designs) and hence not of first-line importance compared to the weapons already available for this task.

Hand grenades were used against the enemy as,

"... judicious employment of a few hand grenades for the scattering of cavalry or the stampeding of led-horses ... may ... amply justify the use of such a device."

This device was, of course, very handy and portable and required a low-level attack to be of sensible use. Not so the steel darts, of approximately 1 oz. weight. They were dropped from a few thousand feet on to the cavalry and men on the march, and reached a maximum velocity of 400–500 ft./s. with a penetration equivalent to "... several inches of spruce planking." For the future, the darts were not accepted to be of much use and there were difficulties allied to bomb-aiming and consequent accuracy on target. In short, Lanchester considers that the gun would reign supreme when developed for a rapid rate of fire and high velocity, but at that time, there were problems relating to mounting a gun satisfactorily in the aeroplane. However, before leaving this subject it should be noted that he expected a bomb-dropping machine could be used on targets sufficiently large to achieve good accuracy. In addition, the aeroplane needed to have a fighting capability with a capacity for rapid-fire from its guns or it would need a supporting fighting squadron to protect it. Again, such tactics were relevant in the latter part of the Second World War when fighters had sufficient range to escort the bombers well into enemy territory and back.

At this stage, Lanchester considers the rôle of aircraft within the Royal Navy and the differing requirements which result in an effectively independent service. The use of aeroplane guns against the enemy's ships is assumed ineffective but they are needed for self-preservation and for attack against hostile aircraft. Against ships, bomb dropping is not as efficient as the use of torpedoes – although the dirigible could carry such weapons, it is itself liable to destruction by enemy aeroplanes. Lanchester supplies an example of a Zeppelin used at very low altitude to launch a torpedo of 21 in. diameter and weighing one ton approximately, with a range approaching 2 miles. In the case of aeroplanes carrying and delivering torpedoes, Lanchester suggests tactics whereby the attack is carried out at specific times appropriate to local camouflage issues. Such periods occur in the haze of the early morning or at dusk when,

"... an aeroplane at the distance in question (i.e. less than 2 miles) is quite invisible."

Furthermore, depending upon the direction of the sun's rays, if an aeroplane is "... assisted by suitable protective colourings ..." or "... approaches against a landscape background, it may reach its target undetected." He also considers night attack as a possibility but also states that,

"... the absence of light may be a greater hindrance than help to the aeronaut."

Scouting and reconnaissance rôles are the prime duty of naval aircraft. Generally land-based aeroplanes would have a range of say 400 miles for coastal, patrol work and away from shore. In this case dirigibles could be used in a similar manner but only where a series of suitable landing points would be available along the coastline. Hence the logical extension for naval aeroplanes to fit floats resulting in the hydro-aeroplane, (currently called a seaplane or flying boat), continued with a floating mothership with a winch-on/winch-off capability to house such machines. Lanchester develops this latter concept (i.e. an aircraft carrier as we know it) further as "... an ocean going aeroplane pontoon base ...", compared with the use of a "normal" ship modified to carry flying boats. Whichever type is used, he maintains again that large numbers of aeroplanes are needed to give the numerical strength necessary to take advantage of the n-square law. Also on that basis the tactics of formation flying would need close study. Lanchester defines an operational and design philosophy for the aircraft carrier which may be seen as very similar to that of later years, with the common requirement of a clear flat-topped area for take-off and landing. He envisages a vessel with a waterline not less than 500 ft. and a maximum beam of 90 ft. Power would be derived from diesel engines (say six) with a power output, totalling 15,000 indicated horse-power. For this design, the funnels would be removed, a situation different from the large present day aircraft carriers.

In case the reader is curious about the interaction between aeroplane and submarine, I would stress here, that Lanchester overall felt that the latter was not a real challenge and could be easily dispatched. This view is based on the use of a reconnaissance strategy and the employment of local bombing tactics by aircraft. The submarine is considered not to have any "power of offence" by which he means that even if fitted with light calibre guns it could obviously only use these after coming to the surface. Lanchester assumes that one or more destroyers or light cruisers would accompany the scouting aircraft and that "... the conning tower of the submarine will be blown away within a few seconds of its appearance." He goes on to say that they would be destroyed in the main by bombs dropped from aircraft at low level (note that this includes the dirigible) even when the submarine was submerged. In view of events in the Second World War for example, I think that such bombing tactics could be effective, but he concludes that, (using what we now know as the Fleet Air Arm and a supporting task force of ships), the enemy submarine would be unable to roam at will or to make unexpected attacks on patrolling vessels. Furthermore, the enemy submarines would need to be protected by a supporting force and without it, they would be "... at the risk of almost certain destruction." Operation at night would be possible but under only limited operational constraints. Such constrictions, I would stress, were not relevant in the Second World War and, particularly in the Atlantic, German wolfpacks roamed near and far, with limited losses from British aircraft and ships in the initial phase of the campaign. Later, Lanchester goes on to say that a strategy of continuous pressure and harassment would be needed with direct attack wherever possible. Furthermore, he believes,

"... that the capacity of aircraft to warn merchantmen of danger will alone be sufficient to render the submarine threat quite ineffective ..."

Inherent in this statement is an assumption that sufficiently numerous aircraft are available to carry out the necessary sweeps over large areas of ocean. However, even though

Lanchester believes that the air service could effectively limit the influence of the submarine, he states that,

"... we can never expect ... that the value ... of the latter will be nullified."

Returning to land based matters, the tactical importance of aircraft tends to be based on a comparison with the cavalry. Opinion of the time inferred that employment of aircraft would not have any noticeable effect on the use of the infantry, cavalry or artillery. These views of course tend to be based on the war experiences of that time. However the point at issue here, is not the encroachment of the aeroplane for example on the the cavalry's duties etc., but the fact that there were few aircraft in service anyway, and hence they were used mainly for specialised duties, such as advance reconnaissance of enemy positions and movements, information that otherwise would not have been obtainable.

Accepting that the strategic advantage lies with the attacker and the tactical with the defender, then the advent of the aircraft has tipped the power scales in the direction of defence. It is accepted of course, that the attacker can use aircraft to report the strength, etc., of defences and apparent points of weakness, but this effect is seen to be small compared to the revelation of the attacker's whole plan. However it is possible that a stalemate situation may be produced due to information gleaned by the aircraft operation.

Considering the future use of aircraft in large numbers by the Aeronautical Arm, then there is a capacity available to assume an effective command of the air with its consequent advantages. This approach could still allow some enemy scouts through, for example, to carry out reconnaissance, but more importantly he will be unable to use his aircraft to direct his long-range guns to maximum effect, his cavalry will be harassed by machine gun and grenade, the safety of his communications such as rail, convoy, will be jeopardised, requiring a large proportion of his aircraft to be used in guarding his own resources. Similarly, Lanchester sees an extension of aircraft duties towards "... raiding of a kind and with a scope not hitherto known in warfare". For the future he perceives extended air raids into enemy territory as the aeroplane is able to move further afield and operate further from its base, and believes that, if an enemy was to lose command of his airspace then there would be a very strong possibility of his ultimate defeat. This strategic use of aircraft on a large scale would need continuous unrelenting attack on important targets. Then the enemy's communications would be destroyed, cutting him off from his supplies and thus his eventual surrender would become unavoidable. Lanchester sees this strategy as a means of "... compelling a bloodless victory." Such strategies and tactics can be reconciled with the events of the Second World War, by taking into account the relevant performance and armament up-dating; however the result was hardly a bloodless victory!

He was of course, looking ahead at a time when, in fact, the Flying Corps was "... scarcely more than necessary to constitute an armed reconnaissance service." Having said this, the tactics recommended at that time for aircraft combat, were to ensure for a given skilled pilot, that speed, climbing power and armament, were maximised, coupled with numerical strength. It was felt that tactical advantages would be gained through engaging a small number of the enemy by a large number of the attacking force "... in order to reap the advantage of the n-square law." It should be borne in mind throughout his work that much of Lanchester's thinking was biased towards the results of naval tactical experiences with ships. Furthermore, an airfleet should be manoeuvred,

"... to defeat the enemy in detail, and if his numbers are superior, to prevent him from bringing his whole concentrated fire to bear by the adroit handling of the weaker numbers ... to neutralise ... his numerically superior force."

Relevant comments indeed when viewed in the light of the Battle of Britain.

The future air force comprising squadrons or units would need to be formed as an independent airfleet, to operate free of routine or set duties. Aeroplane types would be based on the fast strategic scout, the slower tactical scout, possibly lightly armed and protected, the military support machine with armour and multiple machine guns, bomb-dropping machines, and specialised naval types. As already mentioned, superiority in speed and climbing power is of great importance, to engage and destroy the enemy. This of course, reduces to a question of sufficient horse power output from the engines and requires the provision of suitable armour plating to protect the attack from above. Lanchester feels that this protection would be minimal since it could affect the climbing performance necessary for the aeroplane to attain the tactically important advantage of height above the enemy. This latter advantage was very much relevant within air combat during the Second World War. As a rider it is interesting to note that he considers the height gain sufficiently important to point out the advantage of operating aeroplanes from airfields situated at altitudes of 6,000–8,000 ft.

The composition of the air force must be homogeneous in that it consists of units with the same speed, climbing power and armament, i.e., ideally of one design. The consequent strength lies in numerical superiority and the fleet must operate in unison upon the numerically inferior enemy. In this way the n-square law is used to fullest advantage. Having assembled such units, then future tactics would need to be developed with reference to formation flying, particularly as the number of aeroplanes becomes large within the raiding system. This comment compares with the evolution of the thousand bomber raids on Germany in the Second World War. Operations such as this demanded careful planning, organisation and operation, as did the large fighter wings evolved for tactical purposes by the R.A.F. At this stage, it is relevant to draw the attention of the reader to typical formation layouts. It is fairly well-known that over long distances, migrating birds often fly together with a flight-line in the shape of a "V", moving with its point forwards. In short, the bird flying behind and slightly to the left or right of the leader should be operating in a beneficial airflow where the net motion is upwards i.e., an upwash, which gives a better support for the following bird. Overall, this can result in a net saving in the work done by the formation and sometimes the leader of the "V" may change throughout the flight duration. By comparison, the formation with consecutive birds in line, has the disadvantage of the following bird suffering from a downward airflow (or downwash) set up by the bird in front. Hence formations tend towards a "V" format or a single diagonal line.

Further discussion on this subject is developed by Lanchester in 1917 (1917b, c, d) and also for example, at the beginning of the Second World War (1940). In the latter case he deals with the details of induced drag reduction available from formation flying. Furthermore, it is of interest to note that in his article "The Defence of London" (1917c), Lanchester summarises the advantages (and disadvantages) relative to flying in formation under operational conditions. As already mentioned, flying in formation results in a power saving and fuel economy. Tactically the aircraft have a concentration of fire power by acting as a single unit both in attack and defence. In the same vein the leader of the formation is able to direct his attack, having selected a particular target, at the last minute for example, and hence utilise the combined bombing power of all the machines. Alternatively if a small number of aeroplanes are used as bombers, then they can be protected by a formation of fighters, or scouts and the larger the formation the more difficult it becomes to attack and disperse. Lanchester believes that it would take an enemy formation to attack and effectively disperse another formation i.e., individual attacking aircraft would have little influence. Moving on to attack from the ground, then a formation would have the advantage of presenting a target to anti-aircraft fire, for a minimum time only. In addition, such a formation would make range finding very difficult by the observers since the techniques used at that time, relied on sightings of a *single* machine. Finally, Lanchester notes the disadvantages, such as difficulty of flight in poor weather, particularly for keeping station in cloud, and the obvious problem of a large number of aircraft in formation being easily spotted by the enemy.

Continuing with his book, he again suggests that (at that time) there was no overall advantage to be gained by designing larger aeroplanes – rather, to develop the small single man type as a stable flying platform. Also, he feels that for bomb-dropping it would be better not to have a second man to assist, but to utilise his weight in extra bomb load. Obviously design and tactics have moved on significantly since then, with the advent of the large Lancaster, Halifax and B-17 Fortress bombers for example, as used in the Second World War and then further to even larger machines as utilised during the last thirty or forty years in Korea and Vietnam. In the general buildup to operations with larger number of aircraft and their influence, Lanchester does not wish us to lose sight of the restrictions on potential air superiority, particularly within a worldwide context. Even during this First World War writers were predicting that command of the air could be comparable with, or more important than, command of the sea. He did not argue with this based on his conclusions at the time, mainly due to the fact that even allowing for future developments, the maximum radius of action of aircraft would be less than 1,000 miles and that without payload: whilst the Navy could operate around the world. This leads to his suggestion for a strategic policy, whereby in peacetime any one of the Great Powers could initially build up and secure its own local (territorial) limits, with appropriate bases on frontiers and coastlines. Then in time of course, it would defend itself, but also extend relevant forays from these bases in order to support the Army and the Navy wherever operating.

By implication, international problems could arise from the movement of aircraft and this raises the potential guidelines for flight over neutral territory. Such deliberations remain inconclusive, insofar as a certain amount of common sense should prevail. Still in a state of flux, the law could be interpreted through the ordinary law of trespass,

"... the owner is entitled to turn the trespasser off using only such force as is necessary and can claim damages only on account of actual injury sustained."

Again, conjecture reigns with regard to requirements for markings on aircraft to declare their nationality etc., but history shows that the flying of a flag does not necessarily mean that this refers to the nation concerned. Certainly in the air warfare stakes, neutral aircraft would most likely be shot down if entering hostile skies, there being no time to ask questions and clarify nationalities; in any case the markings shown could well be false. Landing an aircraft in a neutral country may well result, as is most likely, in it being interned in that country. Similarly, aircraft taking refuge in a given territory should not strictly be using any facilities for repair, refuelling, etc. hence, international convention needs to be fully established in this matter and any modifications required would result "... as the natural outgrowth from experience in warfare." Lanchester, throughout his work, indicates a poor opinion of the process and delineation of international law and convention. Typically, with regard to the necessity for the defusing of situations in neutral territories where conflict can arise, he quotes that this,

"... is almost lost sight of in a quagmire of dangerous and namby-pamby sentimentality."

Such comment is indicative of the then current politics relative to methods of destruction within the framework of warfare.

Looking forward to peacetime there would be a requirement for a large organisation with access to airfields, training schools, workshops etc., necessary to form the background to the training of air personnel. Lanchester voices his concern over the casualties arising from military flying, both then and potentially for future training, and stresses the need for better airfields so that accidents could be reduced, at least in Great Britain. Arguments of the day which defended the rough fields, suggested they gave useful experience for pilots and also further confirmed the strength of the aircraft itself. This latter test could possibly lead to a reduced reliability in the battlefield (the life of an aeroplane at that time was about three to four months on continuous active service). Furthermore, improved airfields were necessary to support night-flying operations, particularly with regard to the future. He also envisaged aeroplanes landing on public roads, possibly cleared and widened to accommodate emergency landings. I feel it worthy of note here, that Lanchester would have been very impressed to see how Sweden had built this facility into its strategic air defence policy particularly since it has been operating with jet aircraft on such roads. He has no shortage of ideas to reduce running costs and suggests that airfields should be used for grazing purposes by cattle and that they would be transferred from time to time, from one section of the field to another. The problems of birdstrikes encountered at many airfields/airports today, pale into insignificance when compared to this situation! Perhaps the whole issue of rough ground is put into context by quoting a subject fundamental to the British way of life, then as now,

"It is a serious reflection on our conduct as a nation that we have so far shown ourselves prepared to spend more money in the provision and upkeep of cricket fields, than we are ready to do for the safety of our flying men and the efficiency of the Aeronautical Arm."

As a rider to the comment on safety, it is worth noting the articles by Lanchester, which consider the influence of the wind on the aeroplane. The contributions on military and commercial matters (1917g, h) indicate how the wind affects the flight economy of the aeroplane, under conditions of fixed or variable flight speeds. Furthermore, the effect of the wind could have been detrimental to the interests of the British aeroplane (1917f) for example, in northern France where the prevalent winds are normally from the west and southwest. This means that the British aeroplane could be affected on returning from his sortie, (flying against the wind) possibly damaged and low on fuel. In short, Lanchester indicates that in the case of engine failure, the machine need not make it back to base, since the gliding angles were modified by the wind. Hence, the aeroplane could finish up landing behind enemy lines. Conversely the Germans could have an advantage in their gliding range, an advantage as Lanchester states "... often equal to a spare two-gallon can of petrol." Such amounts are akin to our present day automobile fuel tank reserves!

Returning to his book and considering the aeroplanes themselves, the British machines had better aerodynamics resulting in increased speed and climbing power for a given weight and engine output. They also possessed better stability, the latest machines at that time being inherently stable. In addition, they were more robust for landing purposes and had better weatherproof qualities, all points which assisted operation under service conditions. Nevertheless at the *outbreak* of hostilities, Great Britain did not have a successful fighter or gun-carrying aeroplane – it was not found easy to incorporate guns into aeroplanes. However, for reconnaissance purposes, the Air Arm was then well equipped, and later superiority it seemed was due to the "magnificent performance and daring of our flying-men", and a well designed and manufactured product. This latter situation was made possible by the work of private firms and the Government Factory at Farnborough. I would hope that it is fairly well known that the age of flight and aeronautical discovery has been well supported and influenced significantly by British pioneering work. The machinery set up by the Government included the Royal Aircraft Factory at Farnborough, the National Physical Laboratory and the Advisory Committee for Aeronautics, with the naval aspect represented by constructional works and a depôt at Aldershot which took over the Dirigible (balloon) Section. The majority of this framework of expertise thus brought together, has continued through to recent times. Unfortunately, the same cannot be said of the political scene in the more recent past. Great Britain has, over the last forty years for example, had a sorry list of discontinued projects, over-budget design etc., the result of an unhappy ill-programmed aviation policy by the Government. Perhaps only now, in nearing the millennium are we beginning to see a more understanding cohesive liaison between the aerospace industry and Government. However, in fairness, I have to state that these comments should be viewed against a greatly changed background to aviation with a significant movement of the goalposts, and a huge increase in technological development over recent years. Many of the preceding problems could arise due to a lack of awareness of aviation procedures and matters, and Lanchester continued to feel that a fair percentage of the general public did not understand that aeroplane design and construction involved scientific analysis as well as empirical methods.

With some hindsight Lanchester is able to look ahead with a strong recommendation for continuation of effort, particularly to maintain our technical superiority within the aeronautical scene. An adequate secrecy net should also be spread to cover much of the then relevant technical information and this could involve a retention of such material for up to a year before release or publication. However, the Government and public expected to see results from the use of their money which could make life awkward with the delay involved. Of course, the organisation and consequent flow of aircraft design and production is just as important as the efficient use of the aircraft in service. As I mentioned earlier, problems in this respect have plagued the British aircraft industry for many years and Lanchester professed an awareness of the problem in 1916. Accepting that fluctuations in the flow are inevitable we have still been left with cases of Government hesitancy and/or inefficient use of tax payers money, in the Vickers 1000, TSR2, AW681 for example. Lanchester states,

"... that anyone acquainted with the working of Government manufacturing institutions could cite innumerable cases of gross extravagance resulting from so-called Treasury economy."

Problems resulting from this type of thinking had carried through to private industry tasked with the design and production of aircraft. Unfortunately, he assumes that the private sector could absorb these fluctuations better, by fitting in the Government projects with their own, a situation not easily managed in the light of the general demand for aircraft.

Following on from these comments, detailed discussion leads to a suggestion of a Directorate or Board to ensure policy continuity, coupled with responsibility for a construction programme and for the required financial support from the Treasury. Hence the title "Board of Aeronautical Construction" to include the responsible heads of the Army and Navy, and the Head of the Royal Aircraft Factory along with civilian representatives of high standing within key areas. Appropriate cabinet ministers would form ex-officio membership. Lanchester is, of course, seeking a "Utopia" with regard to his proposals for continuity of aircraft design and construction but was particularly clear about the Board's responsibilities. One particular example stands out i.e.,

"... a refusal or a cutting down by the Treasury of the requisitions by the Board, either annual or supplementary, should be rendered next to impossible."

Indeed he felt so strongly about the matter that he looked to a resignation en bloc of the civilian section of the Board if any such thing happened. The reader is referred to additional articles by Lanchester (1916i–o, 1917e, k, l) which add background detail and discussion to the relevance of committees, boards and an Air Ministry, both here and abroad.

At the end of his book Lanchester turns his attention to national defence issues – they are considered as the more far reaching effects of aircraft in warfare. He concludes that Zeppelin raids on England had not paid off but felt that this could change with the advent of better organisation and effective operation on a larger scale. This leads to a warning against not being prepared for the bombing and hence the fire problem of large cities such as London, where administrative headquarters and centres form important targets. At that time, some people believed that attacks in force would not happen for humanitarian reasons, and thus they could not contemplate the destruction by fire of a city of five million peaceable inhabitants. Hence to be effective, a raid on London would knock out the administrative centres etc., thus it would be sensible to move the latter elsewhere beyond attack. Then the enemy, if he were "... destroying private property and murdering civilians ...," would expose himself to reprisal. Again Lanchester indicates his unhappiness with international convention, treaties and suchlike which he believes tie the hands of the defending powers. A reprisal has to be immediate, and conventions can result in delays while consultation takes place. Success by the enemy would be interpreted in the sense of causing a general conflagration due to fire. Defence on this score would be by the use of fire-resistant building construction and sensible town planning schemes to ensure fire localisation.

At the end of 1916, Lanchester was asked by a *Birmingham Daily Mail* reporter for his views on air raids (1916p) an opportunity which allowed him to address the general public on the problem, and in the following years he returned to this humanitarian issue, adding further discussion (1917a) to the contentious matter of air raids and reprisals. It would have been interesting to see the public reaction, had he included a quote from an earlier publication (1916k) which referred to a Government request for *economy* during the war,

> "Reports show an increase of consumption of bread and an increase of consumption of meat per head of population. Our cheap jewellery trade is experiencing a boom, our pianoforte trade cannot procure supplies fast enough, picture palaces are reaping a golden harvest."

The reader should appreciate that during the First World War, the traditions of the Empire and Services were still prevalent in the minds of the English people and thus they were not expecting to be bombed with London as a civilian target. Hence the war should be conducted, as Lanchester puts it "... without recourse to measures which may be looked upon as hitting below the belt." Indeed, he suggests that many Englishmen might have wished to see it carried out along lines "... which may almost be said to have been handed down from the ancient days of chivalry." In other words, Germany was not playing the game and he states "... the profession of arms has never been so degraded as by the German nation in the present war."

I think that Lanchester's comments are best summarised by a list:

i) The German raiders appeared, apparently unmolested over London,
ii) Shell fragments (from anti-aircraft guns) fell all over London causing damage,
iii) London thus smarts under an insult,
iv) The Englishmen are indignant,
v) Parliament goes into secret session,
vi) The man in the street would like to see somebody hanged, and
vii) The insolence of the enemy is paramount, and will be remembered for a very long time, whilst the bombing itself is nearly forgotten.

As in many areas, he felt that the general public showed apathy towards the build-up of an effective Air Service before the war, when the country would then have been better prepared for hostilities. With the relatively limited resources available during the war, he realised that the defence of London would require the use of a disproportionate number of aircraft squadrons to be kept in readiness at strategic points. This would take aircraft away from more important areas of the war with consequent dilution of effort and concentration. Tactics suggested (if these squadrons were available) would be to patrol at high altitude ready for interception. However, they could miss the enemy due to the problem of recognising an aircraft from above in other than ideal weather conditions. Again the prime issue of altitude is important since if the aircraft patrolled at low altitude they would run the risk of losing contact with the enemy during the climb required to intercept. Lanchester does at least concede (but only just), that the anti-aircraft guns might have kept the enemy at high altitude during the London raids, whereas if they had descended to a sufficiently low altitude they would have been better situated to drop their bombs with greater accuracy. He feels that the defence measures necessary for a city should be dependent upon the importance of the targets within it, i.e., the defence would be proportional to the incentive to attack by the enemy, as already mentioned. The latter of course would be reduced if the targets were moved elsewhere, out of bombing range, for example to the Belfast or Clyde area.

At that time, the ensuing public outcry raised the question of retaliation (1917i) and Lanchester feels that they (the public) should be made more aware of the operational issues involved, particularly in view of an announcement in the Press of 4th October 1917,

"... that in reply to the recent air raids on London, a vigorous policy of reprisals on German towns (my underlining) will henceforth be adopted."

He had often put forward his view of the Press in no uncertain terms hence the comments,

"... the bulk of what is written in the public Press, is no more than an unreasoned interpretation frequently to be heard when a short-tempered man gets an unexpected slap in the face ..."

Initial reaction to bombing was of course at the minimum one of indignation, as discussed earlier, (also see 1916g, h) but at that time compared to knowledge and strategy in the Second World War, little was known publicly about the operation of the "artillery of the air" as the aircraft of the Air Service had been called. The radius of action capabilities of the available aeroplanes then, could be up to 600 miles but this with almost negligible payload: 400 miles for a bombing squadron would be a better estimate. Hence any retaliatory raiding of German towns, in the extreme Berlin, could result in only one cwt of bombs being dropped, whereas ten cwt might be dropped on Essen, Dusseldorf or Cologne. The main strategy at that time was to drop substantial bomb loads on the Germans in Belgium, i.e., on military establishments and aerodromes. Such is the rate of progress in aviation that Bomber Command and the American Air Force had the aircraft and the capacity to deliver huge bomb-loads on vital German targets in the Second World War. Here was a deliberate strategy of saturation bombing in the main, within the enemy territory.

Returning to his book, Lanchester suggests that moving vital resources out of bombing range could be looked upon as a form of passive defence and he believed at that time, that they would remain safe from attack,

"... so long as we assume the motive power engine subject to its known restrictions as a form of heat engine, it may be regarded as safe for all time."

Assumedly, he considers this as a limit on aeroplane performance. He suggests also, that naval dockyards, shipyards and depôts should be built in "out of range" regions, along with production centres for aeroplanes and sea-planes. Furthermore, the Admiralty should take charge of Britain's defending air forces, rather than the War Office, since such defence would be closely linked to the distribution and operation of naval bases. In the future the Aeronautical Arm, with the greatest potential for development, would become of national importance. It would be available strategically during war and peace and be able to act decisively "... within a few hours of the outbreak of hostilities." In retrospect, he considered that Great Britain had the upper hand in military aeronautics due more to technical achievement than a state of national readiness for war. Looking ahead, I would think that Lanchester's latter comment on "readiness" might be viewed as particularly relevant in the prelude to the Second World War. In the case of Germany and its Zeppelins, he decided that they "... backed the wrong horse ..." in "... devoting altogether disproportionate attention to the large dirigibles. The Zeppelin from the military standpoint has proved a complete failure."

Strategically the upper hand in aeronautical armament in the future would be vested in the relative manufacturing resources available to the warring factions. However, much later on, Lanchester still leaves the question open, relative to commercial travel, in an article (1937) which refers to the demise of the Hindenberg. After some discussion, he is led to a final comment that,

"... the writer feels very doubtful whether the airship has any prospects of a commercial future at all, ..."

a statement which is still of relevance in this day and age!

In writing this book, Lanchester was carefully building up a case for a series of actions to be initiated then and for the future. I consider it worthwhile to quote these actions in their entirety to allow a direct comparison with the situation and policies leading up to and during the Second World War. He advocates (ostensibly in a wartime situation) "... an immediate and thoroughgoing overhaul of our programme and administration as touching the future of the Aeronautical Arm ..."

Consequently he urges:

"1.   That in view of the potentialities of the Aeronautical Arm, a comprehensive scheme of construction should be forthwith prepared, in which provision shall be made for organising, utilising and developing every available source of manufacture and supply.

2.   That if possible certain of our present type of aeroplane be virtually "adopted for the duration of the war", and existing manufacturing facilities should be utilised for their uninterrupted production to the utmost of their capacity.

3.   That where it is decided that new types are required, new sources of production should so far as possible be tapped or new works equipped, in order that output should not be made to suffer. In other words, the policy should tend in the direction of establishing each new type with the factory for its production as a complete proposition.

4.   That more adequate provision be made for the development of improved models and new types, both as regards initial manufacturing facilities and finance.

5.   That a Board of Aeronautical Construction (under the Presidency of a responsible Minister), be formed on the lines adumbrated in the present work, to deal with the needs of the Services and to settle specifications and approve the designs for new types, and generally to assume control and responsibility for our National Aeronautical Programme, both as to sufficiency and otherwise."

The reader should note that in action (2) Lanchester is advocating the continuation of a particular fixed aircraft design *without* improvements being added during manufacture. He argues that this is detrimental to output, especially when working under high pressure conditions. Furthermore, a given type of aeroplane becomes well known by the mechanics even though it "... may not be the best we know how to make."

Even towards the end of the war he continues to show interest in future policy and attempts to describe the building bricks necessary for this to happen (1917m), in the light of potential prejudices within the system. In this latter publication he still maintains his forthright approach to his writing and I quote (referring to fields of research),

"... the mathematician, however useful his work may be, is utterly incompetent so far as specifying or designing an aeroplane is concerned; his is actually the narrowest field, although nonetheless important. Unfortunately the mathematician is frequently a man with an intellect of childlike simplicity, and he is often unable to understand why anything beyond his symbols (which he usually keeps within easy reach of the end of his nose) should be allowed to count."

He goes further to describe that the,

> "... work of the physicist is vitally important, but this sphere is only one degree less narrow than that of the mathematician. For the physicist to invade territory outside his own ken would certainly not lead to satisfactory results."

Following on with the engineer whose job he sees,

> "... is frequently that of a buffer. He has to listen both to the mathematician and the physicist; he has to keep one ear open to the pilot, ... Above all he has to keep one eye on futurity."

The pilot, "... has to have his say ... His title to express an opinion cannot be gainsaid; his life is at stake." Finally, "... we have our military and naval authorities, whose requirements have definitely to be met." I suspect that Lanchester perhaps felt it was time to stir up the hornet's nest and liven up proceedings even more, as comments such as the preceding could cause some interesting discussion in the present day Press!

There is no doubt that he appreciated the need for teamwork with appropriate leadership and as he says,

> "It would be difficult to do justice to the keenness and enthusiasm which has characterised the work of Aeronautical development in all branches ..."

Returning to policy matters he maintains that past difficulties,

> "... have been due to two causes, one a failure to define properly the spheres or fields of activity of the different contributory factors; the other an absence of foresight and far-sighted policy, both in detail and in the gross."

He later adds that

> "Personally, my great fear for the future is that we shall witness a period of unwholesome political interference in place of sound political support. Therein lies the greatest danger of the future."

In retrospect I feel that much of my account reflects a world in which Lanchester encountered for example, the influence of vested interests, an uneducated, uninformed Press, a public with a lack of awareness and thinking still closely linked to the British Empire with its attendant rigidity and way of life. All this within the context of warfare strategy previously conducted along naval lines and with its entrenched influence on the future, allied to the dawn of military aviation and its inherent potential.

Hence I would close with a final quote from Fred Lanchester taken from a publication (1916k) which again discusses an independent air service i.e.,

> "One has only to reflect on the enormous accumulation of experience and data which has been necessary to render possible the organisation of a modern European army or fleet, to realise that the task of forming an independent air service, if eventually it should come to achievement, will be an affair of decades than years, and can only be considered as proven when it has emerged successfully from a first-class European War."

Indeed with reference to his book and work as described herein, his preceding comments form a concise backcloth to the "Dawn of the Fourth Arm" and of course a perceptive insight, by Lanchester, into the future.

THE INESTIMABLE SCOUT.

Aviator (who has landed badly):—"Thank you, sonny, I'm all right; but you might see if that cow needs any attention, and then I want you to take my card to the Rector and say I'm awfully sorry about the spire. After that you will please send off this wire, and ask the local bicycle plumber to come here—and above all, don't forget to bring some cigarettes! I'll stay and beat off the souvenir hunters."

[With acknowledgement to reference 1917m]

References

Please note that all of the following references (except one) are attributed to F.W.Lanchester either as written publications or as reports on his papers and personal views.

1894: The soaring of birds and the possibilities of mechanical flight, *General Meeting (19th June), Report on Meeting* (October), Birmingham Natural History and Philosophical Society

1907: *Aerodynamics*, Constable & Co. Ltd., London

1908: *Aerodonetics*, Constable & Co. Ltd., London

1909a: The Wright and Voisin types of flying machines: a comparison, *Aeronautical Journal*, Vol. 13, Jan., pp.4–12

1909b: The flight of birds, *The Engineer*, 19th Feb., pp.198–9, and 26th Feb., pp.225–6

1909c: Mechanical flight, *Times Engineering Supplement*, 3rd March

1909d: Mechanical flight, *Times Engineering Supplement*, 7th April

1915: Report on high altitude flying and the development and improvement of the aeronautical motor, Advisory Committee for Aeronautics, R. and M., No. 220

1916a: *Aircraft in warfare: the dawn of the fourth arm*, Constable & Co. Ltd., London

1916b: Britain's aeroplane policy, *Land & Water*, 10th Feb., pp.13–14

1916c: Britain's aeroplane policy, *Land & Water*, 17th Feb., pp.15–16

1916d: Development of the military aeroplane: the question of size, *Engineering*, 3rd March, pp.212–14

1916e: *Anon.* The question of size of military aeroplanes, *Engineering*, 31st March, p.296

1916f: The capabilities of an aeroplane in relation to its size, Advisory Committee for Aeronautics, Sub-Committee Note T.673/1, March

1916g: Aircraft policy and the Zeppelin menace from the national standpoint, *Land & Water*, 23rd March, pp.17–18

1916h: Aircraft policy, and the Zeppelin menace from the national standpoint, *Land & Water*, 30th March, pp.13–14

1916i: Air defence problems and fallacies: The failure of the Derby committee, *Land & Water*, 20th April, pp.12–14

1916j: Air defence problems and fallacies: Air Ministry or Board of Aeronautics, *Land & Water*, 27th April, pp.13–14

1916k: Air problems and fallacies: Air Ministry or Board of Aeronautics, *Land & Water*, 4th May, pp.13–14

1916l: Rise and fall of the French Air Ministry, *Land & Water*, 11th May, pp.10–12

1916m: The so-called "air-muddle", and some of its exponents, *Land & Water*, 18th May, pp.17–18

1916n: The Air Board, *Land & Water*, 15th June, pp.17–18

1916o: The sea and the air: the future of the Air Board: a lesson from naval history, *Land & Water*, 29th June, pp.13–14

1916p: Air raid precautions: why they are not yet available: Mr. Lanchester's views, *The Birmingham Daily Mail*, 12th Feb., p.6

1917a: The air raid of July 7th, *Flying*, 18th July, pp.479–80

1917b: Formation flying, *Flying*, 25th July, pp.3–4

1917c: The defence of London: II. Formation flying, *Flying*, 1st Aug., pp.19–20

1917d: Formation flying as taught by nature, *Flying*, 8th Aug., pp.35–37

1917e: A campaign of slander: a few facts, *Flying*, 15th Aug., pp.51–54

1917f: The handicap of the wind, *Flying*, 29th Aug., pp.83–84

1917g: Flying as affected by the wind: the economic aspect, *Flying*, 12th Sept., pp.115–16

1917h: Flying as affected by the wind, *Flying*, 26th Sept., pp.147–8

1917i: The military aspects of reprisals, *Flying*, 17th Oct., p.200

1917j: Vertical climb or direct lift: the chaser of the future, *Flying*, 14th Nov., pp.259–60

1917k: The Air Ministry, *Flying*, 21st Nov., pp.275–6

1917l: The Air Ministry, *Flying*, 5th Dec., pp.307–8

1917m: The foundation stones, *Flying*, 19th Dec., pp.354–6

1937: The future of the airship, *Engineering*, Vol. 143, No. 3724, 28th May, pp.613–14

1940: Formation flying, *Flight*, Vol. 37, no. 1631, 28th March, pp.286e–286f, and pp.287–9

Chapter Nine

MATHEMATICS IN WARFARE: LANCHESTER THEORY

by Prof. K.C. Bowen and K.R. McNaught

Introduction

Just prior to the First World War, F.W.Lanchester had some ideas about how to examine certain aspects of battle, primarily those concerned with concentration of force. These ideas were documented firstly in a series of articles in *Engineering* (Lanchester, 1914), and later in a book (Lanchester, 1916), on the impact of air power on warfare. As a result, his name has become one to be conjured with in a field far from his mainstream activities.

Although Operational Research (O.R.) was not born until another twenty years had passed, and did not become a formal discipline until after the Second World War, one area of military O.R. bears Lanchester's name. Mathematical models of battle take many forms, but a very large group are governed by Lanchester Theory, which enables the conditions for winning or losing to be established in principle. Lanchester is also immortalised through The Johns Hopkins University Lanchester Prize, an annual award, since 1955, by the O.R. Society of America for notable contributions to Operational Research.

Nothing much seems to have been applied directly until after the Second World War, although Lanchester's work was studied and developed by U.S. defence scientists, mainly those in the Operations Research Group of the U.S. Navy. Their work (Morse and Kimball, 1963 which makes special mention of Koopman, 1943) was openly published in 1951 and most post-war workers "discovered" Lanchester from this interesting account. Few have, in fact, seen Lanchester's 1916 book! However, his chapter on *Mathematics in warfare* became more easily available in 1956. It was reprinted in *The World of mathematics* (Newman, 1960), described as "a small library of the literature of mathematics from A'h-mosé the Scribe to Albert Einstein presented with commentaries and notes by James R.Newman". In volume four, Lanchester is in glorious company with G.H.Hardy, Henri Poincaré, John van Neumann, A.M.Turing, Claude Shannon, Sir James Jeans, Augustus de Morgan and many others (even Stephen Leacock and Lewis Carroll are there!)

Lanchester was not the first to produce equations of battle. Equivalent ones were developed by Admiral Fiske of the U.S. Navy eleven years previously (Weiss, 1962; Engel, 1963a). Fiske's ideas, however, were less well developed and it was Lanchester's thorough consideration of the implications of very simple representations of battle that eventually caught the attention of military analysts in many countries.

The main purpose of this contribution is to give an account of what has happened to Lanchester Theory over the last fifty years. It will be far from complete. In the English language alone, there are at least five hundred published papers and books in the open literature. In the restricted literature of the defence world there must be far more, even not counting work in such countries as Russia*, Japan, Germany, Israel, Egypt, Sweden, France and so on, where Lanchester Theory has certainly not been ignored.

So, a flavour of Lanchester's impact is all that can be offered. It is necessary to start with a brief summary of Lanchester's twenty pages that set it all going. Those who wish to know

---

* Osipov (1915) discovered the square law independently. A translation has recently been published. It is not known what influence this had on Russian studies of warfare.

more should read his own account, a splendid piece of writing, concise, incisive and creative: the mathematics is comparatively straightforward and full understanding demands only elementary calculus.

Concentration and the "n-square" law

Lanchester discusses two different situations of battle. Ancient warfare demanded a series of duels, one man against one man at any moment, so that concentration in any part of the field of battle yielded little or no advantage. Only if special circumstances allowed concentration of "fire-power", as when an army broke and ran or when disciplined archers could dominate areas of an enemy's advance, would there be a departure from the condition that equal losses would be suffered if the combatants' individual fighting values were equal.

Modern warfare, he argues, can be modelled by the assumption that casualties per unit time are proportional to the numerical strength of the opposition. Under these circumstances, concentration of force is shown to have major effects on the outcome. Lanchester gives numerical examples based on two equations which describe the rates of loss of Blue and Red forces:

$$\frac{db}{dt} = -r \times c, \quad \frac{dr}{dt} = -b \times k,$$

where b and r represent numerical strengths at time t, and c and k are constants (individual kill rates).

Lanchester deduces that such a representation of battle implies that if M and N (proportional to k and c) represent the efficiencies, or fighting values, of individual units of Blue and Red, there is then a condition of equality of battle given, at all times t, by

$$Nr^2 = Mb^2.$$

This is Lanchester's square law. He interprets it as saying that *the fighting strength* of a force may be broadly defined as proportional *to the square of its numerical strength multiplied by the fighting value of its individual units* (his italics). The law is called the "n-square law" in the 1916 book, although it was the "n^2 law" in the 1914 articles. n is nowhere defined, but, since N is used to represent fighting strength, it makes most sense to interpret "n-square" as being the fighting strength of an individual unit (n) multiplied by the square of the number of units.

Clearly, if two forces of 1000 each, say, and of equal fighting value per unit meet, they are of equal fighting strength. Unless one side can bring its full force to bear successively on separate parts of the enemy force, the result is a "draw", with equal losses on each side at any time. However, if, for example, 1000 can first engage with 500, they will eliminate them for a loss of only 134. The remaining 866 will then engage the other 500 and annihilate them for a further loss of 159, leaving them victorious with 707 remaining. The figure below, illustrating this, is Lanchester's.

With simple examples of this sort, accompanied by graphical illustrations of the progress of battle, Lanchester examines a range of assumed or historical circumstances, looking at various sizes and concentrations of force and differing unit fighting values. He also discusses different formulations of battle: for instance, long-range fire is seen to imply a rate of loss which is also proportional to the size of the force attacked (a greater density of targets).

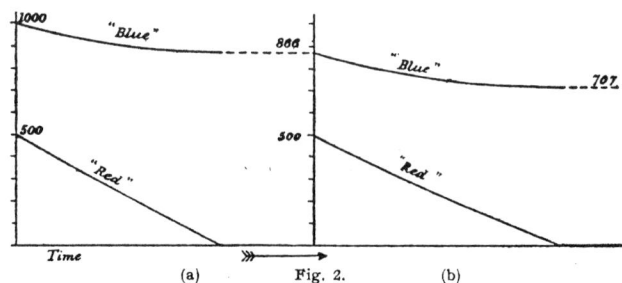

Fig. 2.

This particular case is shown to result in a linear law, albeit for different reasons than those which posit the linear law of "ancient warfare"; the former is the linear law that will be referred to later.

Importantly, Lanchester stresses the limitations of simple models as well as their value. In showing that Nelson's tactical skills implied that he acted as if he were aware of the square law, he comments that Nelson might well have had some similar mental model (based perhaps on experience and common sense). In general, Lanchester is clear that models of warfare can only offer general guidance and cannot generate scientific truths about the consequences of battle, whatever broad agreement might be found between "theory" and practice.

Indeed, at an early stage in his essay, Lanchester addresses the fact that his square-law model assumes a continuous process of attrition, whereas, in practice, integer values only are involved, losses occurring at discrete intervals. (There are senses in which a fraction of a unit lost could be regarded as a reduction in its efficiency, but this is not generally an easy concept to model.) Consequently, the square-law solution is seen as not being satisfactory at later stages of a battle; indeed, in an evenly balanced battle, the equations imply that it continues ad infinitum, whereas, in practice, the last man on one side is eventually killed.

Many of the points which Lanchester makes have been addressed in considerable mathematical depth over subsequent years. Studies of historical battles and different formulations of battle, generalised rather than specific or probabilistic rather than deterministic, have led to new insights into military problems. Yet, in many cases, Lanchester's square-law and other simple models have been sufficient, despite the inherent unrealities to which Lanchester drew attention.

Comment on the post-Lanchester period

In the sections that follow, no judgement will be made on the "operational" relevance of individual "advances" in Lanchester Theory. It is apparent, however, that the military and their analysts have devoted a large effort to achieve greater sophistication and realism in modelling for particular studies of strategy, tactics and equipment development. It is fair to assume that, overall, value for money has been gained.

Nor will any judgement be made on the past or future relevance of the many ingenious mathematical excursions that solving variations of Lanchester's equations have encouraged. But although, inevitably, some of the challenges faced have undoubtedly attracted mathematicians to Lanchester Theory for purely mathematical reasons, most of these have been defence scientists or academics working on defence contracts and so constrained to practical ends.

What is of importance here is the enormous scope and the potential value of work that has been done in furthering defence interests. Although papers referenced are primarily concerned with conventional warfare, it is reasonable to infer that many aspects of deterrence would have been studied by similar modelling. Whether similar work would have been done without Lanchester is unknowable, but in simple fact what exists today has stemmed from the brief but brilliant commentary of one man.

Lanchester as force in history

The heading of this section is the title of a paper by Willard (1963), who examined land battles over the period 1618–1905. Although he found that, for many reasons, simple Lanchester equations or slightly modified versions, gave only a moderate or poor fit to the outcomes of these battles, they did provide a valuable starting point for his inquiries. Among the issues which seemed to influence discrepancies were the need for a stochastic treatment (use of earlier work by Brown (1955), of whom more later, gave a much improved fit between theory and observation), and the absence of any functions of command and control. On this latter aspect Willard says that "the equations seem to be averaging over dehumanized events".

Helmbold (1961a and b) had done similar studies of ninety-two battles and, although Willard was worried by Helmbold's concentration on an assumption of square-law type battles, it appears that the two findings are complementary. Willard also noted that others (Snow, 1948; Weiss, 1957) had questioned the appropriateness of square or linear laws in most battles.

At the other extreme, Engel (1954) obtained a remarkable verification of Lanchester's Law as a description of the Second World War action that captured the island of Iwo Jima. He added one feature only, namely the rate at which friendly forces entered combat. The Japanese forces on the island had no support from outside and fought to total attrition. It was perhaps a unique situation. In later private correspondence, Engel (1981) made some important comments on his published paper.

a. He had suggested in that paper that conclusions of more general interest and value could be obtained by looking at a large number of specific and largely comparable combat situations. He believed that such work was in fact done and much of value had been gained, although little had been published.

b. The good fit does not mean that alternative models should not have been tested and he referred to work on "guerrilla warfare" generalisations of Lanchester's work.

c. Accordingly, the paper *did not* prove that the results of the battle of Iwo Jima were generated by processes inherent in the Lanchester formulations used.

There are a number of papers that could be possible candidates for Engel's a. and b. above. In particular, Karns (1954) used the Iwo Jima data as an example of amphibious assault and examined the theoretical effects of variations in the build-up rates of U.S. forces, and Varma (1957) applied a similar model to an assault on Breskens Pocket. Among the guerrilla warfare papers that combine theory and historical data are those of Deitchman (1962) and of Schaffer (1967). To these can certainly be added Engel's own paper (1963b) on the Second World War invasion of Crete, although this was not a final version of the research under way.

What is referenced above can be only a small sample of the work undertaken to gain from the past. There was comparable work in almost all areas of military interest seeking to move from data, to models, and thence to recommendations for future operations and future equipment development. Other authors have used Lanchester formulations as a small part of much wider studies of military history.

To assess realistically what came from it all would need reference back to those who are mentioned and to many others. They were all, of course, engaged in a wide range of defence studies and the knowledge or ideas generated from their Lanchester work on history may have coloured their many endeavours in useful but indefinable ways. Direct experience of warfare is not generally part of an analyst's expertise: playing in artificial battles is not a bad substitute. How good it can be made is the subject of the next two sections.

Generalised Lanchester equations

Many of the models already referred to, including one studied by Lanchester, depart in small ways from the square-law formulation. However, they remain analytically simple. It has long been understood by operational researchers that, in dealing with complicated situations, simple models that provide useful insights are very often to be preferred to models that get so close to the real world that the mysteries of the world they intend to unravel are repeated in the model and remain mysteries! Lanchester's models score heavily here.

Brackney (1959) concentrated on basic issues of warfare and produced what he called a "theoretically simple combat dynamics" which paralleled the known facts of warfare. Both in the simplicity and the style of treatment, this paper might well have been Lanchester's second offering had he sought to push his ideas further. Brackney made comparisons between eight different combat situations. One was the "linear law" situation of which Lanchester had been aware: another was defined by a "mixed law" with the two sides differently disposed, e.g. the defender hidden and the attacker exposed. Deitchman (1962) and Schaffer (1967) used the latter concept in their guerrilla warfare studies cited above. Karr (1983) has drawn attention to the many sets of assumptions and interpretations that can lead to the various laws. Other examples of simple models that push Lanchester's formulations a little further are given by Bowen (1973). One describes a U.S. Air Force study of the value of aircraft shelters: this was probably a part of wider studies of the survival of deterrent forces against pre-emptive strike, and comparable studies of missiles vs missiles were also carried out. A second example examines submarine vs escort forces, as a stage in estimating the probability of survival of a protected aircraft carrier. Bram (1962) developed comparable models for various naval situations: his work has a special interest because the stochastic nature of the battles was also treated – computers were beginning to open up possibilities.

Very early on, Morse and Kimball had drawn attention to the ways in which Lanchester equations could be generalised to deal also with reinforcement (or withdrawal) and "operational attrition" (losses not due to enemy fire). They too did not take generalisation beyond the power of closed analytical solutions. Dolansky (1964), summarising the state of Lanchester Theory ten years later went a little further by dealing with area fire and heterogeneous forces.

Helmbold (1965) provided a generalisation of which the square law and the linear law were special cases, so also was the logarithmic law (Peterson, 1967) which could have relevance to the early stages of combat when losses may be proportional to a force's size only. Helmbold's primary concern was to deal with reduction in the relative efficiency of a larger force as observed in historical data (see earlier references (Helmbold, 1961a and b)). More recently, Helmbold (1994) has restated his belief that "fighting strength" requires a function of size of force that gives much less weight to size than the Lanchester square law suggests.

Bonder provided several models aiming to predict weapon performance in battle. In one of these (Bonder, 1963), he stresses the aim to provide analytical expressions for the performance parameters. Not only does he deal with kill-rate probability distributions, but also with mobility and degrees of damage. This tendency to push mathematical ingenuity to the limit was, despite the undoubted insights gained by the analysts, beginning to have serious shortcomings in communication with the military.

How far could one and should one go? There was no doubt about the instructive use of quite complex formulations in differential equation form; even if solutions in closed analytical form are elusive, numerical integration is possible. However, for combinatorial reasons, relationships between variables and sensitivity to changes in parameters will be harder to establish. Appendix A (pp. 151–152) is a brief indication of the scope for inquiry: it is a statement of the mathematical expressions that summarised a study of the essential parameters of military environments (Bowen, 1962). It mentions one paper, (Clark et al, 1960), in which mobility treatment was very detailed.

But increasingly, there was concern that few of the models paid attention to the probabilistic nature of battle and to the fact that such chance effects occurred at discrete intervals and not continuously. While some models did introduce this stochastic element, this was often tailored to retain analytical treatment and solution. The following section looks at both early ideas and at a particular extensive treatment that took place in the 1970s.

The probability distributions of battle states

Morse and Kimball defined P(m,n,t) as "the chance that at time t there are m Red units and n Blue units still unhit". They then considered "an elementary engagement lasting a time dt" and produced the following model of the battle, where they used E as the "exchange rate", the ratio of the individual attrition rates on the two sides, adjusting the time scale appropriately:

$$\frac{d}{dt}P(m,n,t) = m\sqrt{E}[P(m,n+1,t) - P(m,n,t)]$$

$$+ \frac{n}{\sqrt{E}}[P(m+1,n,t) - P(m,n,t)], \quad m,n \neq 0;$$

$$\frac{d}{dt}P(m,0,t) = m\sqrt{E} \ P(m,1,t);$$

$$\frac{d}{dt}P(0,n,t) = \frac{n}{\sqrt{E}}P(1,n,t);$$

$$P(m_0,n_0,0) = 1; \text{ and}$$

$$P(m,n,0) = 0, \ m \neq m_0 \text{ or } n \neq n_0.$$

m_0 and n_0 are the m and n values at $t = 0$.

They then say that these can be solved although *the calculations are tedious* for large m_0, n_0. Simple examples are given showing solutions to be polynomials in e^{-t}. What is meant by *large* m_0, n_0 is not made clear, but the number of variables in these equations is $[(m_0 + 1)(n_0 + 1) - 1]$. Clearly, direct analytical solutions and their evaluation very soon became impossible. Computers were in their infancy when Morse and Kimball produced this model, and, even for $m_0 = n_0 = 10$, handling the equations with 120 variables would have needed arithmeticians with the patience and will-power that Napier showed in producing logarithmic tables in the early seventeenth century. But there were important early insights.

Morse and Kimball have also observed that the solutions of the equivalent Lanchester square-law equations will correspond closely to the "expected values" obtained from the probability analysis as long as Lanchester's equations are not "pressed too hard (such as by going to the limit of annihilation of one force)". This necessarily vague restriction is commented on further in Appendix B (pp. 152–153), based on work by Bowen (1975).

Brown (1955) looked at the probability of winning, and provided a very useful approximate analytical solution. He also developed an integral form for the solution to Morse and Kimball's model, but the integrand was too complicated, in general, for any practical uses.

Brooks (1965) examined large battles and showed that Lanchester square law and linear law models, when expressed in stochastic (probabilistic) form, were "stochastically determined", that is that the casualties to each side showed a very small spread compared to the initial numbers involved. In these cases, the deterministic models gave useful approximations.

Smith (1965; 1967) provided treatments of the probability distributions of survivors as a basis for establishing satisfactory rules for attrition in war games. This was probably the first extensive study of Lanchester-type stochastic models linked directly to practical assessment work, and certainly the first in the UK. He drew on work by Brackney (1959) and Brown (1955) which has been cited earlier.

Bowen (1967) used a somewhat cumbersome process to evaluate the first and second moments of the probability distributions of survivors, but in the process provided two conclusions which were later confirmed by Weale (1971):

a) the stochastic model gives expected values close to those of the deterministic Lanchester model at all times at which the probability of either side being annihilated is small.
b) there is a *square-law* relationship in the stochastic model that connects the expected values of the squares of the number of survivors, involving also the expected values of the numbers of survivors.

These results are proved in Appendix B using a more direct method than that used in the references above.

Weale's work has been extensive and provides full numerical solutions to the stochastic square-law battle and many extensions of this. Many thousands of battle-states can be handled efficiently. The Fortran programs, using the Taylor series method, are fully documented. The first of his six papers, (Weale, 1971), describes an efficient model for solving the Morse and Kimball equations for probabilities of the states of battle. Bowen (1973) gives a brief account of this and provides a number of diagrams illustrating the complex nature* of the bivariate distributions of survivors and the probability distributions of battle termination (shown only for annihilation of one side or the other, although Weale's programs allow various definitions of battle termination). Gye and Lewis (1974; 1976) developed analytical forms for the terminal distributions, when termination implies annihilation. Goldie (1977) gives analytical forms for both the transient and terminal distributions as more generally treated by Weale.

Weale's other papers cover calculation of the moments of the probability distributions of battle states (Weale, 1972; 1977) (a very useful way of describing distributions); distributions using more general attrition functions (Weale, 1976a; 1977); the distribution of the duration of battle, (Weale, 1976b); and battles with several types of weapon system on each side (Weale, 1978). The interesting feature of all this work was that it was mathematical research with a strong operational flavour, i.e. it was driven by what was seen to be needed to construct useful operational models. Earlier extensive work, for example that by Bonder to which one reference has been made earlier, or that by Taylor and his researchers (e.g. Taylor and Brown, 1974; Taylor and Comstock, 1976) at the Naval Postgraduate School, Monterey, California, was generally operational research with a strong mathematical flavour, i.e. it was driven by specific operational needs. Weale's work is unique as a sustained and coherent mathematical inquiry. Whether it or the more specific problem-by-problem acquisition of new approaches prove more influential in the long-run will be interesting.

Related approaches

The assumptions usually associated with Lanchester theory (particularly the deterministic variety) make it most suitable for battles involving large force sizes. Ancker and Gafarian (1987) provide a useful discussion of these assumptions. This limitation led some researchers to develop the theory of stochastic duels, thus permitting a mathematical treatment of very small engagements.

In this context, stochastic refers to the random times between shots fired by each combatant. An early introduction to a variety of one-on-one stochastic duels is given by Williams and Ancker (1963). Here they emphasise the importance of single shot kill probabilities and rates of fire as the parameters of fundamental importance in a stochastic duel. They describe and solve the "fundamental duel" and the "classical duel" before going on to consider duels involving surprise.

In the fundamental duel, both combatants enter the duel simultaneously with unloaded weapons. Their firing processes are independent and each combatant's inter-firing times follow an appropriate probability distribution. The single shot kill probabilities are assumed to remain constant over time. In the classical duel, both combatants begin with loaded weapons and fire their first rounds simultaneously, thus permitting a draw. The classical duel then continues in the same way as the fundamental duel. In the duel with random initial surprise, one of the duellists is allowed to fire at his opponent for a random period of time before any shots are returned and the duel then returns to a normal pattern.

---

* This is one reason why computer simulations do not provide efficient solutions unless a very large number of runs of the battle are analysed.

Numerous additions to the theory of one-on-one duels have been made by a number of contributors over the years, such as the effects of limited ammunition, limited time-duration, projectile flight times and non-constant single shot kill probabilities. Wand, Humble and Wilson (1993), using a different method of solution, explicitly incorporated detection within a one-on-one stochastic duel. Williams (1963) established connections between duels and Lanchester's square and linear laws. An early history of stochastic duel theory and its place in relation to Lanchester theory and game theoretic approaches is due to Ancker (1966). More recently, Ancker (1982) has given an updated account of the theory, including a compendium of research results and an annotated bibliography of research papers.

Groves (1964) provided a solution to the fundamental duel with fixed firing times using a Markov chain approach. Barfoot (1974 and 1989) also utilised the theory of Markov chains to investigate one-on-one duels where the single shot kill probabilities depend on the outcome of the last round fired and firing times are assumed to be discrete. He later extended this work to allow for exponentially distributed inter-firing times.

Slightly larger engagements have also been tackled. An early approach by Williams (1965) only considered fixed firing times. More recently, Gafarian and Ancker (1984) have provided an analytic solution to the two-on-one engagement with stochastic firing times while Gafarian and Manion (1989) have extended this approach to the two-on-two situation. Kress (1987, 1991 and 1992) considered a many-on-one stochastic duel, a two-on-one stochastic duel with manoeuvre and a many-on-many stochastic duel model for a mountain battle. (N.B.: "duel" is often used as a synonym for engagement in this field, regardless of the number of combatants involved.)

Another area related to both Lanchester theory and the theory of duels is volley fire methodology. A volley is a set of shots fired in parallel, without adjustment, by a group of weapons at a target array. Of interest is the resulting attrition to the target array. Obvious important applications include artillery and ballistic missile exchange. Helmbold (1992) has recently gathered together the previously fragmented work in this field, concentrating on the single volley. He remarks that "... the current state of the theory does not provide a very satisfying treatment of multiple or counter volleys", nonetheless single volleys "... can be chained together in various (often ad hoc) ways to estimate the effects of successive volleys, if so desired."

Some recent developments

Over the last ten years or so, military O.R. contributions to refereed journals have been primarily in the areas of game theory, logistics, search theory and stochastic duels. Work on Lanchester theory has by no means ended, however, and the following examples show how it is still being extended in various directions.

We have already mentioned Schaffer's work which was effectively an attempt to include some of the effects of information in a model of battle. Indeed, it is the first such attempt known to the authors. More recently, Conolly and Pierce (1988) have taken this idea much further. Their model of battle consists of three differential equations – one for each side's loss rate and one for the rate of change of information. In fact, only one side's information gathering is included in the model – if it were to be included for both sides, four differential equations would be required. The amount of information in existence at any particular time represents knowledge of the enemy's target locations at that time. It increases at a rate proportional to the information gatherer's force size and decays if information gathering stops. In addition, both sides are reinforced at constant rates. In this model it is assumed that the side not gathering information only uses unaimed fire, while the other side uses a greater proportion of aimed fire as its information about enemy target locations grows. As Conolly and Pierce point out, however, such behaviour is only sensible if the aimed fire is sufficiently effective in relation to the unaimed fire.

After both analytic and numerical analysis of these equations, Conolly and Pierce come to the following general conclusions. First of all, stalemates can occur when a small force with good target information holds a much larger force in stable equilibrium. The added production of information can compensate for initially unfavourable force ratios. Also, there appears to be an "information production threshold", around which small changes in the information production rate can cause dramatic changes in the final results.

Given the ever-increasing sophistication and importance of C^3I on the battlefield*, work such as this is both necessary and important. Moreover, it emphasises once again what a powerful and flexible platform, for basing an analysis of the factors important in a battle, Lanchester's original equations provide.

Protopopescu, Santoro and Dockery (1989) have partly considered some of the spatial aspects of battle. By replacing Lanchester's ordinary differential equations with partial differential equations, one-dimensional spatial effects have been modelled. Such equations are required since the force levels are no longer just functions of time, but functions of a one-dimensional space as well. Their intention is to extend this approach to two spatial dimensions which should permit an envelopment to be directly represented.

In a series of articles, Woodcock and Dockery attempt to integrate combat theory with catastrophe theory. They are particularly concerned with trying to embed the effects of command and control in a combat model within the framework of catastrophe theory. They show that the loss rate equations of several Lanchester-type models, including the square, linear, logarithmic and mixed law models can be generated from a structure known as a double cusp catastrophe (1988a). In another paper, using a system dynamics approach, they look at command and control in low-level insurgency warfare (1988b). Drawing on an analogy between nearest-neighbour interactions on the battlefield and the interactions between cellular automata, Woodcock, Cobb and Dockery (1989) developed a combat simulation with temporal and spatial aspects, based on cellular automata mathematics. Applying this model to a variety of scenarios including combat penetration and envelopment, they showed that the results from at least some of these simulations were still fairly well fitted by Lanchester-type models.

Barr, Weir and Hoffman (1993) have developed an "indicator of combat success" based on Lanchester theory. This is a measure of combat effectiveness they call the "battle trace", which varies during the course of a battle and shows how well the two sides are performing relative to one another. It highlights the occurrence of important events and as such could be a useful tool in post-wargame and post-exercise analysis.

Roberts and Conolly (1992) considered an extension of the Lanchester square law to a heterogeneous battle where the attacking side had two weapon types compared to the defending side's one. They went on to use their results to tackle an optimisation problem – minimising the resources needed by the attacker to secure annihilation of the defender.

Another active research area is the aggregation and disaggregation inherent in combat models. Hillestad and Juncosa (1995) explore various approaches to the problem of aggregating several dissimilar weapon system types in order to estimate the strength of one side in a battle with a view to calculating attrition based on a Lanchester square law model. This then leads to the problem of disaggregating the losses over the various weapon types. Davis and Huber (1991) emphasise the continuing need for combat models at a variety of resolutions to support decision making and problem solving at different levels. They advocate the development of an integrated hierarchical family of models. Hillestad, Owen and Blumenthal (1995) describe some experiments they carried out with two simple combat models of different resolution. They show that seemingly intuitively reasonable aggregations can be misleading. They also try to fit a two-stage constant coefficient Lanchester model to their results. McNaught (1994) shows how significantly different outcomes are produced when a battle is broken up into several smaller engagements before being resolved using a stochastic Lanchester model. These smaller engagements are known as mini-battles or fire fights as suggested by Rowland (1984), Bathe (1984) and Ancker and Gafarian (1991). The

* Command, Control, Communications and Intelligence.

degree of battle splitting and the scale of the battle are both shown to be important to the probability of winning and other measures of effectiveness.

Epidemiology – a diversion

In the 1960s, there was a passing interest in the mathematics of epidemiology. There seemed to be a possibility that something akin to a battle was being studied. However, the process considered (Bailey, 1960) was different, being concerned not with the elimination of agents responsible for disease and death, but with the removal from the susceptible population, by isolation or death, of those infected. The language of warfare was not appropriate.

Nevertheless, the mathematical formulations of epidemics did have similarities to those of battle, as represented by generalised Lanchester equations (see Appendix A). Two aspects are noteworthy. Firstly, equations treating the progress of an epidemic as deterministic, defined by rates of infection and rates of removal of infectives, were, in general, not amenable to analytic solutions; numerical analysis did however, provide solutions for the less complicated models.

Secondly, there was dissatisfaction with all deterministic models, unless the populations involved were very large, since what was ideally required were solutions for the probabilities, at any given time, that various states of the epidemic would be reached.

With present-day computer facilities, the methods of solution are very similar to those used in battle model studies. Griffiths and Wheeler (1991) give an example of a deterministic model, while Griffiths, Smedley and Weale (1987; 1989) deal with stochastic solutions. In the latter cases, the similarities to Weale's work on Lanchester Theory in the 1970s (Weale, (1992), is a more readily available reference to those cited earlier) are perhaps not surprising!

Because of the way that ideas beget ideas, a subject that started with the merits of simplicity, and provided insights of great value, can grow into great complexity. The models that are developed need solution-methods of increasing sophistication and the solutions themselves may be of uncertain applicability. It is not inappropriate therefore to draw attention to a paper that looked at the transmission of ideas as an epidemic (Goffman and Newill, 1964). This work assumed the spread of ideas to be benevolent, given the availability of well-designed information-retrieval systems to cope with the problem of identifying which papers are relevant for what purposes.

To a limited extent, the earlier sections have offered some clues of the relevance of some of the literature for which Lanchester was the initial infective, reinforced by those who, during and immediately after the Second World War, spread his ideas widely. The penultimate section deals briefly with the size of the epidemic as indicated by available bibliographies.

Bibliographies

The first published bibliography of Lanchester papers, mainly those in English, came, characteristically, from F.O.A., Sweden (Wrigge and Wigg, 1975). It contains 230 entries although some of these are peripheral. It also draws attention to papers which compare the combat models with the reality they are supposed to describe. Interestingly, it includes references to models of interacting biological populations.

Five years later, Haysman and Mortagy (1980) produced a listing of 180 papers which they described as "important open references in the field of deterministic and stochastic Lanchester models". This was part of a research study which expanded when, in 1981, Haysman and Daly formed the U.K. Lanchester Study Group (Daly, 1985). This group, brought together, informally, those in the U.K. who were Lanchester "experts", and several in the United States and elsewhere who wished to be corresponding members.

The output of this group was an increasingly comprehensive listing of Lanchester references, 465 in 1982, growing to 485 in 1985. Listings are available (Program WAR-TEXT) from the addresses given at the end of the list of references to this chapter.

It is to be noted that papers presented at the N.A.T.O. Conference on Recent Developments in Lanchester Theory, Munich, Germany, July 1967 are included, e.g. Schaffer, 1967; Bowen, 1967. Unfortunately, the Proceedings of this conference were not published: the edited material was mislaid and never recovered. The papers, or most of them, probably exist in several private or research institution holdings, e.g. most are known to have been placed in the library of the then Defence Operational Analysis Establishment, Ministry of Defence, U.K. They are all unclassified.

James G.Taylor's prolific contributions to the development of Lanchester theory in order to make it more directly applicable to operational problems deserve a special mention, in particular, his development of variable attrition coefficient solutions and the use of Lanchester models in optimisation problems. A full account of these and other developments in both the deterministic and stochastic theory can be found in two sizeable volumes by Taylor (1983). A comprehensive list of references is provided at the end of each chapter. Although many are repeated, there are at least 500 different references in all. A much slimmer account of the main features of deterministic Lanchester theory, also by Taylor (1981), makes an excellent and concise introduction to the subject.

Final comment

Over the last fifty years, a veritable army of mathematicians has created a body of theory describing the process of battle. The authors of this account of that theory, Lanchester Theory, have concentrated on what has most influenced their own research. Of the papers referenced, not all will be universally acclaimed and some major contributions to the field will inevitably have been overlooked.

We make no apologies for this. Our purpose has not been to honour those who have followed Lanchester, but to indicate something of the variety, ingenuity and sheer size of the inquiries that his seminal ideas have stimulated. These inquiries are continuing. They may well provide new material for future festschrifts to honour a man whose contributions covered so many years and so many applications.

APPENDIX A   Generalised Lanchester equations

Weapon systems $W_1, \ldots, W_n$ consist of $w_1(t), \ldots, w_n(t)$ units of firepower at time t.
Opposing systems $W_1', \ldots, W_n'$ consist of $w_1'(t), \ldots w_n'(t)$ units.
Against W_s', W_r has a "standard" effectiveness k_{rs}, i.e. each unit of firepower of W_r can nullify k_{rs} units of W_s' per unit time. "Standard" refers to defined environmental conditions of battle and there will be a modifying term, l_{rs}, whose components depend on

a. local conditions applying to W_r against W_s' (weather, terrain, radio propagation, etc);
b. external support and information available to W_r;
c. measures taken by W_s' to mitigate the effectiveness of attack by W_r; and
d. special vulnerability of W_s' to attack by W_r (e.g. morale, lack of replacement personnel, etc.) and similar special effects on the efficiency of W_r, not taken into account in a. and b. above.

Note that k_{rs} may be a function of w_s': l_{rs} may be similarly dependent on w_s' in cases c. and d. above.

Firepower allocation may be allowed for by λ_{rs}, the proportion of w_r devoted to reducing W_s'. Here, $\sum_{r=1}^{n} \lambda_{rs} \leqq 1, 0 \leqq \lambda_{rs} \leqq 1$ for all r,s. The choice of λ_{rs} may affect k_{rs}, l_{rs} through problems of control, and effects on morale and logistics.

There is wear and tear of battle independent of any direct effects of opposing weapon systems. This relates to a multitude of considerations of training, leadership, maintenance, supply and so on. For simplicity, let the rates of change per unit of firepower be μ_r, μ_s'.

Steady reinforcement, replacement (or withdrawal) can be allowed for by terms c_r, c_s'. Finally, supporting systems must be included; these are regarded as weapon systems with no offensive capability. All the above leads to equations of combat,

$$-\frac{dw_s'}{dt} = \sum_{r=1}^{n} \lambda_{rs}(k_{rs}+l_{rs})w_r+\mu_s'w_s'+c_s', \quad s=1,\dots n', \text{ and}$$

$$-\frac{dw_r}{dt} = \sum_{s=1}^{n'} \lambda'_{sr}(k'_{sr}+l'_{sr})w_s'+\mu_r w_r+c_r, \quad r=1,\dots,n.$$

Time-dependencies of the parameters, and such events as discontinuities in units engaged, provide immense potential complications. Mobility for example, may involve movement that affects weapon effectiveness and, as for many other aspects of battle, necessitates a redefinition of the set of equations for future time periods. There may also be a dependence of k_{rs} on range (defined in terms of time and speed of movement). Clark et al (1960) have taken this further in small land battles, in which k_{rs} was also dependent on dw_r/dt, since W_r would have to adjust the range in the light of the effect of opposing firepower: they used step-by-step numerical methods to obtain solutions. The variations possible seem endless, limited only by simplifying assumptions to make the equations amenable to solutions, with accompanying adjustments to such conclusions that can be safely drawn.

APPENDIX B   "Equivalent" stochastic battle models and the Bowen square-law

Consider a battle between two sides, M and N. Let $p(\mu, \nu; t)$ be the probability of μ and ν units, respectively, remaining at time t. In a small interval dt, let the probabilities of the loss of a unit by M be $\beta\nu dt$ and by N be $\alpha\mu dt$, provided that each side has at least one unit remaining. The probability of losses being incurred by both sides or more than a single loss by either side are taken to be of the order of $(dt)^2$.
The mean values at t are

$$E(\mu) = \sum_{\mu=1}^{M_0} \sum_{\nu=0}^{N_0} p(\mu, \nu; t).\mu \text{ and}$$

so that

$$E(\nu) = \sum_{\mu=0}^{M_0} \sum_{\nu=1}^{N_0} p(\mu, \nu; t).\nu,$$

where M_0, N_0 are the initial numbers of units and $E(.)$ is an expected value in the usual statistical sense.
The decrement in $E(\mu)$ over the interval dt is

$$dE(\mu) = - \sum_{\mu=1}^{M_0} \sum_{\nu=1}^{N_0} p(\mu, \nu; t)\beta\nu dt,$$

so that

$$\frac{dE(\mu)}{dt} = - \sum_{\mu=1}^{M_0} \sum_{\nu=1}^{N_0} p(\mu, \nu; t)\beta\nu$$

$$= - \beta[E(\nu) - E(\nu|\mu = 0)\Pr(\mu = 0)].$$

Similarly,

$$\frac{dE(\nu)}{dt} = - \alpha[E(\mu) - E(\mu|\nu = 0)\Pr(\nu = 0)].$$

Provided that $\Pr(\mu = 0)$ and $\Pr(\nu = 0)$ are small, the mean values are given by the Lanchester square-law equations based on attrition rates of $\beta\nu$ and $\alpha\mu$. In that sense, the battle model considered is the stochastic "equivalent" of the Lanchester square-law model. However, it can be shown that the time scale of the Lanchester model is greater than that of the stochastic model, and this seems to be generally the case in so far as "equivalence" can be defined (Bowen, 1975).

It may also be shown that

$$E(\mu^2) + dE(\mu^2) = \sum_{\mu=1}^{M_0} \sum_{\nu=1}^{N_0} p(\mu, \nu; t)[(\mu-1)^2 \beta \nu dt + \mu^2(1 - \beta \nu dt)] + \sum_{\mu=1}^{M_0} p(\mu, 0; t)\mu^2$$

$$= E(\mu^2) - 2\sum_{\mu=1}^{M_0} \sum_{\nu=1}^{N_0} p(\mu, \nu; t)\beta \mu \nu dt + \sum_{\mu=1}^{M_0} \sum_{\nu=1}^{N_0} p(\mu, \nu; t)\beta \nu dt,$$

so that
$$\frac{dE(\mu^2)}{dt} + \frac{dE(\mu)}{dt} = -2\sum_{\mu=1}^{M_0} \sum_{\nu=1}^{N_0} p(\mu, \nu; t)\beta \mu \nu = -2\beta E(\mu \nu).$$

There is a similar equation with $E(\nu^2)$ and $E(\nu)$, and it follows that

$$a[E(\mu^2) + E(\mu)] - \beta[E(\nu^2) + E(\nu)]$$

remains constant throughout the battle. Thus, the fighting strength of M is, in Lanchester's terms, proportional to $a[E(\mu^2) + E(\mu)]$.

This is Bowen's stochastic square law. It was first derived in a paper (Bowen, 1967) presented to a N.A.T.O. Conference on Lanchester Theory in 1967. It was confirmed in a more elegant and more general treatment by Weale (1971). The derivation above is essentially the same as that of Weale, but it is more direct: it has not been published in this form before.

References

Ancker, C.J. Jnr. (1966): The status of developments in the theory of stochastic duels – II, *Opns. Res.*, **15**, 388–406

Ancker, C.J. Jnr. (1982): One-on-one stochastic duels. Research monograph published by the Military Applications Section, Operations Res. Soc. of America

Ancker, C.J. Jnr. and Gafarian, A.V. (1987): The validity of assumptions underlying current uses of Lanchester attrition rates, *Naval Res. Log.*, **34**, 505–34

Ancker, C.J. Jnr. and Gafarian, A.V. (1991): An axiom set for a theory of combat. A paper presented to the T.I.M.S./O.R.S.A. Conference, Nashville, U.S.A.

Bailey, N.T.J. (1960): *The Mathematical theory of epidemics*, Griffin & Co., London. (The second edition, 1975, is entitled *The Mathematical theory of infectious diseases*.)

Barfoot, C.B. (1974): Markov duels, *Opns. Res.*, **22**, 318–30

Barfoot, C.B. (1989): Continuous-time Markov duels: theory and application, *Naval Res. Log.*, **36**, 243–53

Barr, D., Weir, M. and Hoffman, J. (1993): An indicator of combat success, *Naval Res. Log.*, **40**, 755–68

Bathe, M.R. (1984): Modelling combat as a series of minibattles. TRASANA Report LR-14–84, White Sands Missile Range, New Mexico, U.S.A.

Bonder, S. (1965): A generalized Lanchester model to predict weapon performance in dynamic combat systems. Research Group Report No. RF-573 TR65–1. The Ohio State University, Columbus, Ohio

Bowen, K.C. (1962): An examination of the essential parameters of a military environment relevant to gaming or analytical studies of limited war. D.O.R. Note 17/63. Department of Operational Research, Admiralty, Ministry of Defence, London

Bowen, K.C. (1967): A Lanchester-type stochastic model related to the operational evaluation of A.S.W. weapon systems. A paper presented to the N.A.T.O. Conference on Recent Developments in Lanchester Theory. Defence Operational Analysis Establishment, Ministry of Defence, U.K.

Bowen, K.C. (1973): Mathematical battles, *IMA Bulletin* **9**, 310–315

Bowen, K.C. (1975): A comparison of the duration of a deterministic battle and the mean duration of an "equivalent" stochastic battle. D.O.A.E. Research Working Paper L4. Defence Operational Analysis Establishment, Ministry of Defence, U.K.

Brackney, H. (1959): The dynamics of military combat. *Opns. Res.*, **7**, 30–44

Bram, J. (1962): A Lanchester-type model for combat between submarines, carrier task groups and hunter-killer groups. IRM-22, Operations Evaluation Group, Washington, D.C.

Brooks, F.C. (1965): The stochastic properties of large battle models. *Opns. Res.*, Jan.–Feb. 1965, 1–17

Brown, R.H. (1955): A Stochastic analysis of Lanchester's theory of combat. Technical Memorandum ORO-T-323. Operations Research Office. The Johns Hopkins University, Bethesda, Maryland

Clark, D.K., Keefer, L.E. and Walton, W.W. Jr. (1960): FOE, a model representing company actions. Technical Paper ORO-TP-17. Operations Research Office, The Johns Hopkins University, Bethesda, Maryland

Conolly, B.W. and Pierce, J.G. (1988): *Information mechanics*, Ellis Horwood, Chichester

Daly, F. (1985): The U.K. Lanchester Study Group, *OMEGA, Int. J. of Mgmt Sci.*, **13**, 131–3

Davis, P.K. and Huber, R.K. (1991): Variable-resolution combat modelling: motivations, issues and principles. RAND Report N-3400-DARPA, RAND Corporation, Santa Monica, U.S.A.

Deitchman, S.J. (1962): A Lanchester model of guerrilla warfare, *Opns. Res.*, **10**, 818–27

Dolansky, L. (1964): Present state of the Lanchester theory of combat, *Opns. Res.*, **12**, 344–58

Engel, J.H. (1954): A verification of Lanchester's law, *Opns. Res.*, **2**, 163–71

Engel, J.H. (1963a): Comments on a paper by H.K.Weiss, *Opns. Res.*, **11**, 147–50

Engel, J.H. (1963b): Combat effectiveness of Allied and German troops in the World War II invasion of Crete. Interim Research Memorandum, IRM-35. Operations Evaluation Group, Washington, D.C.

Engel, J.H. (1981): Private communication to K.C.Bowen

Gafarian, A.V. and Ancker, C.J. Jnr. (1984): The two-on-one stochastic duel, *Naval Res. Log. Quart.*, **31**, 309–24

Gafarian, A.V. and Manion, K.R. (1989): Some two-on-two homogeneous stochastic combats, *Naval Res. Log.*, **36**, 721–64

Goffman, W. and Newill, V.A. (1964): Generalisation of epidemic theory: an application to the transmission of ideas, *Nature*, **204**, 225–8

Goldie, C.M. (1977): Lanchester square-law battles: transient and terminal distributions, *J. Appl. Prob.*, *14*, 604–10

Griffiths, J.D., Smedley, J.K. and Weale, T.G. (1987): Terminal distributions along a "knight's line" for a stochastic epidemic, *IMA J. of Maths. Applied in Medicine and Biology*, **4**, 69–79

Griffiths, J.D., Smedley, J.K. and Weale, T.G. (1989): Modelling the spread of an epidemic. In *Applications and Modelling in Learning and Teaching Mathematics*, Ellis Horwood, London, pp.300–5

Griffiths, J.D. and Wheeler, K.A. (1991): Modelling the spread of AIDS. In Dunstan, F. and Pickles, J. (eds) *Statistics in Medicine*, Clarendon Press, Oxford, pp.65–82

Groves, A.D. (1964): The mathematical analysis of a simple duel, Ballistic Research Laboratories Report 1261, Aberdeen Proving Ground, Md, U.S.A.

Gye, R. and Lewis, T. (1974): Some new results relating to Lanchester's square law. Research Report. Department of Mathematical Statistics and Sub-department of O.R., University of Hull

Gye, R. and Lewis, T. (1974): Lanchester's equations: mathematics and the art of war, *Math. Scientist*, **1**, 107–19

Haysman, P.J. and Mortagy, B.E. (1980): References to the Lanchester theory of combat to 1980. Department of Management Sciences Working Paper OR/WP/6. R.M.C.S., Shrivenham

Helmbold, R.L. (1961a): Lanchester parameters for some battles of the last two hundred years. Combat Operations Research Group, CORG-SP-122

Helmbold, R.L. (1961b): Historical data and Lanchester's theory of combat. Combat Operations Research Group, CORG-SP-128

Helmbold, R.L. (1965): A modification of Lanchester's equations, *Opns. Res.*, **13**, 857–9

Helmbold, R.L. (1992): Foundations of the general theory of volley fire. U.S. Army Concepts Analysis Agency Report CAA-RP-92-1, Bethesda, Md, U.S.A.

Helmbold, R.L. (1994): The constant fallacy: a persistent logical flaw in applications of Lanchester's equations, *Euro. J. of O.R.*, **75**, 647–58

Hillestad, R. and Juncosa, M.L. (1995): Cutting some trees to see the forest: on aggregation and disaggregation in combat models. *Naval Res. Log.*, **42**, 183–208

Hillestad, R., Owen, J. and Blumenthal, D. (1995): Experiments in variable resolution combat modeling. *Naval Res. Log.*, **42**, 209–232

Karns, C.W. (1954): Lanchester's equations applied to an amphibious assault. Operations Research

Group, M.I.T., Office of Naval Research, Washington, D.C. (presented at the Chicago meetings of O.R.S.A., May, 1954)

Karr, A.F. (1983): *The Mathematics of conflict.* Systems and Control Series, 6. North Holland, Amsterdam

Koopman, B.D. (1943): see Morse and Kimball (1963) for details

Kress, M. (1987): The many-on-one stochastic duel, *Naval Res. Log,* **34,** 713–20

Kress, M. (1991): A two-on-one stochastic duel with maneuvering and fire allocation tactics, *Naval Res. Log.,* **38,** 303–13

Kress, M. (1992): A many-on-many stochastic duel model for a mountain battle, *Naval Res. Log.,* **39,** 437–46

Lanchester, F.W. (1914): Aircraft in warfare: the dawn of the fourth arm, parts V and VI – the principle of concentration, *Engineering,* **98,** 422–3 and 452–4 (October 2 and October 9)

Lanchester, F.W. (1916): *Aircraft in warfare: the dawn of the fourth arm,* Constable and Co., London. [Apart from minor editing, this reproduces a series of sixteen articles in *Engineering,* **98,** September 4–December 25, 1914. Importantly, a section giving a proof of statements deduced from the equations of battle is added to the October 2 article.]

McNaught, K.R. (1994): The effects of nodal structure on stochastic exponential Lanchester battles. A paper presented at the T.I.M.S./O.R.S.A. Conference, Boston, U.S.A.

Morse, P.M. and Kimball, G.E. (1963): *Methods of operations research* (pages 63–67). The M.I.T. Press, Cambridge, Massachusetts. (This is a revised edition, ninth printing, of the original 1951 book previously available as an internal defence document, 1946.)

Newman, J.R. (ed.) (1960): *The World of mathematics, Volume Four* (pages 2138–57). George Allen and Unwin Ltd., London. (This is the first edition to be published in Great Britain – first publication 1956.)

Osipov, M. (1915): The influence of the numerical strength of engaged sides on their losses (in Russian). Translation by Helmbold, R.L. and Rehm, A.S. (1995) in *Naval Res. Log.,* **42,** 435–90

Peterson, R.H. (1967): On the "logarithmic law" of attrition and its application to tank combat, *Opns. Res,.* **15,** 557–8

Protopopescu, V., Santoro, R.T. and Dockery, J.T. (1989): Combat modelling with partial differential equations, *Euro. J. of O.R.,* **38,** 178–83

Roberts, D.M. and Conolly, B.W. (1992): An extension of the Lanchester square law to inhomogeneous forces with an application to force allocation methodology, *J. Opl Res. Soc.,* **43,** 741–52

Rowland, D. (1984): Field trials and modelling. D.O.A.E. Report, Ministry of Defence, West Byfleet

Schaffer, M.B. (1967): Lanchester models of guerrilla engagements (*Opns. Res.,* **16,** 1968), presented at the N.A.T.O. Conference on recent developments in Lanchester Theory, Munich, 1967

Smith, D.G. (1965): The probability distribution of the number of survivors in a two-sided combat situation, *Op. Res. Q.,* **16,** 429–37

Smith, D.G. (1967): The probability distribution of the number of survivors in a two-sided combat situation: comparison with experimental data. A paper presented to the N.A.T.O. Conference on Recent Developments in Lanchester Theory, Munich 1967. R.A.R.D.E., Ministry of Defence, U.K.

Snow, R.N. (1948): Contributions to Lanchester attrition theory. RA-15078. RAND Corporation, Santa Monica, California

Taylor, J.G. and Brown, G.C. (1974): A mathematical theory for variable-coefficient Lanchester-type equations of "modern warfare". NPS-55TW 74111, Naval Postgraduate School, Monterey, California

Taylor, J.G. (1976): Optimal fire-support strategies, NPS-55TW 76021, Naval Postgraduate School, Monterey, California

Taylor, J.G. and Comstock, C. (1976): Force-annihilation conditions for variable-coefficient Lanchester-type equations of modern warfare – I. Mathematical Theory NPS-55TW 76081, Naval Postgraduate School, Monterey, California

Taylor, J.G. (1981): Force-on-force attrition modeling. Research monograph published by the Military Applications Section of the Opns. Res. Soc. of America, Arlington, Virginia

Taylor, J.G. (1983): *Lanchester models of warfare* (two volumes). Military Applications Section, O.R. Society of America, Washington, D.C.

Varma, R.S. (1957): Lanchester's equations applied to assault on Breskens. *Defence Science Journal*, January 1957, Defence Science Laboratory, New Delhi

Wand, K., Humble, S. and Wilson, R.J.T. (1993): Explicit modelling of detection within a stochastic duel, *Naval Res. Log.*, **40**, 431–50

Weale, T.G. (1971): The Mathematics of Battle – I. A bivariate probability distribution. D.O.A.E. Memorandum 7129 Defence Operational Analysis Establishment, Ministry of Defence, U.K.

Weale, T.G. (1972): The Mathematics of Battle – II. The moments of the distribution of battle states. D.O.A.E. Memorandum 7130 (as above)

Weale, T.G. (1976a): The Mathematics of Battle – V. Homogeneous battles with general attrition functions. D.O.A.E. Memorandum 7511 (as above)

Weale, T.G. (1976b): The Mathematics of Battle – VI. The distribution of the duration of battle. D.O.A.E. Memorandum 76126 (as above)

Weale, T.G. (1977): The mathematics of Battle – VII. Moments of the distribution of states for a battle with general attrition functions. D.O.A.E. Memorandum 77105 (as above)

Weale, T.G. (1978): The Mathematics of Battle – VIII. The Heterogeneous Battle Model. D.O.A.E. Memorandum 78106 (as above)

Weale, T.G. (1992): Two numerical methods for computing the probability of outcome of a battle of Lanchester type, *J. Opl Res. Soc.*, **43**, 797–807

Weiss, H.K. (1957): Lanchester-type models of warfare. In *Proceedings of the First International Conference on Operational Research*, O.R. Society of America, Baltimore

Weiss, H.K. (1962): The Fiske model of warfare, *Opns. Res.*, **10**, 569–71

Willard, D. (1963): Lanchester as force in history: an analysis of land battles of the years 1618–1905. Technical paper RAC-TP-74 Research Analysis Corporation, Bethesda, Maryland

Williams, T. (1963): Stochastic duels – II. System Development Corporation Report SP-1017/003/00, Santa Monica, California, U.S.A.

Williams, T. and Ancker, C.J. Jnr. (1963): Stochastic duels, *Opns. Res.*, **11**, 803–17

Williams, T. (1965): Some discrete processes in the theory of stochastic duels, *Opns. Res.*, **13**, 202–16

Woodcock, A.E.R. and Dockery, J.T. (1988a): Models of combat with embedded C^2 [Command and Control] – I. Catastrophe theory and the Lanchester equations, *Int. CIS J.*, **2**, 34–62

Woodcock, A.E.R. and Dockery, J.T. (1988b): Models of combat with embedded C^2 – III. Recruitment, disaffection and the tactical control of insurgents, *Int. CIS J.*, **3**, 5–38

Woodcock, A.E.R., Cobb, L. and Dockery, J.T. (1989): Models of combat with embedded C^2 – V. Cellular automata, *Int. CIS J.*, **4**, 5–44

Wrigge, S. and Wigg, L. (1975): Lanchesterteori och andra matematiska beskrivningar av strid: en bibliografi. F.O.A. Rapport C 10035-M6 Forsvarets Forskningsanstalt, Stockholm, Sweden

WAR-TEXT bibliographical listings can be obtained from
  National Technical Information Services
  U.S. Department of Commerce
  5285 Port Royal Road
  Springfield, VA22151, U.S.A.
or from
  University Microfilms International
  P.O. Box 1764
  Ann Arbor, Michigan 489106, U.S.A.

Chapter Ten

APPLICATION OF n-SQUARE LAW TO BUSINESS

by Yoichi Takeda

I. Lanchester's principles

The key to success in business is to increase profits and win a battle against many competitors. While it is difficult, if not impossible, to invent an effective managerial strategy that guarantees victory, Lanchester's principles provided extremely useful insights into a successful strategy. Triggered by the need to suggest effective military strategies for the First World War that broke out on 28th July 1914, F.W.Lanchester came up with the following two principles on 2nd October.

> The first principle (known as the "linear law"):
> Fighting strength = the number of units × fighting value of each unit
> The second principle (known as the "square law"):
> Fighting strength = square of the number of units × fighting value of each unit

The purpose of this paper is to discuss possible applications of these two principles to business and examine how effectively they may help businesses to increase their sales and profits.

1 *The first principle*

The first principle is valid when it is applied to a close combat in which short range weapons are used. Provided that both sides possess the weapons and soldiers with equal capacity and efficiency, the fighting strength is proportional to the numerical strength multiplied by the fighting value of its individual units. In the ancient wars, in which only swords and spears were used as weapons, the soldiers first needed to go sufficiently close to their enemies, and then engage in one-on-one fight, since the fighting range of these weapons was limited to only two metres or so. Thus the first principle is sometimes referred to as a principle of man-to-man fight.

When the soldiers used spears, they formed one line across the field on either side, and dashed themselves against the enemies. In fact such a spear-to-spear fight is often called "a suicide attack" or "a crash fight". In such a fight the death toll is the same on either side regardless of the initial number of soldiers. Another example that illustrates this principle is when a fighter plane crashes into the other in an air battle. The amount of loss is the same on either side, since both planes crash.

Case No. 1: Force A possesses one hundred men and Force B has sixty. If the fight continues until all sixty men in Force B are killed, there will be the same amount of loss on either side, since Force A also loses sixty men.

Case No. 2: Force A has two hundred men and Force B sixty. When they fight until entire Force B is wiped out, the amount of loss is still the same, as sixty men are also dead in Force A.

In one-to-one combats with short range weapons, either force suffers the same amount of damage regardless of the initial number of units. The exceptions to this principle are when a large force completely surrounds a small force, and when troops are attacked off guard. When one looks into the combat fatalities in wars in Japan before guns were first used, one will find that approximately the same number of soldiers were killed on either side. Records of ancient wars in England and Europe, if available, would show the same result as in the wars in Japan. This is the summary of the first principle.

2 *The second principle*

We will now turn to Lanchester's second principle. This principle is valid only when long range weapons are used, and the forces are far apart from one another. Provided that there is not a great difference in the efficiency of weapons or soldiers, the fighting strength is proportional to the square of its numerical strength multiplied by the fighting value of its individual units. How does the distance between the forces account for the difference of the square numerical strength?

Suppose Force A consisting of five soldiers equipped with rifles and Force B with two men with rifles engage in a combat, keeping the maximum range distance. Each soldier of Force A targets either one of the soldiers of Force B. There is, therefore, the probability of a soldier in Force A killing a soldier in Force B is one out of two. Since there are five men in Force A, the expected damage calculated for Force B is: $1/2 \times 5 = 5/2$.

Force B also strikes back. Each soldier targets one of the five men in Force A, and the probability of a soldier in Force B killing a soldier in Force A is one out of five. The expected damage of Force A is: $1/5 \times 2 = 2/5$. The rate of damage expected to be inflicted on the two forces is, therefore, four (Force A) to twenty-five (Force B).

Since the fighting strength is the reverse of the value of damage, it is said that Force A possesses fighting strength of twenty-five, as opposed to Force B possessing fighting strength of four.

To further illustrate, the formula to calculate Force A's damage is: the square root of the difference between the square of A's numerical strength and the square of B's numerical strength. If A has one hundred men and B has fifty, and they fight until all fifty soldiers of Force B are killed, there will be thirteen casualties and eighty-seven survivors in Force A.

Now how will the result change, if the rifles with a longer shooting range are used and there is a greater distance between the forces?

Case No. 1:   How many soldiers will survive and how many will be killed, when Force A with one hundred men and Force B with sixty fight until all the sixty soldiers in Force B are killed.

The numbers of survivors and casualties can be calculated by finding out the square root of the value gained by subtracting the square of sixty from the square of one hundred. Twenty men will be killed and eighty will survive. In other words when all sixty men of Force B, the inferior side, are killed, there will be twenty casualties in Force A, the superior. This means that there will be three times as much damage in the inferior force as in the superior force.

Case No. 2:   The superior force has two hundred men, and the inferior force has sixty. The difference between the square of two hundred, i.e., 40,000 and the square of sixty, i.e., 3,600, is 36,400. The square root of 36,400 is 190. This means that 190 men will survive and ten lives will be lost, and that the damage to the inferior force is six as opposed to one to the superior force, a six-fold difference.

Furthermore if there is a ten-fold difference in the numerical strength between two forces, then the ratio of damages is as much as one to twenty.

All of these examples point to the fact that the greater the difference in numerical strength between forces in space fight, i.e., keeping distance from one another, the more disadvantage the inferior force faces.

The ratio of numerical strengths			The ratio of damages		
A		B	A		B
100	vs	60	1	to	3.0
150	vs	60	1	to	3.2
200	vs	60	1	to	6.5
300	vs	60	1	to	9.9
400	vs	60	1	to	13.2
600	vs	60	1	to	20.0

Table 1   Sample Calculations of Damage Ratios

3 *Case studies in wars*

When Japan and the United States first went to war with each other, the loss was greater on the American side, as the capability of the Japanese Zero fighters was superior. The U.S. State Department ordered a project team consisting of mathematicians and physicists to study to find a strategy that would help the U.S. military to gain superiority over the Japanese.

The project team discovered a promising theory by applying the Lanchester principles and game theory. The essence of the theory was to triple the fighting strength of the U.S., i.e., to use a three-plane force against one Zero fighter. The U.S. Air Force adopted this theory and tripled the number of air fighters. The ratio of fighting strength in this case can be found by squaring the numerical strength of each side: square of one to square of three equals one to nine. In the beginning of 1945 the U.S. Air Force began to use four planes against one Japanese fighter, and the difference in fighting strength was as much as sixteen times.

The American airplanes had more horsepower, which enabled them to carry nearly twice as much ammunition as the Japanese planes. Despite a superb fighting capability of the Zero fighters, the Japanese had little possibility to win the war. Eventually the ratio of damage was one for the U.S. against ten for the Japanese. Air fight clearly proves the validity of the n-square law.

The difference in fighting strength in ground battle can also be calculated by applying the n-square law. The U.S. military adopted the theory in their battle in Okinawa, site of the greatest battle during the Pacific War. The number of U.S. troops that landed in Okinawa was 2.73 times as many as their Japanese counterparts.

Furthermore a convoy of battleships and aircraft carriers lent support to the landing troops. Including the support from the sea, the difference in numerical strength was four times altogether, making the ratio of fighting strength one to sixteen. The U.S. also used many times more ammunition than Japan. As a result there were ten times more casualties on the Japanese side than the U.S., which again proves that the n-square law can explain the effect of numerical difference in fighting forces.

The United States used the same strategy in virtually every battle against Japan during the war, and as a result Japan lost 1.5 million lives whereas the number of U.S. casualties was 150,000, or one-tenth of that of Japan. Use of long range weapons put the n-square law into effect, and made it impossible for the Japanese to rely solely on their spiritual strength and willpower. There was only the physical and numerical difference in fighting strength between the two forces. This is a characteristic of a modern war, and the many lost lives have helped to demonstrate the validity of Lanchester's principles.

The same principle was adopted when the British attacked General Rommel in North Africa. They used three times as many troops as their counterparts. Data concerning European wars or their detailed analyses are not available, but we can imagine that the same principles can be applied to account for the damages.

II. Application to business management

We will now see how the n-square law may be applied to actual business management. Some observable differences available for a comparison of power in business are the salesroom space of a department store or a supermarket and their gross sales. When comparing two nearby department stores or supermarkets that are competing against one another, we can use the n-square law to find out the relationship between the salesroom space and sales. The sales are proportional to the square of salesroom space. As department stores attract their customers from a large commercial area, the principle used to explain the effect of long range weapons can be used, i.e., the n-square law.

To illustrate, suppose one department store with salesroom space of 3,000 square metres and the other with 1,500 square metres are in direct competition. The ratios of salesroom space is two to one, but the ratio of fighting strength is four to one, which means that there should be a four-fold difference in total sales.

While the running cost such as personnel expenditure, air conditioning, lighting, and rent per square metre should be all the same in each store, the sales per square metre is twice as much in the larger store. The larger the salesroom is, therefore, the more the sales. Conversely the smaller store has a great difficulty winning the competition.

We can use some data from several department stores in Hiroshima, Japan, as specific examples to demonstrate the validity of the n-square law applied to business. A recent newspaper article has featured the general trends of department stores in the centre of the city. The article listed Sogo, Fukuya, Tenmaya, and Mitsukoshi with their salesroom space and annual sales. Mitsukoshi may be known in England also as it has a branch store in London.

A quick look at the chart indicates that the sales are proportional to the square of the salesroom space of the respective store. Sogo Department Store that has the largest salesroom space is overwhelming the rest, and Mitsukoshi with the smallest salesroom is having the most difficult time.

Another example, which I used in my lecture to the Junior Chamber of Commerce in Kumamoto, my neighbouring town, illustrates the same principle. The top department store in Kumamoto is Tsuruya, and the runner up is Iwataya Isetan. The ratio of the salesroom space is 10 to 7.35. When squared, it is 10 to 5.4, and it is amazingly the exact ratio of net sales between the two stores.

When two stores are far apart from each other, or when there is a large river, valley, or mountain separating the two, the margin of error becomes great. The closer the two stores are, the narrower the margin of error and the more valid the n-square law.

When a sales representative sells door-to-door, the profit before tax is proportional to the number of representatives. The profit before tax is gained by subtracting personnel and other business expenses, and income from the gross sales. When Company A has thirty sales representatives, and Company B has fifteen in the same territory, the ratio of numerical strengths is two to one. The profit before tax per representative is four to one. These examples come from Japan, but I am certain that the same can be seen in England as well.

Stores	Salesroom space (square metres)	simple ratio	squared ratio	sales (hundred million)	sales ratio
1 Sogo	34,700	1.00	1.00	680	1.00
2 Fukuya	30,500	0.88	0.77	480	0.71
3 Tenmaya	22,000	0.63	0.40	228	0.34
4 Mitsukoshi	16,400	0.47	0.22	143	0.21

Table 2 Ratios between salesroom space and net sales: The case of department stores in Hiroshima

Stores	Salesroom space (square metres)	simple ratio	squared ratio	sales (hundred million)	sales ratio
1 Tsuruya	33,698	1.00	1.00	650	1.000
2 Iwataya	24,756	0.73	0.54	350	0.538

Table 3 Ratios between salesroom space and net sales: The case of department stores in Kumamoto

1 *Effective strategies for superior and inferior forces*

Lanchester's principles and their application to sales management have been so far discussed in this paper. It is important to reiterate the core concepts of the principles.

In a man-to-man battle between two forces with one hundred men and sixty men, respectively, each force will lose the same number of men. The ratio of damage is, therefore, one to one. When one force has two hundred soldiers and the other sixty, the result is still the same.

On the other hand, in a space fight using long-range weapons between the force of one hundred men and a force of sixty, the damage ratio is one to three. Furthermore in a fight between two forces of two hundred and sixty each, the ratio is one to six.

The Lanchester principles have thus far proved that the degree of damage in a battle is largely influenced not so much by the numerical strength as the kind of weapons used and the strategy followed in the battle.

If you were the commander in chief of the superior force, what kind of weapons would you let your men carry? The choice is between short range weapons for a man-to-man fight, and long range weapons. It would be wise of you to order your force to carry long range weapons and engage in space fights against the enemy.

Conversely if you were to command an inferior force with fewer soldiers, how would you change your strategy? You would immediately let your men attack the enemy in a man-to-man manner with short range weapons. This is because in a short distance fight both forces will lose equal number of soldiers, and the superior power will have as hard a time as the inferior force in controlling the field. The Communist Viet Congs and Afghan Guerrillas used this tactic to win their battles.

To sum up, a superior force is better off following Lanchester's second principle, taking advantage of the effect of n-square law, and an inferior force may find it advantageous to respect the first principle. For this reason the first principle is called a strategy for the inferior, and the second a strategy for the superior. This principle leads to the following set of implications:

A superior force with more numerical strength should:
1. choose long range weapons that are effective in space fights;
2. engage in a fight in an open field with high visibility; and
3. keep a long distance from the enemy.

The force with more numerical strength can put the n-square law in effect by observing these principles. It is the commander's responsibility to "manoeuvre the battle" into such a favourable condition. Such a tactic was named "a strategy for the superior".

An inferior force, on the other hand, should:
1. choose short range weapons that are effective in man-to-man fights;
2. engage in a fight in such places as a valley, and a basin where there are mountains, rivers and other obstacles that make it difficult for the larger force to move around freely; and
3. fight the enemy by keeping the distance as short as possible.

By observing these principles, the inferior force can prevent the n-square law from taking effect, and may be able to overcome its numerical disadvantage. Of course the commander has a great responsibility to create such a favourable condition on his own will. This tactic was named "a strategy for the inferior".

Lanchester's principles have led to effective strategies for both the strong and the weak. The strategies are naturally quite different from each other. This is the most critical core of the Lanchester strategy, and consequently the most important.

These principles are just like any other new achievement that seems impossible until it has been actually tried and easily accomplished. Though each individual's skills and mental factors had been regarded as important, it was Lanchester who first helped design a strategy to win a competition in a pure mathematical fashion by excluding the mental aspects.

2 *Application of Lanchester's principles to business management*

Applying the principles that have been discussed in the previous section to management, I can suggest that the superior company should:

1. choose to sell the merchandise for broad based competition against the competitors;
2. choose to focus on the market area where they can take advantage of the square law; and
3. choose the strategies that will multiply the effect of the square law.

These are called the "managerial strategies for the superior".

The inferior company, on the other hand, should:
1. choose to sell the merchandise for direct competition;
2. choose to focus on the market area where they can take advantage of the linear law; and
3. choose the strategies that will multiply the effect of the linear law.

These concepts describe the "managerial strategies for the inferior".

Once the corporate leaders understand the fundamental concepts and their differences, they can simply apply them to their actual management and anticipate immediate effects.

3 *Distinction between the superior and the inferior by market share*

Effective business strategies clearly differ between the superior and the inferior. In war the superiority is determined by the number of troops, whereas in business it is determined by the market share.

Before the Second World War began, the U.S. State Department had formed a project team. Bernard Koopman, a member of the team and mathematician, designed a model of Lanchester strategy by combining the Lanchester principles and game theory. The Koopman model calculated the long-term fighting strength.

The fighting strength at a battle field is made up of the fighting capacity available on site and the strength that is supplied and supplemented from the home country. Yet the quantity of the strength constantly changes. Even if a force is superior at one point, it could consume its own strength at each battle and eventually lose the war without sufficient supply supplemented by the home country. To win the war, a force must obtain and maintain more supply than its counterpart.

The quantitative fighting strength, including soldiers, weapons, and food decreases with the passage of time. The long-term fighting strength is thus secured by the reserve troops and material shipped from the home country. The relationship between time and quantitative change in the fighting strength was first mathematically demonstrated by Koopman.

The extent to which a company may receive supplement is shown by the profits gained through sales of goods and services. The supplementary power of a force translates into the extent to which the company has monopolised the source of profits, i.e., customers. Simply put, it is the company's market share.

If a company has a larger market share than its competitors, its capital power becomes stronger. The strong capital power enables the company to hire more personnel, expand the production capacity, etc., further increasing the company's fighting strength.

Koopman's calculation contains complex differential calculus, and also it is so long that I will extract only its conclusions. For full detail one may consult the book on operation research published in the U.S. in 1943. According to Koopman there are three conditions that a company must meet in order to be superior.

1. The company must have the largest sales among the competitors.
2. It must secure 26.1% or more market share.
3. In comparison to the runner-up there must be at least a difference of ten to six.

In reality a large number of companies compete against one another to expand their market share, attempting to draw more customers to their own goods and services. In such a "dog-eat-dog competition" each company struggles to increase the probability to monopolise the market by expanding their share as much as possible.

If the top company has gained 26.1% or more market share, essentially monopolising one third of the market, its power to control the market increases. These are, however, the minimal conditions that a company must meet in order to be a superior power, and if one of them is not met, then the company is not regarded as being superior.

Above the superior level is a super superior power. The three conditions to be met are:

1. The company must have the largest sales among the competitors;
2. It must secure 41.7% or more market share; and
3. In comparison to the runner-up there must be a difference of at least ten to six.

When a company meets all these conditions, it is called a "relative monopoliser". By monopolising 41.7% or more market share, the company practically controls 50% of the market.

If a company becomes a relative monopoliser of 41.7% or more, the capital gain dramatically increases. The amount of profit before tax per employee is four to six times more than the average in the same industry, and two or three times more than the runner-up. The reason for the large difference is accounted for by the n-square law. Without the n-square law effect, profit per employee would be the same, even though there should be a difference between a large company with many employees and a small company that has fewer employees.

When one company owns 73.88% of the market, it is called an ultra superior company. With such a high percentage of the share, the company essentially controls 100% of the market, and the runner-up has very little chance to reverse the situation.

The degree of market share is thus divided into three levels: superior, super superior with relative monopolisation of the market, and ultra superior that has the absolute monopolisation of the market. These levels are marked by three percentage points: 26.1%, 41.7%, and 73.88%, respectively. Although such accurate figures may not be available in actual market surveys, the fact is clear that a company with a market share above a certain point can enjoy a great advantage over its competitors.

In the next sections I will discuss the business strategies for the superior and the inferior.

III. Business strategies for the superior

1 Merchandise strategies

a. The superior company should choose merchandise that is likely to appeal to massive public as potential customers, and merchandise that is likely to bring a large order in each transaction.

A long range weapon translates into merchandise that can be easily transported over a long distance. In order to be easily portable, the merchandise needs to be relatively light, carried in cardboard boxes, and fairly easily stored.

In order to take advantage of the n-square effect, the superior force in war also needs to increase the number of shots. In business this means merchandise that is frequently used by the customers. The merchandise that meets these two conditions, i.e., easy portability and frequent use, is generally consumer goods for an indefinite number of potential customers. Instant foods, snacks, powder soap, underwear, etc. that are necessary in people's daily lives are all examples of merchandise in this category. Since these goods are used by many people, their potential market is quite large, and also their sales at one transaction is potentially very large.

b. The superior company should expand the line of merchandise by increasing the range of goods.

The superior force in war uses many different kinds of weapons to attack the enemy from different angles in order to promote the n-square effect. Applied to business, this strategy means expansion of business by selling many different kinds of merchandise. This tactic can help the company not only gain control over the market, but at the same time keep the inferior competitors from growing strong.

Henry Ford of the United States developed a system of mass production by using large scale assembly lines. Ford had become an ultra superior company. But since the company failed to expand its range of merchandise, its sales declined until General Motors took a leading position and Ford almost went bankrupt. An example of success is Matsushita Electric of Japan that has kept manufacturing a large selection of appliances.

The strategies for the superior can be summarised into two tactics: to choose merchandise that appeals to a large number of people as target customers, and to expand the range of merchandise. Facility for mass production of goods, construction of warehouses that enable the company to make the flow of merchandise from the manufacturer to customers efficient, and also employing sufficient sales and other staff all cost a large sum of money. These strategies, therefore, can be successfully adopted only by companies with sufficient capital.

2 *Market strategies*

a. The superior company should focus on large cities.

In battle a superior force must choose a flat battleground without any steep hills or valleys to implement their strategies successfully. The force can concentrate their fighting strength on destruction of the enemy.

Applying this principle to market strategies, we can see that a superior company needs to target big cities with large population. The company sets up a number of branches in the big city and stations sufficient sales representatives in each branch. The strong company can thus put the n-square law in effect in the entire city and increase sales.

b. The superior company should expand the market and leave no area unattended.

A superior force in war disperses its power over a large area, restricting the inferior force's military movement. An application of this principle to business indicates an expansion of the market area. This strategy helps not only to increase the sales but to keep the competitors from becoming strong, since it leaves no blind spots in the market.

The companies that are known to have the largest markets in the world are Coca Cola and IBM of the United States. The Japanese companies that have used the same strategy are Matsushita Electric or better known as Panasonic, Sony and Honda.

3 *Distribution strategies*

A superior force uses multiple layers of weapons: guns with long shooting ranges, guns with medium shooting ranges, tanks, and infantry. Given such a variety of forces, the superior can concentrate their strength on their attack on the enemy, thus causing the n-square law to take its effect.

Translated into business language, this strategy means multiple layers of distribution systems. The distribution systems may be classified into main and supplementary systems. The main system is for customers that place large scale orders, and the supplementary system is for individual customers that may be large in number but can not be counted on for large scale transactions. The superior company puts a large number of sales staff to control the main distribution system, and at the same time uses a sufficient number of personnel to dominate the entire market through individual sales.

4 *Sales strategies*

a. The superior company should value indirect sales.

A superior force uses long range weapons. Their n-square effect is clear, particularly when there is a sufficient distance from the enemy.

An implication of this principle to business is that the superior company should keep an adequate distance from the customers. In other words, the company should have multiple layers of distributors such as wholesale dealers and retail stores, which altogether make the sales of merchandise indirect. If the merchandise is distributed to more than one chain of wholesale dealers, the number of retail stores increases, which helps to increase their power of control over the market.

The multiple layers of distributors contribute not only to an increase in sales, but are a safeguard against the competitors.

b. The superior company should value the effect of advertisement.

A superior force attacks the enemy from many different angles such as ground, air, and sea.

The multi-angled strategy in business is the use of advertisements. The superior company promotes its sales through various means of advertisement such as neon signs, billboards, newspaper ads, and radio and TV commercials. The sales promotion through multi-channel ads covers the entire market, leaving very few blind spots, and helps increase the sales.

5 Counterattack strategies

An inferior company attempts to fight against the superior by innovating new products and ideas. If the superior company leaves the new ideas and merchandise unattended, the inferior power is likely to gain strength and become a serious competitor. The superior company should immediately counterattack the competitor by using the same ideas, merchandise, strategies, etc., and equalising the sales conditions.

Evidence of failure in this strategy can be seen in the U.S. automobile industry. The U.S. "big three" had known that small cars had been exported by Japanese and German car makers. Since the margin of profit was naturally greater for a large car than a small one, the U.S. companies neglected to develop small cars. When the U.S. companies began to build small cars, the Japanese makers had already taken 25% of the U.S. market, and it took them fifteen years to regain competitive power.

This section has summarised the strategies for the superior, focusing on merchandise, market, and sales tactics. Lanchester's principles can be applied to finance and personnel management as well.

IV. Business strategies for the inferior

As previously mentioned, the profit before tax per employee is proportional to the square of the percentage of market share. The company that has the largest share is more advantageous than it appears. The aim in business is, therefore, either to concentrate on one area where the company may win a larger percentage of revenue, i.e., customers, than the competitors, or to manufacture goods that win the most popularity in any market area.

While the capital in such forms as cash and properties are valuable assets for any company, priority market, where the company makes more sales than the competitors, and popular merchandise are equally valuable assets. Selling in markets where there is no possibility of winning a favourable revenue, or manufacturing merchandise that does not sell is of no value to the company.

It is, however, not easy to be number one in sales in a particular area or of a particular merchandise with limited fighting strength. That is precisely the reason why a company that is numerically weak must concentrate its power on developing a single premier market and manufacturing and selling a number one product. This principle is called a "partial winning strategy".

In the following sections are the specific tactics for a numerically inferior company to develop a premier market, and to develop and sell a popular product.

1 Avoid direct competition against a superior power

The actual difference of fighting strength in any competition is proportional to the square of the difference in numerical strength. The difference between a force of ten men and a force of seven may appear to be insignificant, but the ratio in an actual fight becomes one hundred to forty-nine, a two-fold difference. The force that has only seven men has no realistic chance to win the fight. Even though the inferior force might appear to be courageous to confront the superior opponent, it is the most unskilful tactic since the inferior force is sure to lose badly.

The same argument applies to business. If a small company challenges face-to-face a larger company, following the same strategy, the difference in actual strength is multiplied, and it will likely suffer more serious damage than initially anticipated. If, on the other hand, a small company competes against a smaller company, then it can enjoy the advantage, because of the n-square law effect.

From these principles one can see that a small company must clearly differentiate between an ideal goal in competition, i.e., a superior company, and an actual target which is a smaller company that it aims to beat. It is important that a small company does not choose a stronger company as an actual target.

2 *Use different strategies than those used by others*

If the inferior force used exactly the same strategies as the superior power, the n-square law would be in effect and there is no chance for the inferior to win the competition. When the numerical ratio between the two forces is 1 to 0.5, the actual ratio of strength is 1 to 0.25, which means that even if a force has an input of 0.5, the output that is available will be only 0.25, only one half of the potential power.

Similarly, given the numerical ratio of 1 to 0.33, the difference in fighting strength will be shown as 1 to 0.01, meaning that the inferior power can put out only one third of its numerical strength. In the business world if such a discrepancy continues, the company will not be able to operate for a long time and its name will soon disappear. In order to avoid such unfavourable situations, the inferior force must devise different methods from those of the superior, and this is called a "differentiation of strategies".

The only way for a numerically weaker company to differentiate their strategies from those used by their competitor is to examine sample success cases and learn from them.

3 *Discover the most effective strategies through detailed observations and analyses of the market*

Sun-tzu, a Chinese philosopher said 2,500 years ago that the best way to win is to win without fighting. The same principle applies to business. An effective strategy for a numerically weak company to win a competition is to discover merchandise or a service that no other company is yet selling, and concentrate all of its power on the sales. This is the key to success for a small business and such merchandise is likely to gain popularity among consumers by "filling the gap" of service provided by major companies.

A company needs to analyse the characteristics of merchandise or a service that may potentially sell, by minutely classifying them into such categories as usefulness, prices, and prospective customers' expected taste, and closely analysing them from a number of angles. Such detailed analysis is likely to help stimulate the company to come up with ideas for a new product and service that other companies may have overlooked.

Sun-tzu also advised choosing an opponent that you will most likely beat. Even when a force confronts a strong opponent and it is improbable that the inferior force will win, they may win battles if they concentrate their power on the weak spots of the superior power. It is thus important for the inferior to find weak spots by analysing the opponent by dividing it into many components. This tactic was put into practice by Nelson in the Battle of Trafalgar. It was also tested on the ground by Napoleon.

The effectiveness of the strategy applies to business as well. A minor company can find itself in the leading position if it successfully discovers the major company's weak points by conducting detailed analyses of various categories of merchandise characteristics such as the manner of use, prices, sizes, and how it may be liked by prospective customers.

The characteristics of a market area may also be analysed in the same fashion. If the area is divided into some regions by mountains and rivers, the inferior company should be able to find out in what regions the superior company is weak by carefully analysing each. Such microscopic analyses are important to inferior forces.

The specific case studies detailing methods of analyses are available, and owners of business are encouraged to consult them.

4 *Maintain light gear*

The weaker company needs to keep their structure, and decision making systems as simple and flexible as possible so that they can quickly respond to the demands of the market, and implement new strategies that a large company has not yet come up with. It is also important for a company not to cling to old traditions and customers, but to be flexible enough to adopt new and effective alternatives quickly.

5 *Concentrate on merchandise with a relatively small potential market*

In military terms when an inferior force engages in a man-to-man fight using short range weapons, it can prevent the n-square law from taking effect. The competition may develop more favourably for the inferior power than if the n-square law is in effect. A typical weapon for a short-range fight is spears. Long range machine guns, on the other hand, enable a force to attack a large number of wide spread enemies simultaneously. Spears, of course, can only be used on a single enemy that a soldier confronts face-to-face. In other words the range of possible attack with spears is quite limited.

Applied to business, this principle implies sales of special merchandise that has limited use for highly technical purposes. Special cameras that are used by civil engineers and constructors in Japan, for example, have this attribute. The on-site engineers need to take a large number of photographs of the construction, file them, and present them as records to the customers, who may be government organisations or private companies. The physical conditions that confront the engineers when taking such photographs are often not favourable with the presence of many possible obstacles such as high humidity, dusty air, rough and unstable terrain. The cameras may often be damaged through rough handling such as banging and dropping, which could cause the pictures to come out blurry, and entirely useless.

A Japanese camera maker developed a water and dust resistant camera that has a protector around it. Named an "On-the-Job Superintendent", the camera sold quite well. The company has maintained the leading position as a manufacturer of the special kind of camera for use at construction sites although the size of the market is considerably smaller than that of regular cameras. The company had aimed at the special market with special merchandise.

Another example is seen in the video camera market. Sony and Matsushita Electric had a fierce competition against each other. As the video cameras have continued to become smaller every year, the monitors attached to the cameras have also become smaller to the extent that they were very difficult to see. The consumers, particularly the elderly with vision problems, complained about the small monitors. Casio, one of the large electric companies, developed a video camera with a five-inch monitor, and stressed it in their advertisement. The camera quickly became a popular item, and it won 30% of the market share.

The company that is in an inferior position can thus develop number one merchandise by specialising in a narrowly targeted group of prospective customers. If the company can further increase the number of such special items, it will surely grow to be a superior power.

Pricing is another important issue that an inferior company needs to consider carefully. Since medium priced merchandise is popular among general consumers and it has a large market, many big companies participate in the competition of its sales. The more companies compete, the less chance for a weaker company to win.

An alternative for an inferior company is to focus on fairly to very expensive merchandise. The gross sales of such goods may be limited in quantity, but the company could quickly lead the competition because the number of competitors is likely to be very small. Rolls-Royce of England and Mercedes of Germany are typical examples. When English companies export merchandise to Japan, it is important that they follow this principle to be successful.

Although it may be generally believed that even a small company could raise its sales and profits by developing a product that helps expand the market, this is a greedy idea.

An old Chinese saying teaches us that the head of a chicken is worth many times more than the rear end of a cow. Chickens used to be regarded as a much lower animal than cows. Everyone would naturally seek to have a cow rather than a chicken. One must be satisfied and proud to have the head, the most important part of any animal, rather than a rear end, even of a highly valuable animal. An equivalent saying in English is: "A large fish in a small pond". In other words it is more meaningful to be number one in a small market than to be at the bottom position in a large market.

6 *Keep the range of merchandise narrow*

The leading company attempts to control the entire market by diversifying the line of product it sells. If, on the other hand, a weak company increases the kinds of merchandise it sells, the already limited fighting strength is spread thinly on each item and the company's competitive power will further decline. The weak company needs to prevent this by limiting the number of products it sells, and concentrating all the power on the small number of items. Once the company is in a leading position with its premier merchandise, then it can gradually expand the line of products.

This principle is true particularly when a new company joins a competitive market with many strong companies. The new and weak company can not hope for any success unless it follows the principle of narrow range of merchandise. The rule for a new company to follow when participating in a competition is that it must bring to the market a premier product. If the company succeeds with the premier product, then it can start developing one or two more products. This strategy is analogous to the "hop, step, and jump". In order to jump far, one must make sure to set the first step in the right direction with a sufficient distance.

The two principles that a numerically inferior business must adopt are: to focus on special merchandise whose prospective market is small and limited, and to keep the range of merchandise narrow.

7 *Market strategies for the inferior*

a. Focus on the market areas where man-to-man fight is the most appropriate tactic.

If a numerically inferior force engages in a fight against a superior power at a flat battle field with a good visibility, the n-square law will take effect, causing serious damage on the inferior. Furthermore there is a good chance that they may be completely surrounded and wiped out by the enemy.

If, on the other hand, the inferior force chooses a battle field that is divided into many areas by mountains, rivers, and other natural obstacles, it is difficult for the strong force to use long range weapons. Also such natural obstacles are likely to prevent the inferior force from allowing the stronger to surround them. The strong force can not take advantage of the n-square law. The commander of the inferior force must deliberately choose such a geographical location as a battlefield. This principle of natural obstacles explains why the Romans could not effectively attack Scotland. Applied to business management, this principle refers to an accurate choice of market area.

The primary strategy for the inferior business is to choose and focus on a market area that is more appropriate to man-to-man fight. It translates into avoiding large cities with big population and choosing instead the market area that is further divided into small local areas by natural obstacles. A small but independent commercial area is also a good choice. Scotland, Wales, the Exeter area in the west, and various regions in the southern part of England are good examples.

The strong business often neglects to focus their strength on these regions. The smaller companies can thus use the geographical features to their advantage, and prevent the n-square effect. A numerically inferior company must first learn how to become a leader in a small market area, and use the experience in other areas so that they gradually increase the number of market areas where they can become the leader.

The secondary market strategy for the weaker company is to choose suburban areas that are commercially independent, being divided into regions by hills, rivers and other markers such as railroad tracks. Another urban market where the company may be able to compete favourably against the strong companies, is an area toward the centre of the city that is clearly identified as independent by such landmarks as rivers, railroad lines, and parks.

The numerically inferior company should identify the competitors' weak spots by dividing the large market area into small units and analysing them in detail. Once they have identified the market area that has high potential, they should send a larger number of sales staff there so that they can be a leader in that area. This is an application of the Nelson Touch. The company should first become a market leader in one area, and attempt to increase the number of market areas where they can be the leader gradually.

When a company makes its way into an entirely unfamiliar market, it should regard itself as the weakest of the weak, and carefully follow the principle that has been just laid out. Otherwise the company will not be successful in running its business. When Honda made their way to England with motorcycles, and Canon started selling photocopying machines, they both concentrated their sales power in the Wales and Exeter regions. Only after they became well known in these local areas, they gradually moved toward London. Coca Cola used the same strategy when they first came to Japan. They practised the urban local strategy, and became number one in the soft drink industry in Japan in ten years.

When a company goes into an area where they have no foundation for sales, they may think that the larger the population in the area, the more sales and profits they will make. This is only a greedy illusion. Many companies from across the world are trying to sell their products and services in Tokyo, making the competition quite fierce. It is very difficult to bring the sales up to a certain level. British companies considering opening business in Japan are well advised not to start in Tokyo.

The key to success in business in Japan is not to choose such big cities as Tokyo and Osaka, but instead go to smaller cities such as Sapporo, Sendai, Hiroshima, and Fukuoka. These areas provide the companies with excellent opportunities for successful sales as well as educational and productive experiments that will be later helpful.

b. Concentrate the fighting strength on a small market.

If a numerically inferior force engaged in a conflict at a large battle field, its power would be dispersed and the strength of each unit would be further limited. As Lanchester's principles have so far demonstrated, the larger the difference in numerical strength is between the two forces, the greater the amount of damage done to the inferior force. In order to minimise the damage, the inferior force needs to keep the battle field as small as possible.

Applied to business, this principle indicates maintenance of a small market. If the number of sales representatives is limited, the company must avoid enlarging the market and thinly spreading its strength. Instead the company should identify a market whose conditions are favourable, and concentrate all the power available on the market by sending a larger number of people than the competitor. This is the key to gaining a leading position in the given market.

The most important market strategies for an inferior company are: to concentrate on the market where they can engage in close fight to their advantage, and to concentrate on a small market.

8 The sales strategies for the inferior

Every force sets up a camp during any battle. If the numerically inferior force sets up camp far away from the superior force, and engages in a space fight using long range weapons, the n-square law is in effect, and they will suffer significant damage. If they fight more closely, however, they can prevent the n-square effect and hold an advantageous position over their counterpart.

Equivalent to a war camp is a sales structure. When a manufacturer sells their product to a wholesale dealer, it is called an indirect sale. If a weak company engages in an indirect sale, the same effect as in a space fight is expected. In other words the n-square law will be in effect, and there is no chance for them to win the competition against larger companies.

The only way to avoid the n-square effect is to follow the principle of man-to-man fight and sell the products to the retail stores, or directly to customers. A good example that shows the effect of this method is "Amway Business" that continues to grow across the world.

An example of failure is an American company that began doing business in Japan, first setting up its headquarters in Tokyo and concentrating its limited power on the Tokyo area. They also spent a large sum of money on TV commercials. Within six months they failed and withdrew from Japan. They had used the strategies that would be good only for a numerically strong business. The president of the company was reported to have attributed the failure to the Japanese people's unfair psychological barrier and their deliberate and collective avoidance of his product. The real reason for the failure is the use of the wrong strategies rather than what the president called the "closed Japanese market".

When foreign companies come to Japan, they often tend to assign only one company as a representative distributor and give all rights to that company. This is not an effective business strategy. Japan is a mountainous country surrounded by the sea. It is divided into an uncountable number of independent commercial areas by rivers, mountains and volcanoes, which may be difficult to imagine for the people coming from England where there is no mountain over 1,500 metres high. The geographical attribute of Japan accounts for the existence of a number of wholesale dealers established in each commercial area.

This is the reason why one distributor in the whole country would never be able to conduct sales of any merchandise, but must distribute the goods to local wholesale dealers. When an English company comes to Japan, they find themselves selling their merchandise through a five-step distribution system: 1) manufacturer, 2) general distributor, 3) local distributor, 4) retail store, and 5) customer. The route of distribution is so long and difficult to control that the foreign company may not be successful in doing business in Japan.

Some alternatives to the long distribution system are to use many local distributors in each area, to find and sell directly to powerful retail stores, or to sell through human networks as in the Amway. By using one of these strategies, the company can take advantage of close fight strategies and prevent the n-square effect. The company staff may also be able to understand directly the customers' needs and demands, which will help further promote the sales.

This is an effective sales strategy for a numerically inferior company to be successful in business. The same principle can be applied to many other areas such as finance, personnel management, or life in general.

As shown in this paper Dr. F.W.Lanchester's principles of competition have been widely applied in Japanese business. Over two million copies of books on Lanchester's business strategies have been sold, and the majority of people in managerial positions in Japanese companies are very familiar with the principles.

I firmly believe that there are more people who are familiar with the name and work of Dr. F.W.Lanchester in Japan than in any other country. It is interesting, however, that Lanchester's career as an automobile company president is not so well known.

Reference

Koopman, B.O. A quantitative aspect of combat. OEMsr-1007, *Applied Mathematics Panel*, Note 6. AMG-Columbia, August 1943

Chapter Eleven

DEFENCE OF GIBRALTAR

by Dr. F.W. Lanchester

Editor's Note

The typescript of the full article, and of this summary are in the Lanchester Collection, Lanchester Library, Coventry University. In publishing the summary for the first time in 1996, I have consciously not edited any of the wording or emphases of the original.

The references to "Clausewitz" are probably to:

Karl von Clausewitz: *Vom Kriege*, 1830, with the first English translation being published in 1853

The problem of defence

That which follows is an abstract of an article written by the undersigned in 1939, but which for reasons explained below has not hitherto seen the light of day. It was offered to the Editor of "Engineering" on the 17th April 1939, but was not accepted for publication on the ground that it was not quite in their line. The original Article occupied eleven quarto pages, but a great deal related to conditions which no longer apply, as for example, the probability that France would be with us, has been omitted in view of the present conditions. Also the remaining matter has been condensed and the points made in simpler and more direct language.

No further attempts were made to publish at the time owing to it having become evident that the Government, or those speaking in their behalf, wished the public to believe that the "Rock" was impregnable, there were Articles in several of the London Dailies within a few days of the submission of my article, with accounts of "Special Correspondents" who had been shown round the defences, expatiating on the wonders they had seen, and telling the public that the defences of Gibraltar were secure beyond question. And since that time further articles have appeared to the same effect, and been sent out on the Broadcast. It not being my business to enlighten my fellow countrymen in matters which the Authorities wished to conceal I withdrew my Article entirely. I supplied our M.P. (Sir Patrick Hannon) with a copy, as being confidential through such a channel, but all he could do was to repeat parrot-like what he had been told, "Gibraltar is impregnable." Who the powers that be were trying to fool I do not know, it cannot have been the Germans or the Spaniards: I shrewdly suspect that the visit of "politeness" paid by the German Navy about that time had for its purpose to convoy to Cadiz and Algeceras the big guns and ammunition for the reduction of the "Rock".

GIBRALTAR
The Dardanelles of the Mediterranean
Abstract of Article (not published) April 17th, 1939

(1) *Preliminary.* The Defence of Gibraltar is of vital importance to Britain's position in the Mediterranean and the East; its loss to us could only be regarded as an irreparable disaster, hence, as far as is humanly possible, its defences must be made secure beyond all question or doubt: it must be held at all cost. *The problem is by no means easy.*

(2) *General Considerations.* Although recent experience has shown that the art of defence keeps pace with the technical advance in the means of attack (other things being equal), it can, as a general proposition, no longer be maintained (as in the days of Clausewitz) that the tactical advantage *always* lies with the defenders. We cannot shut our eyes to the fact that when the area held by the defenders is small, effective defence may become extremely difficult or even impossible. A mediaeval castle for example, although impregnable in its day, cannot be defended at all against attack by modern weapons.

(3) *Gibraltar.* Here we touch on the main source of weakness and difficulty: the peninsular on which the Rock of Gibraltar stands, Map Fig. 1, has a total length of less than three miles, and an average width of only half a mile; its area is approximately $1\frac{1}{2}$ sq. miles. Whatever the natural strength of the position may be this is too restricted for the problem of defence to be regarded lightly. The failure of past attacks on the "Rock" means nothing, the weapons of those days were little better than toy cannon and squibs in comparison with the weapons of to-day.

(4) *Physical Features.* The peninsular of Gibraltar runs almost exactly North and South, and is bounded on the East by the Mediterranean (open sea), and on the West by Gibraltar (or Algeciras) Bay (see Map, Fig. 2) the opposite shore being part of the Spanish province of Cadiz.

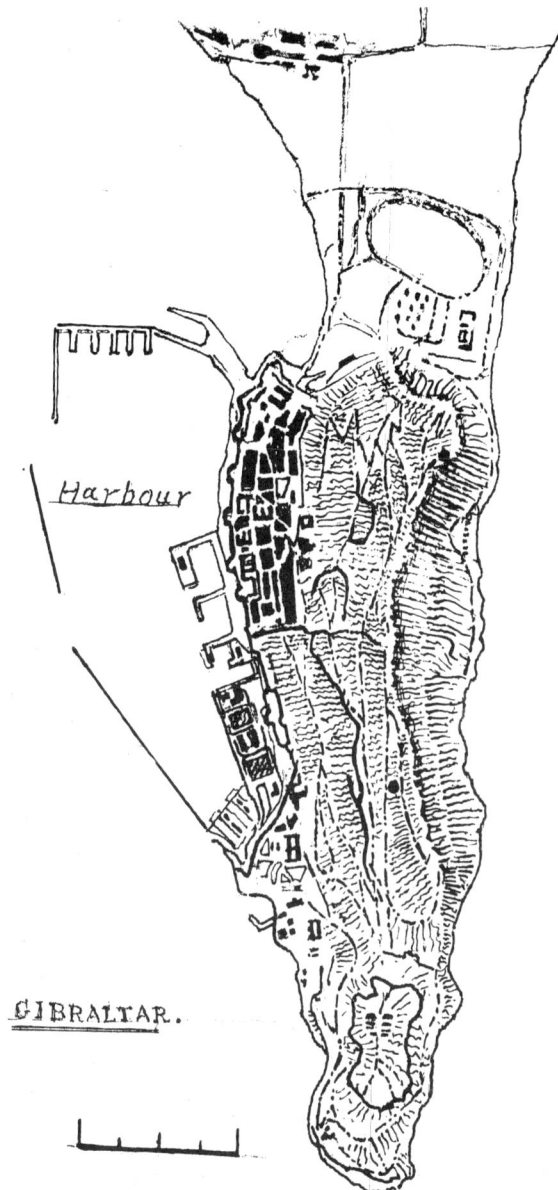

Fig. 1. The Gibraltar peninsula

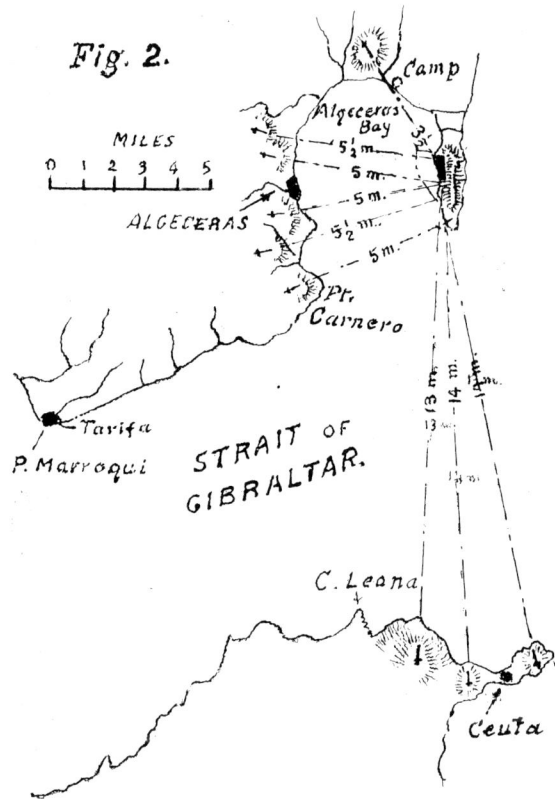

Fig. 2. The Straits of Gibraltar

The Bay has an average width of about 4½ miles or 8,000 yards. From the southern extremity of the Gibraltar peninsular to the nearest point of the African continent across the Strait is 13 miles.

(5) *The Political Situation.*   As the outcome of the civil war the present Government of Spain under General Franco undoubtedly owes its existence to the armed intervention of Germany and Italy known as the Axis Powers. Although Britain nominally stood on one side, in theory, preaching the doctrine of non-intervention, our actual policy viewed from the stand-point of General Franco, and his supporters was marked by ill-will and pin-pricks. If a European War were to break out on a grand scale in which we should find ourselves together with France ranged against the Axis Powers, it is difficult to believe that Spain could preserve her neutrality for long, and we must therefore count her as a potential enemy.

(6) *Comparison with Gallipoli.*   There is a certain analogy, – not very close perhaps, – between the peninsular of Gibraltar and that of Gallipoli. Strategically Gallipoli in the hands of the Turks had every advantage, for not only was the mainland in their hands, ensuring their supplies and evacuation of wounded, but also the Asiatic mainland across the strait was in their undisputed control and the strait – the Dardanelles – is but some four or five miles in breadth, or about one third of that of Gibraltar. In the last Great War our combined forces failed to dislodge the Turks from Gallipoli, but no one need doubt that had the mainland, (Turkey in Europe) and the coastal region on the opposite side of the Dardanelles been in our possession Gallipoli would have fallen.

(7) *Guns, – Artillery, – Attack and Defence.*   The chief weapon and it may be the decisive weapon, which the enemy would bring to bear against the defences of Gibraltar is their Artillery, chiefly guns of heavy and medium calibre. And likewise the only defence possible against this form of attack is Artillery, – guns and still more guns. As is standard practice Aircraft would act in conjunction with Artillery for spotting by both attack and defence, but in this the advantage would rest with the attack, for the peninsular offers very little accommodation for aircraft, and the take-off and alighting grounds, – such as they are, – are under enemy fire. As a subsidiary means of attack the enemy might be expected to use bomber planes to attack shipping in the

harbour and harbour works; but this is not essential for the guns would be capable of smashing everything at less cost. It must be realised that only two things matter where Gibraltar is concerned, these are the *harbour* and the *big guns* which exercise some control over the passage of vessels through the Straits: If the former were rendered untenable, and the latter (the guns) silenced, Gibraltar must be regarded as gone; the holding of the bare rock as a matter of prestige would have no military value.

(8) *Attack and Defence. (contd.)*   We then may visualise operations as opening with a large scale Artillery duel; let us take stock of the situation. Referring to Fig. 2 which is a map of Gibraltar and its environment. It will be observed that the town and harbour are vulnerable to enemy gun fire from all points of the compass from due South through the West to North. From the Northern area batteries of field pieces could be brought to bear from a range little over three miles, (area marked "Camp"). From the West across the bay the Town and harbour would come under the fire of numerous guns of medium calibre from a range of 5 to $5\frac{1}{2}$ miles mounted on mobile road or rail vehicles in the rear of Algeciras, and beyond, that in the hinterland, at ranges up to 8 to 10 miles or more, from North to S.W. heavy naval guns firing from fixed emplacements would be able to plaster the vital area with devastating effect. Moreover from heavy guns of 14" or 15" calibre, mounted in the vicinity of Ceuta the whole position could be enfiladed, and any aircraft or aerodrome equipment on Egypt Point bombarded from a distance of only 13 miles. The question is, – "What have we to show against this?"

(9) *Guns versus Guns.*   The answer to gun-fire is gun-fire. The bomber aeroplane in this combat may be dismissed, since owing to the circumstances above set forth the air strength at the disposal of the defenders is negligible. The enemy in this are better situated, there is nothing beyond the economic limit to the air force they could bring to bear, but we are here discussing defence. There is obviously a limit to the number of guns of heavy or medium calibre which could be mounted on or *in* the Rock; there is virtually no such limit in the number which the enemy might bring to bear. Once the defender's guns have been unmasked their positions are known to the attack, and from distant fixed emplacements they may be bombarded night and day, regardless of visibility. In clear weather the enemy can sustain aircraft aloft spotting for their artillery; the defenders can do little or nothing to obstruct or prevent this or to take equivalent measures themselves. The conditions would hardly allow of the use of an aircraft carrier such as the Ark Royal. Unless the defence can silence the enemy guns which command the harbour they may be deprived of the means of renewing their ammunition, or executing repairs, and the harbour may be rendered untenable by the fire of medium calibre guns, and even field pieces which owing to their mobility are very difficult to locate and destroy, without securing mastery in the air. And the bombers of the enemy may at any time take a hand in this.

(10) *An Application of the N^2 Law.*   At a date prior to the last Great War the writer deduced a generalisation which he termed the N-square Law; it is this, – That under certain well-defined conditions the power or potential strength of a military or naval force is proportional to the square of its numerical strength. This was first published in the columns of "Engineering" on October 3rd, 1914, and republished in Book form in January, 1916, Chapter V. "The principle of Concentration." Applied to an Artillery duel means that *other things being equal* the strengths of the opposed forces are proportional to the square of the numbers of guns they have at their disposal and can bring simultaneously into action. This assumes that the guns are all of one size and capacity; but the principle applies generally although direct comparison may be difficult. In applying this in the present case we may regard some of the lesser guns of the enemy as occupied in demolishing the harbour works, but of the heavy stuff, he is in a position to have at command a great numerical superiority. Let for example his numerical superiority be in the ratio 2:1*, then his strength will be four to one greater than that of the defenders, and their fate is sealed. This is not all, the condition, "Other things being equal" needs consideration. Other things are not equal, for the attack (as already pointed out) has the advantage of being able to control and direct his artillery fire from the Air, which is denied to the defence. Even with gun power equal this alone would be expected to give a winning advantage. And the enemy would possess the additional advantage of being able to draw on the manufacturing

---

* It is of interest to note that the great authority of last century, Clausewitz, pointed out that a 2:1 superiority in numbers, save in exceptional cases could be relied upon to give a winning advantage! And *he* knew nothing of the N-square Law.

resources of his own arsenals (at Barcelona for example – and elsewhere). He would have the resources of the Axis Powers as well as their technicians at their beck and call.

The above abstract contains the main substance of the Article which in all comprised some eleven quarto pp. of MS. The conclusion is clear, Gibraltar could not be defended against the combination of Powers postulated, *from within*. In the original many sister facts were mentioned as illustrating the direction in which things were drifting, and the possibility of defence *from without* was discussed, but as this largely depended upon the assumption that France would be with us that part of the report or rather Article has ceased to have any bearing on the situation as it stands.

The only effective method of defence would appear to be based on an invasion, the landing of an expeditionary force in the region of Cadiz, which to judge from the trend of things and the difficulties of such an undertaking must be ruled out.

The problem then resolves itself into making every effort to keep Spain from entering into the war, by making her fear our blockade and the possible loss of her world possessions. Not by act of so-called "appeasement" which would only engender contempt.

Chapter Twelve

LANCHESTER'S USE OF PETROL-ELECTRIC DRIVES

by Roy L. Thomas

Historical note

In his Presidential Address to the Institution of Automobile Engineers in 1910 F.W. Lanchester discussed the state of the art of motor vehicle design. He suggested that, "The change gear-box is evidently the weakest point in the chain of mechanism. In the gear-box (of the sliding gear type) we have a piece of mechanism that we should most of us like to see abolished". (Lanchester, F.W., 1910–11, p.20). As a suitable alternative to straight mechanical drive with a gear-box he described the use of a petrol electric combination in the form of a direct current dynamo directly coupled to a petrol engine. The dynamo was of the series wound type and could act as a dynamo or motor depending on its speed. A further advantage of the use of a series wound motor is that maximum torque is produced at starting, the torque then falling as the speed rises. This allows a smooth start to be easily achieved.

Lanchester then described a Daimler bus driven by an electric auxiliary system (Lanchester, F.W., 1910–11, pp.23–33). This was known as the K.P.L. bus, from the initials of Knight, originator of the sleeve-valve type of engine used, Pieper, on whose design the electric transmission was based and Lanchester, who improved the overall design of the bus for the Daimler company. A four cylinder petrol engine was direct coupled to a dynamotor and the drive to the propeller shaft was through a magnetic clutch. There were two units, one for each rear wheel, so there was no need for differential gearing. On starting, the dynamotor took current from the battery for acceleration, then as speed rose part of the engine power was used to charge the battery. According to Lanchester only twelve were built.

In 1915 Lanchester took out a patent (No. 7366) for an internal combustion engine, an electric motor and a generator all on one shaft and with all the electrical windings in series. This was not developed. Then in 1916 the chief engineer at Daimler asked him to look into the P.E.T.-Thomas system, a petrol-electric transmission system in which an epicyclic gear was used to connect a petrol engine and two electric motor-generators. Starting, initial acceleration and battery charging were by control of the field strength of the motor-generators and reversal of their functions as motor and generator. At a suitable speed the electric drive was disconnected and the car was driven by the petrol engine. Lanchester made some favourable comments and was asked to design a similar system without an epicyclic gear. A patent taken out in 1917 (No. 108970) was probably the result. In this design a generator was connected to a petrol engine by way of a clutch and gears, and an electric motor was similarly connected to the output shaft. The petrol engine shaft and output shaft were in line and connected by a clutch. With this arrangement the electric devices could run at a higher speed than the main shaft so allowing them to be of a more economical size. An advantage claimed for the system was that the motor and generator could be removed without disturbing the petrol engine.

The Petrelect system

In the early 1920s Fred Lanchester and his brother George became interested in producing a cheap, light car and they designed the series of wooden cars, as described in Volume 1 of

The Lanchester legacy. After a time, although George thought they had gone far enough to go into production with the Mark III motor, Fred decided to use a petrol-electric power unit and so for wooden cars from Mark IV to Mark VIII Fred produced variations on the "Petrelect" engine, striving to satisfy his dream, and causing George to comment some years later, "How often does the genius contrive to improve and develop his design until the original conception is lost." (Lanchester, G.H.). Fred drove the various cars to and from his works and the sight of the strange-looking motor is remembered by some who lived near the Lanchester's Laboratories works in Tyseley.

The final "Petrelect" system was the subject of a patent (No. 241965) taken out in 1925. The patent covered all possibilities, allowing for a four-pole or a six-pole dynamotor and including the controller as well as the general arrangement of the petrol engine and dynamotor. The patent specification says that the invention has for its objects, "to provide a dynamotor having a greater torque and to allow of the better control of the motor and charging regimes and to render the charging regime under better control in order to prevent excessive currents entering the battery and to avoid the necessity of a clutch and enable the motor to start both engine and car from standing".

The principal improvement over previous systems was that the internal combustion engine and the electric dynamotor were directly coupled and the armature of the dynamotor was the flywheel of the engine. A controller was provided which allowed the driver to select forward or reverse electric drive, in which only field windings in series with the armature winding were used, or a battery-charging regime in which field windings in parallel with the armature winding were also in operation, with the petrol engine driving the car. When acting as a dynamo the current in the shunt winding reduced the magnetic field and so limited the magnitude of the charging current.

The field magnet poles were positioned to allow as much ground clearance as possible, and in the four-pole design the poles were arranged at other than 90° to one another thus also shortening the magnetic circuit, so economising on material and energy required for excitation. To reduce resistance, copper brushes were used; but to reduce sparking, carbon brushes were fitted to the leading and trailing edges. The connections between the armature windings and the commutator segments were made truly radial to avoid the risk of failure due to the action of centrifugal force on bent conductors.

Petrelect motors

F.W.Lanchester's notebooks contain general notes on the design of dynamotors and the matching of petrol engine to dynamotor. There are detail calculations on the design of dynamotor with six sizes of armature, from 13 in. diameter and $1\frac{1}{2}$ in. long to 18 in. diameter and 4 in. long. Two of these are still in existence, the $13\frac{1}{2}$ in. diameter, 2 in. long is on a Mark VIII engine and the 16 in. diameter, $3\frac{1}{2}$ in. long on the Mark VII motor car, both at the Birmingham Museum of Science and Industry.

Series wound direct current electric motors, in which the armature is in series with the field coils, give a very high starting torque (called the static torque) but the torque falls with increasing speed until it is virtually zero. The speed at which this occurs is called by Lanchester "cut-in speed" because the dynamotor can then become a generator. To limit the rise in voltage a shunt winding (a field winding in parallel with the armature) is added when running as a generator.

With the object of designing conditions of similarity such as would allow of specification to suit any particular conditions and also to enable performance chart and experimental data to be of universal application, under the heading "Generalisation re Dynamotors for the 'Petrelect' System", Lanchester set down the following assumptions:

"peripheral speed of the armature is the limiting factor, as is piston speed in a petrol motor, and there should be a definite numerical relation between these two limiting speeds;

in order to preserve a relation between the dynamotor and the engine the torque must be proportional to the cube of the armature diameter (or the horse power proportional to the square of the armature diameter);

flux density is constant for any given state: that is, for equivalent states the ratio of flux density to diameter is constant:

there is geometric similarity so far as is compatible." (LC. SB 5, p.32R)

Lanchester also wrote a list of "Factors in Dynamotor Design" in which he states that for a given armature diameter the iron weight is reduced by increasing the number of poles, but too many poles would incur an excessive weight of copper and total weight is no longer saved. On the other hand, increasing the copper in the armature means greater weight of iron, which will increase the flywheel effect. In the note book is the comment in capital letters and underlined, "THIS MUST BE DETERMINED". (LC. SB 5, p.48L)

The size of the armature is set by safety considerations. The maximum diameter is set by the bursting speed, which should be twice the maximum running speed. The length, while depending on the pole area required, must not exceed one quarter of the diameter and, if possible, should be less than one-fifth of the diameter. A limit to the number of poles is set by the radial thickness of the iron required to give sufficient strength to support the field windings. Although six poles are possible, a four pole winding is better.

A summary of his thoughts on the design of a dynamotor is to be found in one of Lanchester's sketch books under the heading of "Commonsense points of design". (LC. SB 5, pp.44R, 45L, 46L). For a given length of armature consideration of the effects of variations in field magnets and armature diameter Lanchester inferred that, "It is vital to employ the maximum diameter consistent with safety. The other vital quantity is the size of the field magnet unit." (LC. SB 5, p.44R). Since for a limited number of poles the weight of the field magnets is constant, increase in diameter (which increases torque) allows increase in torque to weight ratio so long as speed can be increased. But if diameter and speed are at the safety limit and magnet design is altered to increase torque, the torque to weight ratio is approximately constant. The effect of dynamotor proportions on the battery power required and the relation between speed and output was also considered.

Lanchester then set out to assess the best torque to weight ratio. His figures 1–7 are shown on two pages reproduced on page 181 as Figs 1 and 2. Starting with Fig. 1, a four pole machine of unit diameter, weight and torque. Increasing the diameter to 1.5 times with the same field, as Fig. 2, increases the torque by 1.5 times but only half the armature weight has been added. Adding a further pair of poles of the same size to give a six pole machine, as Fig. 3, increases the weight to 1.5 times the original but gives a torque 1.5^2 times the original. Attempting to increase the size of the six poles to 1.5 times to obtain 1.5^4 times the torque for less than 1.5^4 times the weight leads to overcrowding of the poles, Fig. 4, but if one pair of these poles is removed, Fig. 5, a torque of 1.5^3 times the original is obtained for a weight increase of 1.5^3 times. From the figures it might be supposed that Fig. 2 is superior to Fig. 1, since the torque is increased 1.5 times and the weight is increased less than 1.5 times, but due to the limit of peripheral velocity Fig. 2 would have to be geared to run slower in the ratio 1.5 to 1 and the effective "utility" torque would be reduced to the original value, so giving Fig. 3 the better torque to weight ratio. Since in Fig. 2 there is, in effect, wasted weight in the armature due to the field magnets not spanning all of the armature, Fig. 6, consider reducing the copper and iron in the armature and the field magnets to two-thirds and adding a further pair of poles. Then the field is three poles each two-thirds the weight of the original poles (i.e. the same total weight) and the armature is two-thirds the weight of the original (Fig. 7). This gives a six pole machine of the same torque as Fig. 6 but at about 15% less weight. (LC. SB 5, pp.44R–46L)

On a page headed, "Fundamental scale factor in Dynamo design" (LC. SB 5, p.46R), Lanchester also considered the effect of size on the design of a dynamotor and showed that larger machines could be made more efficient by reducing ohmic losses.

Plate 1. Rear view of the "Wooden car", Mark VII, showing engine compartment

Plate 2. The engine of the Mark VII car

Plate 3. The Mark VII car, belonging to the Birmingham Museum of Science and Industry, in the drive of Fred Lanchester's house, Dyott End, Moseley, Birmingham, during the Lanchester Centenary Rally, September 1995

Plate 4. The pendulum accelerometer made by Lanchester in 1904, now on display in the
Library of the Institution of Mechanical Engineers (photograph by permission
of the Institution)

Plate 5. General view of Lanchester rally, Coventry (Lanchester) Polytechnic, 29th April 1984

Plate 6. Mrs. "Steve" Lanchester

Plate 7. "Steve", Mrs George Lanchester, at the grave of George and Rose Lanchester
in the churchyard at Hampton Lucy, Warwickshire

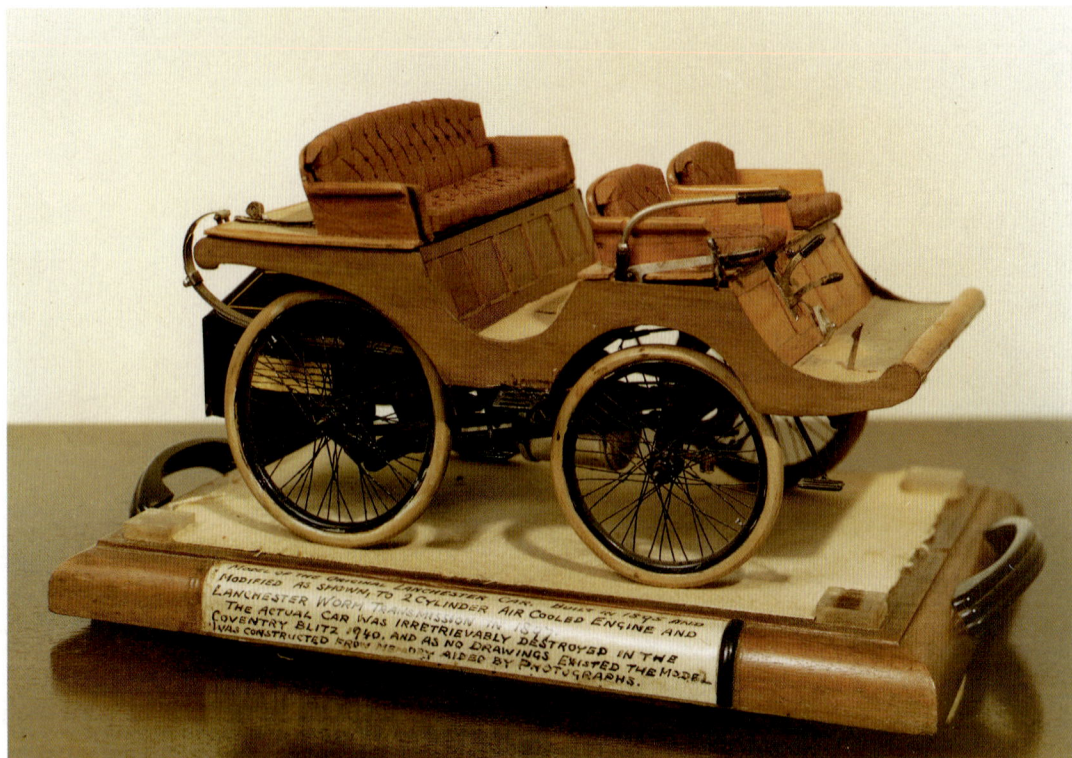

Plate 8. George Lanchester's model of the first Lanchester car

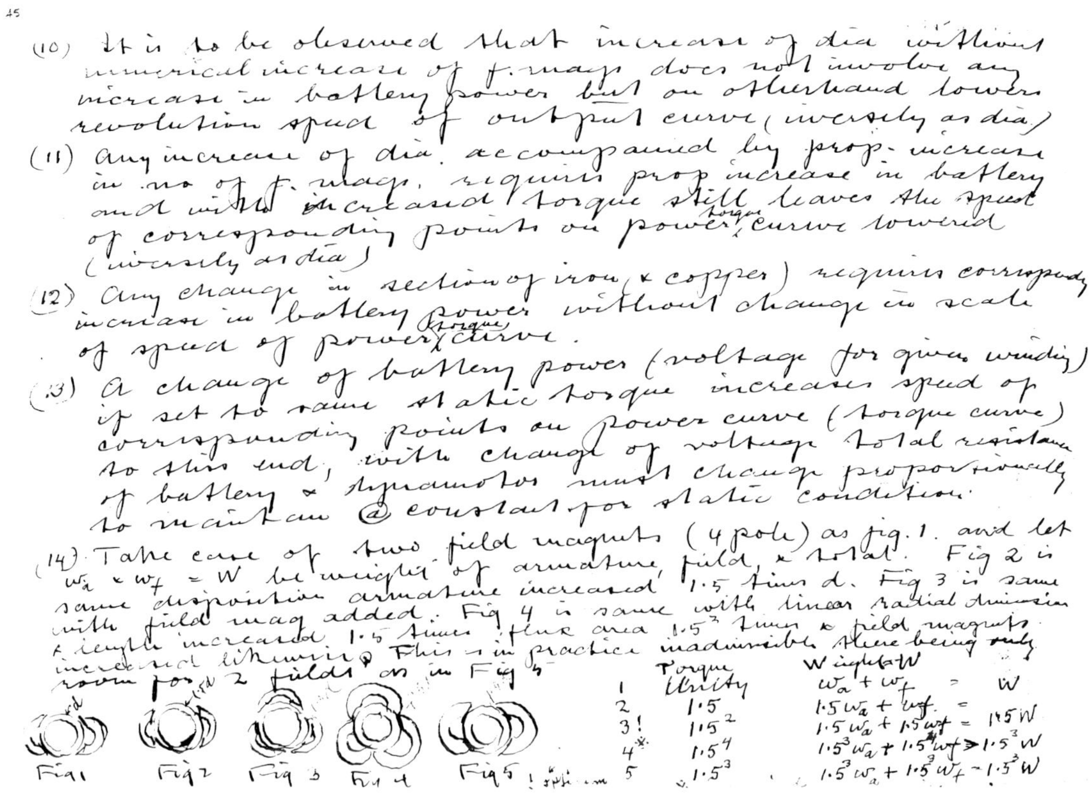

45

(10) It is to be observed that increase of dia without numerical increase of f. mags. does not involve any increase in battery power but on otherhand lower revolution speed of output curve (inversely as dia)

(11) Any increase of dia. accompanied by prop. increase in no. of f. mags. requires prop increase in battery and with increased torque still leaves the speed of corresponding points on power curve lowered (inversely as dia)

(12) Any change in section of iron & copper) requires corresponding increase in battery power without change in scale of speed of power curve.

(13) A change of battery power (voltage for given winding) if set to same static torque increases speed of corresponding points on power curve (torque curve) to this end, with change of voltage total resistance of battery & dynamotor must change proportionally to maintain @ constant for static condition.

(14) Take case of two field magnets (4 pole) as fig. 1. and let $w_a + w_f = W$ be weight of armature, field, & total. Fig 2 is same disposition armature increased 1.5 times d. Fig 3 is same with field mag added. Fig 4 is same with linear radial dimension & length increased 1.5 times, flux area 1.5^2 times & field magnets increased likewise. This in practice inadmissible there being only room for 2 fields as in Fig 4.

	Torque (Utility)	Weight = W	
1	Unity	$w_a + w_f$ =	W
2	1.5	$1.5 w_a + w_f$ =	1.5 W
3	1.5^2	$1.5 w_a + 1.5 w_f$ =	1.5 W
4	1.5^4	$1.5^3 w_a + 1.5^3 w_f >$	$1.5^3 W$
5	1.5^3	$1.5^3 w_a + 1.5^3 w_f$ =	$1.5^3 W$

Fig 1   Fig 2   Fig 3   Fig 4   Fig 5

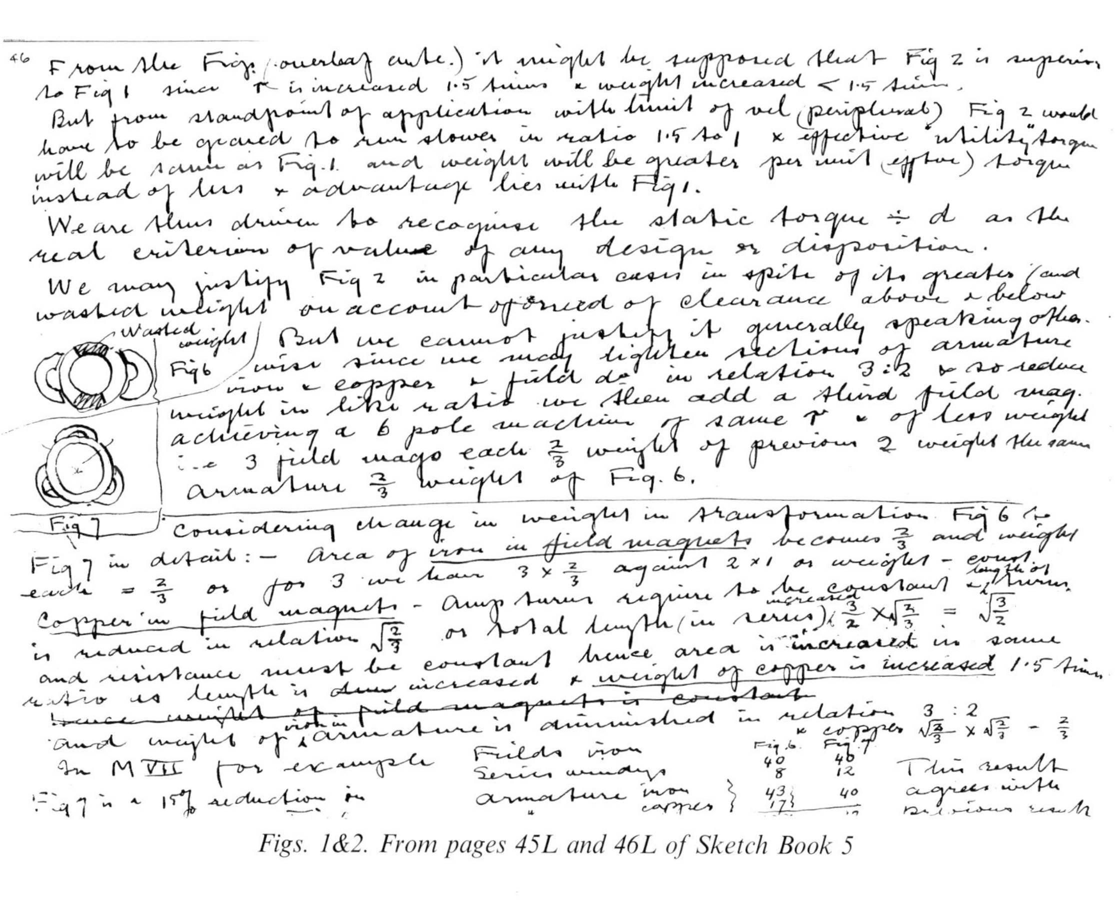

46 From the Figs. (overleaf ante.) it might be supposed that Fig 2 is superior to Fig 1 since τ is increased 1.5 times & weight increased < 1.5 times.

But from standpoint of application with limit of vel (peripheral) Fig 2 would have to be geared to run slower in ratio 1.5 to 1 & effective "utility" torque will be same as Fig 1. and weight will be greater per unit (effective) torque instead of less & advantage lies with Fig 1.

We are thus driven to recognise the static torque ÷ d as the real criterion of value of any design or disposition.

We may justify Fig 2 in particular cases in spite of its greater (and wasted weight) on account of need of clearance above & below. But we cannot justify it generally speaking otherwise since we may lighten sections of armature iron & copper & field do in relation 3:2 & so reduce weight in like ratio we then add a third field mag. achieving a 6 pole machine of same τ & of less weight i.e. 3 field mags each ⅔ weight of previous 2 weight the same armature ⅔ weight of Fig. 6.

Fig 6

Fig 7

Fig 7 in detail:— Area of iron in field magnets becomes ⅔ and weight each = ⅔ or for 3 we have $3 \times \frac{2}{3}$ again 2 x 1 as weights - equal. Copper in field magnets – Amp turns require to be constant & turns is reduced in relation $\sqrt{\frac{2}{3}}$ or total length (in series) $\sqrt{\frac{2}{3}} \times \frac{3}{2} = \sqrt{\frac{3}{2}}$ and resistance must be constant hence area is increased in same ratio as length is increased & weight of copper is increased 1.5 times and weight of armature is diminished in relation 3:2 × copper $\sqrt{\frac{3}{2}} \times \sqrt{\frac{3}{2}} = \frac{3}{2}$

In M VII for example Fig 7 is a 15% reduction in

	Fig 6	Fig 7	
Fields iron Series windings	40 8	40 12	This result agrees with previous result
armature iron copper	43 173	40	

Figs. 1&2. From pages 45L and 46L of Sketch Book 5

In an Internal Report of Lanchester's Laboratories Limited (undated, probably early 1928) entitled "Fundamental Basis of Calculations for the 'Petrelect' System" (LC. Baxter 13–14), Lanchester set out the method of determining the maximum torque from consideration of the heaviest battery allowable (the peak load available being about 50 watts per pound of battery weight) and then deciding the slow running speed of the petrol engine. Determination of the battery charging current is then described. The maximum possible is basically half or less of the ratio of turns in the shunt field winding to turns in the series field winding. The report indicates ways of reducing the charging rate if necessary.

Design of dynamotors

On 16th December 1912 Lanchester wrote some notes headed "Dynamotor in relation to Petrol Engine" (LC. SB 5, p.33R) in which he derived a formula for the relation between the stroke of a petrol engine and the diameter of a dynamotor armature based on the force on the periphery of the armature. This showed that, up to a definite limit of size, the smaller the installation the larger relatively the armature needed to become. Comparison of a two-cylinder engine (as in the Mark VII) and a four-cylinder engine (as in the Mark VIII) designed for the same mean torque showed that for a four-cylinder engine the diameter of the armature should be five times the stroke. But a note says that further investigation is required. The armature for the Mark VII dynamotor, with a two-cylinder engine, was 16 in. diameter and $3\frac{1}{2}$ in. long. For the Mark VIII dynamotor, with a four-cylinder engine, two sizes of armature were considered, 13 in. diameter by $1\frac{1}{2}$ in. long and $13\frac{1}{2}$ in. diameter by 2 in. long.

In order to fill the winding slots more economically, Lanchester found that, using "hydraulic pliers", he could compress round cotton covered wire 0.056 in. diameter over the cotton to 0.040 in. flat over the cotton without damaging the insulation and so fit ten wires instead of seven into each slot. This increase in area of copper in each slot would give a current capacity in the ratio of $(\frac{10}{7})^2$, or about double.

Concern about overheating of the windings led to the calculation that, even stationary and without cooling, the armature would take three minutes to overheat, but at the calculated rate of cooling the electric motor would be safe for 24 minutes: "Battery exhausted long before." (LC. SB 5, p.52R).

The Mark VII Petrelect wooden car

This motor car is preserved in the Birmingham Museum of Science and Industry, its general design has been described in Volume I of *The Lanchester legacy*, and we now note some details of the design of the engine and dynamotor which are to be found in Lanchester's sketch books. (See Colour plates 1, 2 and 3)

The petrol engine had two cylinders and was fitted with an exhaust valve lifter for "reverse and electrical manoeuvring" which also cut out the ignition. A control switch enabled the driver to select conditions for electrical drive in forward or reverse or petrol drive with or without charging the battery.

The dynamotor armature was 16 in. diameter and $3\frac{1}{2}$ in. long. Calculations of the magnetic circuit show that the static torque for 1000 amps was 2.5 times the engine torque and the cut-in speed would be 10 m.p.h. A 100 ampere-hour Exide battery was to be used. Calculations led Lanchester to the conclusion that the series turns were "Probably O.K." In output tests on 6th and 7th January 1926, the maximum torque at 24 volts was 150 lb. ft. and at 36 volts, 130 to 135 lb. ft. (LC. SB 5, pp.86, 87).

Later in the year Lanchester redesigned the 16 in. diameter armature, trying different magnetic field strengths. The final design gave an initial starting torque of 600 lb. ft., equal to $2\frac{1}{2}$ times the petrol engine torque, and the cut-in speed was $8\frac{1}{2}$ m.p.h. (LC. SB 5, pp.77–9).

The Mark VIII Petrelect engine

Although the Mark VIII car was never built two engines were completed, as described

Fig. 3. The Mark VIII engine

in Volume I of *The Lanchester legacy*, and one of these is preserved in the Birmingham Museum of Science and Industry. (See Fig. 3)

Trial calculations for a two-cylinder engine with cylinder diameter of either 2.75 in. or 2.95 in. and a 3 in. stroke showed that the horse power expected at 5,000 r.p.m. with a brake mean effective pressure of 80 lbs. per square inch would be three times the R.A.C. rating of 6 or 7 h.p. respectively. (This is stated to be usual for car engines in Newnes *Mechanical World Year Book, 1944*). However, the engine finally designed for the Mark VIII had four cylinders of $2\frac{3}{4}$ in. bore and $2\frac{3}{4}$ in. stroke giving an R.A.C. rating of 12 h.p. Running on petrol only it was forecast that at maximum torque a speed of 49 m.p.h. would be achieved at a petrol consumption of 45 m.p.g.

The petrol injection system, termed the "carburettor", was operated by a double action pump situated in the petrol tank. Lanchester designed a device for preheating the mixture using exhaust gases. This was a block of metal (Dural) with 97 holes of $\frac{1}{8}$ in. diameter for the air and 48 holes of $\frac{1}{8}$ in. diameter at right angles for the exhaust gases. To obtain 100°F temperature rise it was reckoned that more than 7% of the exhaust gases must be employed. He also designed an electrical preheater in which a dome fitted above the carburettor contained a winding which heated a 3 in. long block containing 127 holes of $\frac{1}{8}$ in. bore. The mixture passed up through an annulus round the heated block, then down through the block into the carburettor.

The four cylinders of the final design were horizontally opposed in pairs. Each cylinder had three valves, two inlet and one exhaust. For running on electric drive decompression would be achieved by a push-pull lever with a locking device on the instrument panel which operated a rod and link motion attached to the exhaust valves.

Notes on the design of the valve gear cams say that lateness of induction closing should not be exaggerated since turbulence suffers and that there should be no overlap as this is inimical to slow running. (LC. SB 6, p.60R) Results of tests of the valve timing as first set up on 24th April, 1929, contain a note that dead centre was determined by a soap film on the spark plug hole. (LC. SB 7, p.119)

Ignition was from the battery using a two coil contact breaker mounted on top of the pump gear box. It was noted that advance was not needed on the distributor.

Cylinder cooling was by copper gills, each "gill system" representing 60 sq. in., not all of which was effective, but sufficient gills were fitted to give about 5 sq. ft. cooling surface per cylinder. The cooling fan was designed to be $13\frac{1}{2}$ in. diameter and was expected to deliver air at a pressure of 2.25 in. water gauge. (LC. SB 6, p.15)

Before settling on the size of armature to use for the Mark VIII motor, Lanchester tried some variations on a 13 in. diameter armature. For various reasons these were unsatisfactory, although differences were not great. The 24 volt battery consisted of two 12 volt batteries, each weighing 97 lbs. After estimating the total weight of the dynamotor, comparison of the weight of the Petrelect design with the same petrol motor with a conventional flywheel, dynamo and starter showed that the Petrelect would be 40 lbs. heavier.

A dynamotor with an armature of $13\frac{1}{2}$ in. diameter and 2 in. in length was finally chosen. For this design a current of 500 amps was needed to provide the required static torque of 100 lb. ft. The 16 volt battery was of 166 ampere-hour capacity. As originally designed the loss in the leads would have been 1,000 watts, almost equal to the armature loss. To reduce this to 360 watts the controller was altered so that the cut-in was operated by a lever through the floor, so needing only 12 feet of cable instead of 32 feet. (LC. SB 6, p.3L)

An electric motor controller was designed which was to be operated by a handle rotating through five positions: Boost – for electric drive; Brake – for regenerative braking; Charge – for charging the battery when the petrol motor had power to spare; Reverse – as required; and Idle – when electric drive was not required but there was insufficient power for charging the battery.

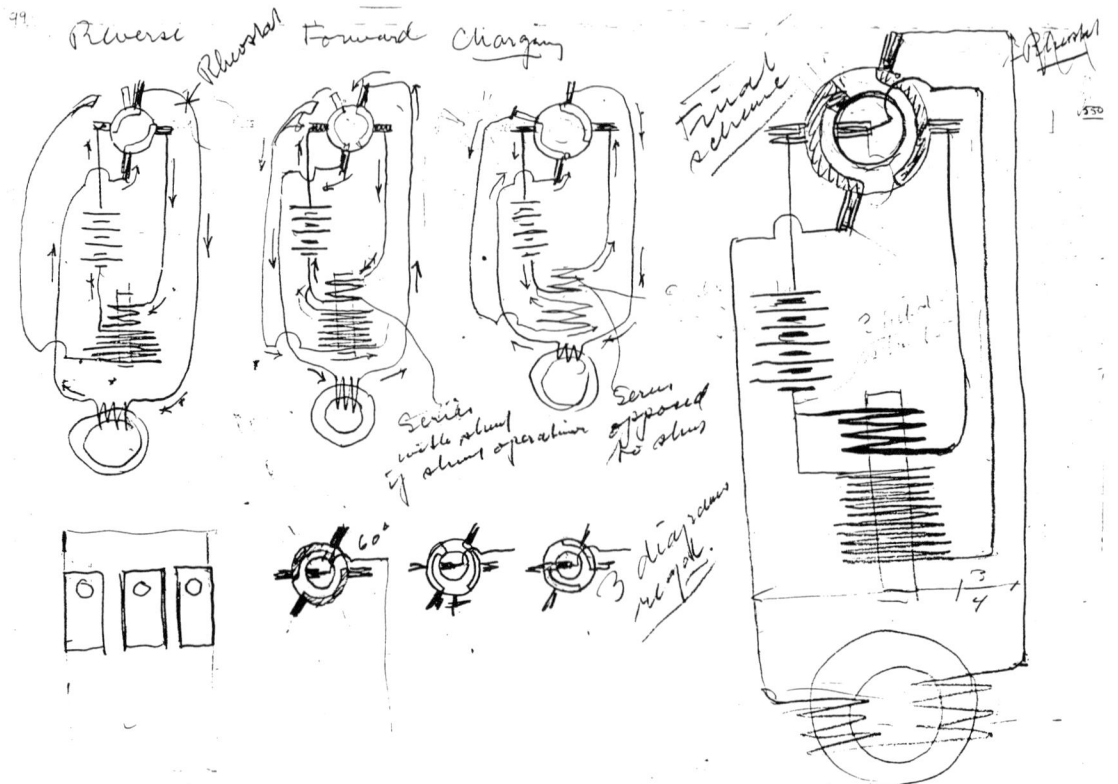

Fig. 4. Sketches of the controller for the Mark VIII motor. From Sketch Book 5, p.99L

References

Lanchester, F.W.: Factors that have contributed to the advance of automobile engineering and which control the development of the self-propelled vehicle, *Proc. I. Auto. E.*, Vol. V, 1910–11, 33–35

Lanchester, G.H.: The Miniature Lanchester car, (B.M.S.I. Historical Records)

Chapter Thirteen

SOUND REPRODUCTION EQUIPMENT AND RADIOS FROM LANCHESTER'S LABORATORIES LIMITED

by Roy L. Thomas

How Lanchester's Laboratories Limited began

Development work on Lanchester's motor car designs was carried out in the production department at the Daimler works, but Fred Lanchester did not consider this satisfactory: he thought that testing was not done as thoroughly as it should. Lanchester's Laboratories Limited was therefore set up with the object of keeping development separate from production. The company was incorporated on 23rd February 1925, and in March 1926 F.W.Lanchester and the Daimler Company each put up £12,000, but Daimler immediately borrowed back £11,000. The "loan" was never returned.

Experimental and development work by Lanchester's Laboratories Limited continued to be done at Coventry. But Lanchester wanted to be separate from the Daimler works and in 1928 he bought land in Birmingham, at Spring Road, Tyseley, and outlined a scheme for Lanchester's Laboratories Limited's headquarters to be built there.

Building plans were submitted to Birmingham City Council in 1928, but early in 1929 Lanchester and Daimler were in dispute over Lanchester's contract with Daimler, and the continuation of Lanchester's Laboratories. Percy Martin, a director of Lanchester's Laboratories Limited and of the Daimler Company, said that Lanchester's Laboratories had been running for some time without appreciable good for anybody. He proposed that Lanchester should either purchase Daimler's shares in Lanchester's Laboratories, or take over all patents and Daimler would take all the cash in Lanchester's Laboratories. If Lanchester did not accept one of these alternatives Martin would advise a veto of active work on Lanchester's Laboratories Limited while retaining rights to all future inventions. At a meeting of the Board of Lanchester's Laboratories Limited on 13th April those directors who were also directors of the Daimler Company resigned. Lanchester was advised by solicitors that it should be possible to show that, by reason of a borrowing, Lanchester's Laboratories Limited could not carry on the work for which it had been formed and that it should be wound up. He did not take this advice, not wishing to become involved in litigation, but decided to carry on with Lanchester's Laboratories, buying Daimler's shares and taking over the loan (which had been interest free) at three per cent per annum. A new board, which met on 15th April, was formed and the premises at Spring Road were built. (LLL; Lanchester, 1935).

Lanchester's Laboratories embarked on the development of high class products for the reproduction of music, and in the course of eighteen months put on the market a number of transformers and loudspeakers. This was a subject in which Fred Lanchester had been interested for some time; he had already patented some improvements in loudspeakers.

The Euterpe-phone demonstrated in Birmingham Town Hall

In January 1929 the *Birmingham Post* carried notices that "In the Town Hall, Birmingham, on Thursday, Jan. 31, at 7.30" there would be a "LANCHESTER EUTERPE-PHONE CONCERT". Perfect electrical reproduction giving full orchestral volume was promised for a programme arranged by Mr. Adrian Boult. Fred Lanchester was to give a short introductory lecture. (See pp. 14–15.)

Fig. 1. The Spring Road factory of Lanchester's Laboratories Ltd.

An interview with Lanchester appeared in the *Birmingham Mail* for 23rd January, 1929, in which a reporter said that at present the equipment was intended for halls, but there was no reason why it could not eventually be for use in the home. Dr. Lanchester had explained that the speaker was constructed with the object of avoiding any directional effect and concentrating on emitting sound laterally, so that everyone would be able to hear direct and not by reflection from the walls. But the reporter was even more interested in the "metatonic amplifier" with its control box to emphasise bass and treble and its specially designed output transformer. When some records were played the reporter found the results extraordinary, from the low notes of a Sousa march to the high notes of a violin or a soprano.

Unfortunately the demonstration in the Town Hall was not the success which had been hoped for. On the following day the *Birmingham Post* reported that, "The first thing to be said about the musical results obtained was that they bore no comparison in point of quality with the reproduction one had been privileged to hear in Dr. Lanchester's own quite large music room." We can only surmise that some component, such as a valve or capacitor, had become faulty; we can be sure that Lanchester had tried everything out beforehand.

The patent for the first Euterpe-phone (Euterpe was the muse of lyric poetry and was usually depicted with a lyre) was No. 317339 of 1929, "An Improved Telephonic Loud-speaker", which describes a pipe of non-resonant material (preferably lead), 3 to 4 ft long (or 8 to 10 ft to cater for the deepest tones of an organ), connecting the magnet and diaphragm to a horn having parallel sides and the top and bottom logarithmically curved. By curving the pipe to fit in a space behind the horn the whole could be made to form a piece of furniture. Circuit diagrams in his sketch books show a "Mystery Box", probably the "control box" in the demonstration equipment.

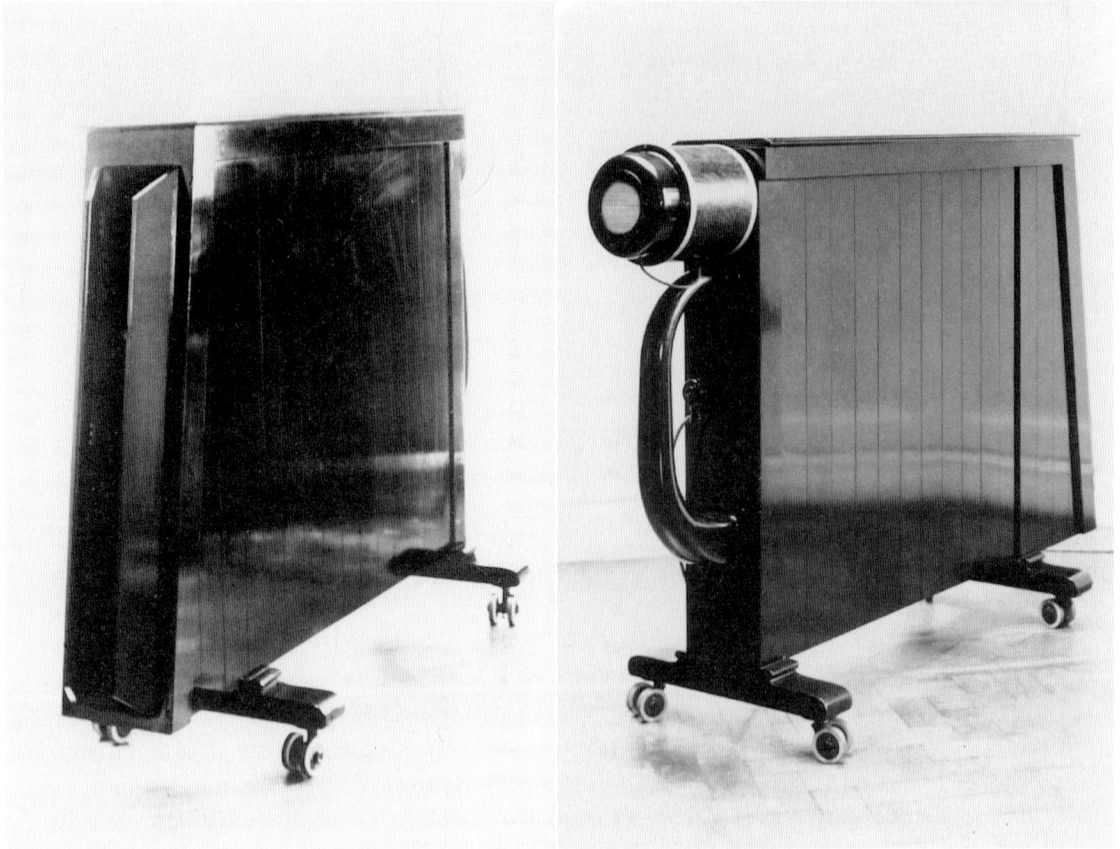

Figs. 2 & 3. The Euterpe-phone for indoor use

Fig. 4. A larger Euterpe-phone fitted with a roof for use outdoors

The Spring Road Works

The plans submitted in 1928 were for a single storey building of five bays stretching back from the entrance in Spring Road. The first bay consisted of two rooms with a corridor between which gave access to a larger room under the second bay; beyond this was the workshop which occupied the rest of the building. In 1929 application was made for an extension consisting of two ground floor offices with drawing office over to be added at the front of the works, but this was never carried out. The building is still in existence, though an extension has been added at the side.

Fred Lanchester was assisted in his experimental and design work by R.H.Pearsall, who had joined Lanchester's Laboratories on motor work at Coventry (described by Lanchester as an excellent laboratory man but not a good manager), and by an apprentice, A.H.Jones, who joined Lanchester's Laboratories at Tyseley at the age of sixteen to begin a career in electronics. We are indebted to Tony Jones for sharing his recollections of those days and suggesting items which should be included in this chapter.

Musical reproduction

Under this heading a Lanchester's Laboratories Limited catalogue of 1930 says:

"We are now able to offer to the public the result of research work carried out in Dr. Lanchester's private laboratory during a number of years. We do not set forth any claim to perfection. Every maker of wireless apparatus claims in his advertisements such perfection that, if true, would render further advance unnecessary, not to say impossible, and generally speaking the worse the outfit the more extravagant the claims made. We take no part in this pandemonium of misrepresentation; we content ourselves by putting before the public the best we can offer and leaving them to form their own opinion."

The catalogue advertised that, "One of our activities is the hire and servicing of Lanchester Musical Reproduction Sets, our terms being based on contracts for the *supply of music*."

Lanchester's Laboratories Limited catalogues and advertisements always carried the slogan, "We sell direct to the public only", and the 1930 catalogue had a reasoned argument on the benefits to the customer of not using dealers because they would want their share of the profits. Notwithstanding this, Fred Lanchester did consider the possibility of supplying the trade when resources permitted an output of sufficient magnitude, and in one of the sketch books he noted the need for twenty "depots" or "stations" as a start, two in central London, eight in Greater London and the remainder scattered round within sixty miles of the city, including one at Brighton. (LC. SB 12, back page).

Lanchester's sketch books have already been mentioned. It is fortunate that many have been preserved because we can see how he made notes on the theory behind the design and use of the items used in the reproduction of sound. As a result of his examination of the theory of loudspeaker performance he was able to produce "rules" or "laws" concerning their design and construction. There are pages of notes on magnetism and the theory of electronics but Lanchester also studied acoustic theory so that he could use the correct procedures to calculate the pressure of the acoustic wave and so determine, among other details, the permissible weight of a diaphragm for best results.

In the sketch books we are able to see some of the designs and calculations Lanchester made in connection with his research. He designed many components from first principles and tested samples of the materials used. One page of a sketch book has samples of paper for loudspeaker diaphragms stuck on, with notes on suitability alongside. In investigating the performance of different types of iron and steel for magnets he considered the effect of hysteresis, a phenomenon of magnetic materials due to the work done when the flux density is alternating. It was the custom to ignore this when dealing with small oscillations, but Lanchester found that the effect was considerable.

Calculations or experiments did not always give sensible results; we see comments such as, "calculation is inconsistent", or, "calculation seems to denote a dilemma". There are many entries headed "Recalculation", sometimes because Lanchester had used a wrong formula: on one occasion the note, "Wrong calculation, used 4π instead of 2π", appears.

Loudspeakers

The public demonstration of Lanchester's first moving-coil loudspeaker, intended for the finest reproduction of gramophone records, the Euterpe-phone, has already been described. When, in 1930, Lanchester's Laboratories Limited went over to the production of musical reproduction equipment it was primarily the moving-coil instrument with which Fred Lanchester was concerned. Although smaller versions of the Euterpe-phone called "Acoustic-tube Type" speakers were made, the principal sales were of moving-coil cone speakers attached to a facia board ready for fitting in a cabinet or to a baffle.

Lanchester calculated the sound energy and pressure required to give a satisfactory volume, and, as a result, decided that a loudspeaker should give one hundred times the output of a voice in normal speech. One of the rules which he developed to aid the design of the speakers related the ratio of the weight of the coil winding to the weight of the diaphragm and the ratio of the resistance of the coil winding to the resistances of the valves and transformer, another concerned the size of gap in the magnet.

Raw materials were sometimes tested at the Spring Road Works, particularly the various steels available for making magnets. Cobalt steel was most often used and Lanchester calculated the cost per joule of energy of using steels with different proportions of cobalt; at 1s.3d. to 3s.11d. per lb of steel the cost per joule was 14s.2d. to 19s.7d. But cobalt steel is difficult to machine, so Lanchester patented a method of capping the pole faces with easily machineable soft iron or mild steel pole pieces kept in register by brass or aluminium, (Patent No. 356246 of 1931).

Lanchester's Laboratories loudspeakers were all based on patents registered by Fred Lanchester. His earliest were improvements to the Euterpe-phone for use in halls. Others concerned the arrangement of a bar magnet and the moving coil and suspension of the diaphragm. One of these, No. 364683 of 1932, was for the use of rubber rings to provide a

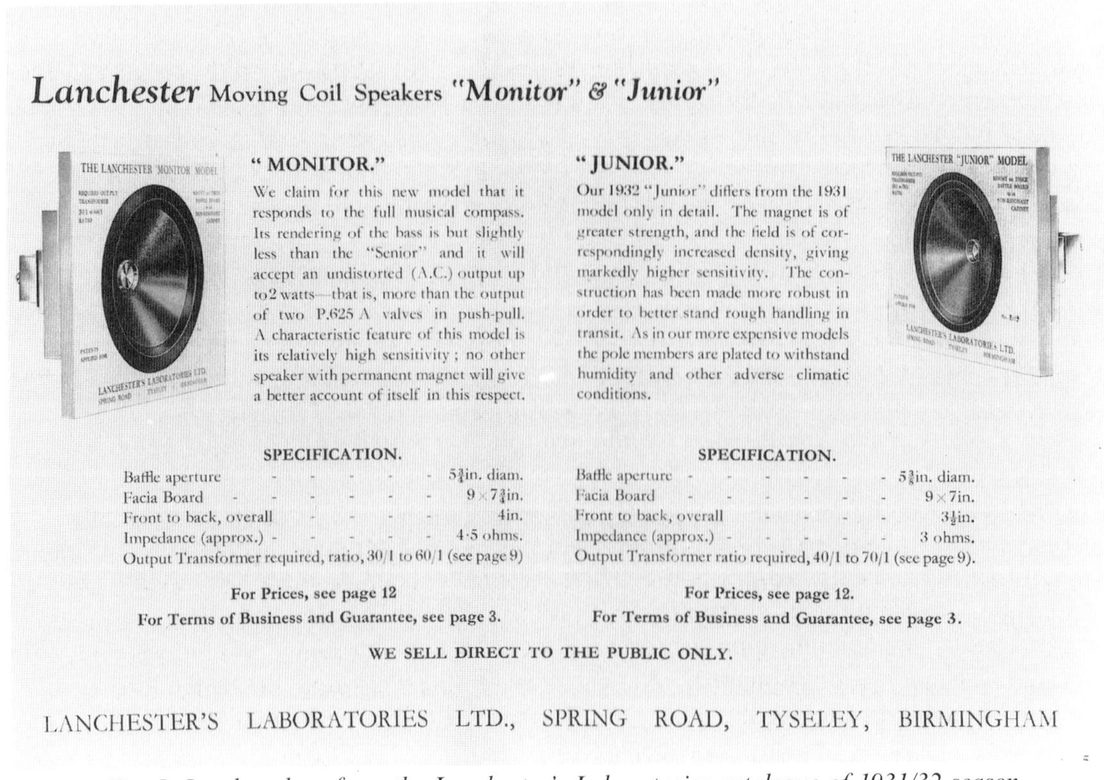

Lanchester Moving Coil Speakers *"Monitor" & "Junior"*

"MONITOR."

We claim for this new model that it responds to the full musical compass. Its rendering of the bass is but slightly less than the "Senior" and it will accept an undistorted (A.C.) output up to 2 watts—that is, more than the output of two P.625 A valves in push-pull. A characteristic feature of this model is its relatively high sensitivity; no other speaker with permanent magnet will give a better account of itself in this respect.

"JUNIOR."

Our 1932 "Junior" differs from the 1931 model only in detail. The magnet is of greater strength, and the field is of correspondingly increased density, giving markedly higher sensitivity. The construction has been made more robust in order to better stand rough handling in transit. As in our more expensive models the pole members are plated to withstand humidity and other adverse climatic conditions.

SPECIFICATION.	
Baffle aperture	5⅜in. diam.
Facia Board	9 × 7⅜in.
Front to back, overall	4in.
Impedance (approx.)	4·5 ohms.
Output Transformer required, ratio, 30/1 to 60/1 (see page 9)	

SPECIFICATION.	
Baffle aperture	5⅜in. diam.
Facia Board	9 × 7in.
Front to back, overall	3½in.
Impedance (approx.)	3 ohms.
Output Transformer ratio required, 40/1 to 70/1 (see page 9).	

For Prices, see page 12

For Prices, see page 12.

For Terms of Business and Guarantee, see page 3.

For Terms of Business and Guarantee, see page 3.

WE SELL DIRECT TO THE PUBLIC ONLY.

LANCHESTER'S LABORATORIES LTD., SPRING ROAD, TYSELEY, BIRMINGHAM

Fig. 5. Loudspeaker, from the Lanchester's Laboratories catalogue of 1931/32 season

mounting permitting free rectilinear movement to the diaphragm and preventing acoustical leakage. A later patent, No. 383376 of 1932, concerned the use of rubber balls to centre the moving coil. The "Junior" and "Monitor" speakers used four ⅛in. diameter rubber balls to centre the diaphragm and the "Senior" speaker used six.

The *Wireless World* reported on tests on Lanchester speakers, always noting the excellent tone because of the very flat frequency response curve. The report in the issue of 13th April 1932, on the Lanchester "Special Senior" moving coil speaker says that:

> "The design of the chassis is similar to that of other Lanchester loud speaker units previously reviewed. The magnet system is, however, of more generous dimensions. The characteristic, which is practically level between 100 and 1,000 cycles actually rises about 5 decibels from 100 down to 50 cycles. This imparts a full round tone to the reproduction, which is unusual in a loudspeaker with a diaphragm diameter of only 5¾in.
>
> The reproduction of speech is of the natural and unforced quality which has come to be associated with Lanchester loud speakers, and the happy balance between bass and treble gives a rendering of music which is entirely satisfying to the ear."

Tony Jones says that the Lanchester speakers' response was pretty flat from 50 to 5,000 cycles, whereas the competition's (at least those he read of) all had resonance "spikes" and some looked like cross-sections of the Alps!

A "Lanchester Speakers" catalogue of 1931–2 lists three sizes of acoustic-tube type loudspeakers and four types of simple cone type with prices ranging from £22.10s.0d. for the largest acoustic-tube speaker to £1.10s.0d. for the Junior speaker. These prices were "Chassis only", a chassis fitted in a cabinet cost £1.2s.0d. extra, and for a further 10s.0d. the cabinet could be of polished mahogany. However the sketch books contain designs for at least three other types which presumably never reached the production stage.

The "LanLab Repeater" speaker was specially designed for use as an extension speaker in a house, wired to the low voltage side of the main speaker transformer so that bell wire could be used. This speaker was also recommended for radio relay services.

Because of the importance of matching the speaker to the set, output transformers were also made, and could be supplied separately or with the loudspeaker. The 1931 catalogue included a short article by Fred Lanchester entitled "The Speaker and the Set", which described the advantages of the moving-coil type over the moving-iron type, and explained the need to match the low impedance of the moving-coil speaker to the high impedance of the output valve. Output transformers cost from 11s.0d. for a two ratio 17 amp design to 55s.0d. for the four ratio "Pentode Master" (described as having a "Massive (5½ lbs.) Stalloy core.").

In the 1931 catalogue were also published valve characteristics plotted "in accordance with the method given in our 'Broadcast No. 3'". These "Broadcasts", also listed in the catalogue, were "short monographs giving information on subjects for which we have found a special demand." They were free, but those who were not customers were asked to enclose a three halfpenny stamp "to prevent us being out of pocket". (LC. Baxter 3–17).

Microphones

Although Fred Lanchester was principally concerned with the design and manufacture of loudspeakers he was interested in all other items required for the reproduction of sound. He realised that a moving coil loudspeaker would work reciprocally as a microphone and, using two loudspeakers and a suitable audio amplifier, he demonstrated this to his apprentice.

As with most of the other items, Lanchester would carefully design a microphone taking account of theory and manufacturer's data on such matters as the magnetic properties of the iron used. When testing a microphone and amplifier Lanchester would sometimes indulge his talent for singing, measuring performance by the swing in anode volts as he sang.

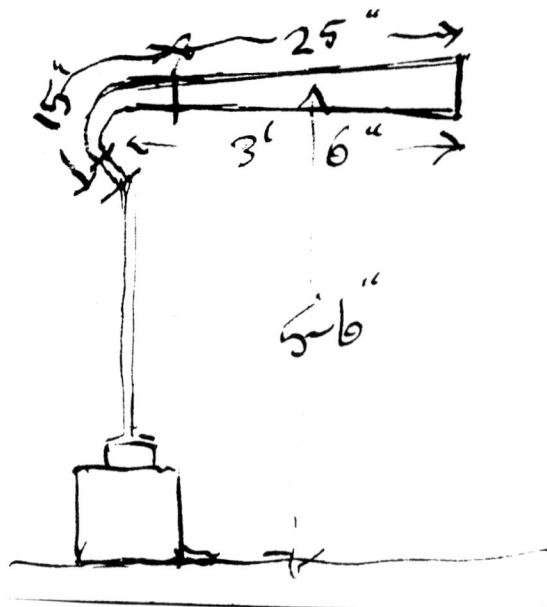

Fig. 6. Lanchester's sketch of his Condenser Microphone, from Sketch book 7, p.173

Lanchester also designed a condenser microphone. To users of tiny modern electret microphones this would be a huge and cumbersome device. A 6 in. square frame had a screen of 10 SWG wires (at about 6 to the inch) soldered across it and a sheet of tinfoil (insulated from the bars by a piece of stretched fine silk) formed the diaphragm. A sketch of the microphone shows the device mounted at the base of a trumpet. The sound entered through the flared mouth and turned through 90° before passing down a long acoustic tube. Lanchester calculated the capacitance, permitted inductance, and work done per vibration for 1% of gap change. Calculation of change in charge indicated that the resistance of the condenser should be 20 megohm, and that of the whole apparatus many times this, or voltage would be lost; alternatively the condenser could be fed through the secondary (lower voltage) of a Tesla coil.

Gramophones and pick-ups

In striving for as near perfection as possible in the reproduction of music, Lanchester investigated the design of gramophones and pick-ups and, although he does not seem to have advertised and sold a full "line" of pick-ups, the article on "Musical Reproduction" in the 1930 catalogue advises the use of a lighter pick-up to save damage to the record and concludes, "We manufacture and can supply a pick-up of either high or low voltage, as may be required."

Lanchester took out a number of patents covering improvements in gramophones. One such was Patent No. 321967 of 1929, "Improvements in Gramophones and like Sound Reproducing Apparatus", concerned with the loss of efficiency occurring due to the limitation of the mass of the pick-up or soundbox, because increase in weight would lead to destruction of the needle and the record. In modern tone arms the vertical and horizontal axes of movement cross, so that the needle is raised by lifting the whole tone arm, but since in a mechanical reproduction gramophone the tone arm is part of the tube connecting the soundbox to the horn, the needle was lifted by rotating the soundbox round the tubular "tone-arm" and the vertical pivot was at the opposite end of the arm. Early electric gramophones continued to separate the horizontal and vertical pivots and Lanchester's invention, while leaving the vertical pivot at the fixed end of the tone arm, provided a crank in the arm at its outer end, so that two pivots could be provided to enable the weight of the soundbox or pick-up to be partially balanced by a counterweight. This allowed the soundbox or pick-up to be easily raised from the record as well as providing a means of reducing the load on the needle.

Another patent, No. 361841 of 1931, entitled "Improvements in Mechanisms for the Reproduction of Sound", was intended to obtain greater volume (using a soundbox) by more direct application of the needle movement to the air column in order to obviate scratch and to amplify the bass notes. The needle operated a piston diaphragm of conical shape. Bass response was improved by the use of a "trombone" tube between the diaphragm and the horn. Calculations of the force on a diaphragm and a diagram in one of the sketch books appear to refer to this patent. (LC. SB 7, p.177).

Three patents applied specifically to electromagnetic pick-ups. The first was No. 337811 of 1930, "Improvements in Mechanisms for the Reproduction of Sound", in which, instead of attaching the needle to a "reed" as in previous electromagnetic pick-ups, the needle became the armature by being held in a rubber bush which was a lining to the coil.

The other two patents had similar titles, being concerned with improvements in apparatus or mechanisms for the reproduction of music, speech and the like. Patent No. 339949, also of 1930, covered an improved electromagnetic pick-up in which the needle and coil were carried at the end of a long cobalt steel bar magnet (8–10 mm diameter, 70–80 mm long) mounted as in patent No. 321967 of 1929, the coil being held between pole pieces in the form of prongs at the end of the magnet. The needle, which could be of soft low hysteresis iron, could be the armature as in patent No. 337811. It was claimed that the construction allowed some irregularities in the record (e.g. warping) without apparent effect on the sound. Patent No. 356367 of 1931 was similar but was particularly intended to avoid unnecessary weight on the needle and to obviate wear. It detailed various forms of tone arm and ways of mounting the pick-up to accommodate open or closed magnetic circuits.

Amplifiers

To demonstrate his improved designs of loudspeaker, Fred Lanchester needed a good amplification system. His sketch books contain circuit diagrams from 1928 showing the development of an amplifier using triode valves with resistance-capacity coupled early stages and a final push-pull stage of six triodes in two sets of three in parallel to obtain the power required. This may have been the amplifier used for the demonstration of the Euterpe-phone in Birmingham Town Hall, as mentioned above.

A good amplifier has some form of tone control. Present-day designs often have a "graphic equaliser" with which almost any form of frequency response can be obtained, but Lanchester was concerned with correcting for the poor response at low and high frequencies of the microphones used in making gramophone records at that time. He and his assistant therefore designed what they called an "integrating circuit" to correct for the loss of tone at low frequencies and a "differentiating circuit" to compensate for the loss at high frequencies. The two together, called the "Mystery Box", were used as two triode circuits in parallel feeding the first stage of an amplifier, and would have been the "control box" used in demonstrating the Euterpe-phone in Birmingham Town Hall.

The earlier amplifiers were battery powered, and indeed the use of batteries was strongly advised in order to avoid the hum introduced by a typical "all mains" set, but Lanchester's Laboratories Limited produced a range of mains eliminators which appear to have ranged from £12.10s.0d. to £3.18s.6d. in price.

Radios

In his "Concise History of My Career" which Lanchester wrote in 1933 he says that Lanchester's Laboratories Limited made "speakers, transformers and, for the first time this season, radios". In fact he had been working on radio reception for some time; a "Domestic Set" designed in October 1929 as a gramophone amplifier had a "radio attachment" which plugged into the first stage in place of the pick-up.

In a sketch book (LC. SB 7, p.148) is a description, dated December 1929, of a balanced

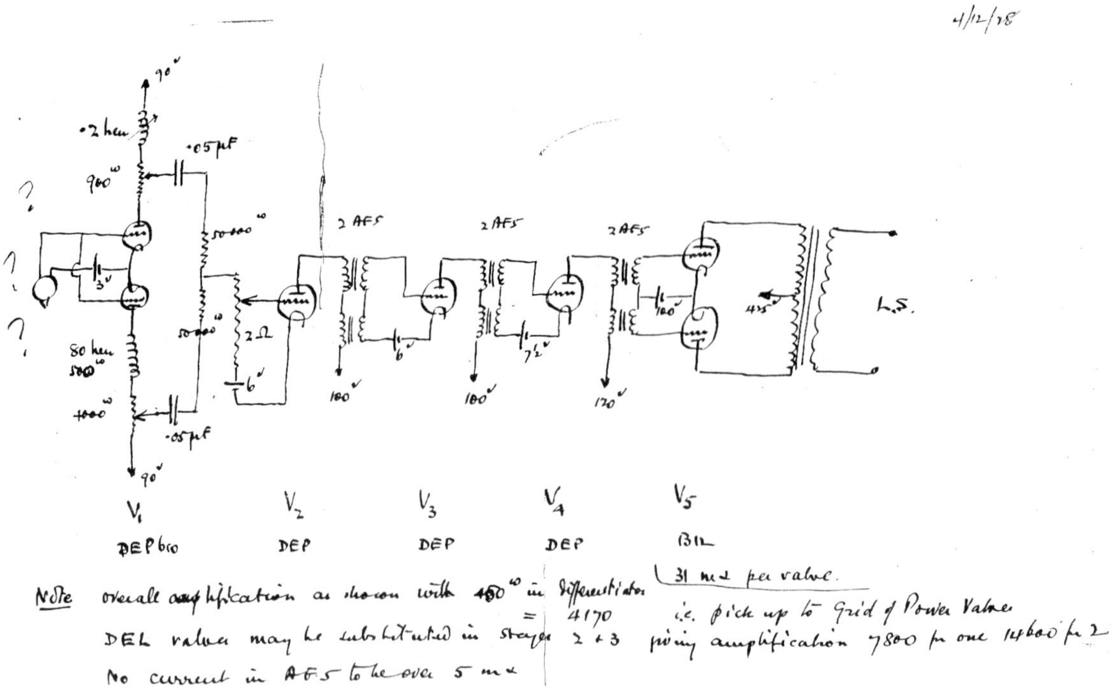

Fig. 7. Amplifier with "Mystery Box" tone control circuit. From Sketch book 7, p.109

selector circuit in which signals from two aerials were fed through two tuned-anode amplifiers, one to each end of a potentiometer which had its slider connected to the grid of a third valve. This was developed and registered by Lanchester in 1931 as Patent No. 341574, "Improvements connected with Wireless Reception and Application therefor" and was described as a means of eliminating inter-station interference, especially that caused by overlapping of sidebands, achieved by connecting two aerials to independent radio frequency and detector circuits. In or after the detector the audio signals were made of opposite sign, either by using plus bias in one detector and anode bend in the other, or by audio frequency reversal, or by using a transformer. The signals were then combined by a potentiometer having one signal applied to each end, the tapping connection leading to headphones or an amplifier. Signals at the potentiometer were of opposite polarity resulting in a tapping point for each station where the signal was zero, not the same point for each station, so allowing removal of the unwanted station.

It is interesting to note how Lanchester used the latest components, other than those he designed and made himself. The screen-grid valve only reached the commercial market in 1927 but Lanchester used one as the R.F. amplifier in his "Home Set". The amplitude of oscillation in grid voltage is limited in screen-grid valves to only about half a volt. Lanchester noted this limitation and proposed to overcome it by using what he called a "retrodyne" circuit. In this circuit the 80 volts supplied to the screen passed through a coil wound over the tuned coil on the anode of the valve in such a way that the screen was made more negative when the anode swung positive and vice-versa, thus enabling the valve to give more amplification. (LC. SB 8, p.64). In 1928 Phillips put on the market the pentode, a valve which had a suppressor grid between the screen-grid and the anode to suppress the negative resistance which occurred when a large signal caused the anode potential to drop below that of the screen-grid. This made a valve which was particularly useful in low frequency stages, and Lanchester made much use of pentodes in later amplifiers and radios. He designed the "Pentode Major" output transformer in 1931 specially for connecting a final stage using pentodes to the Lanchester's Laboratories "Senior" moving-coil loudspeaker.

Fred Lanchester was prepared to give credit to others when appropriate. His apprentice, A.H.Jones, devised a "Quiescent Driver" circuit for a push-pull amplifier in which a rectified signal was fed back from the grid of one of the push-pull valves to the grid of the driver.

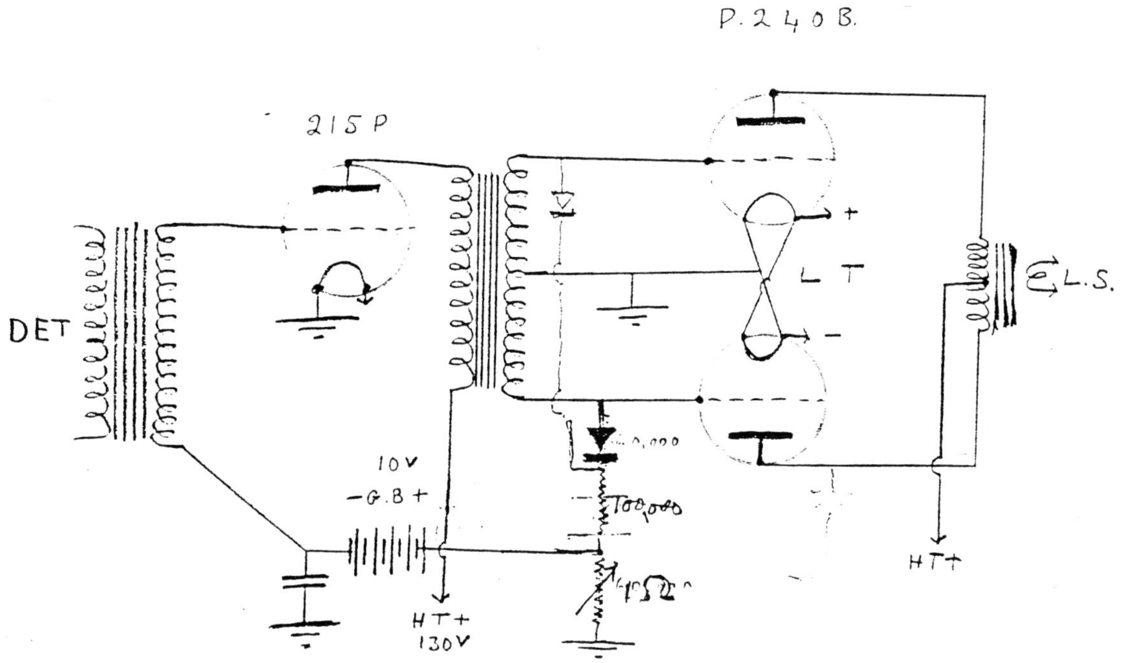

P.240 B.

215 P

DET

10 V
−G.B +

HT+
130V

L.T

HT+

E L.S.

Fig. 8. Tony Jones' original "Quiescent driver" circuit

The circuit is shown, with evidence of some changes before the final design, at Figure 8. In *Practical Wireless* for 9th December 1933, the Quiescent Driver circuit appeared under the title of "The Lanchester-Jones 'Class B' Circuit". The circuit was described as being designed to allow the use of a large amount of bias on the driver valve, so reducing its power consumption, but to provide rectified feed-back from the power circuit to reduce the bias on the driver valve so that a strong signal could be amplified without distortion. A footnote said that the patentees had no intention of asking for royalties from "bona fide amateurs". The circuit diagram (Fig. 9) drawn by Tony Jones, of a portable radio in which the device

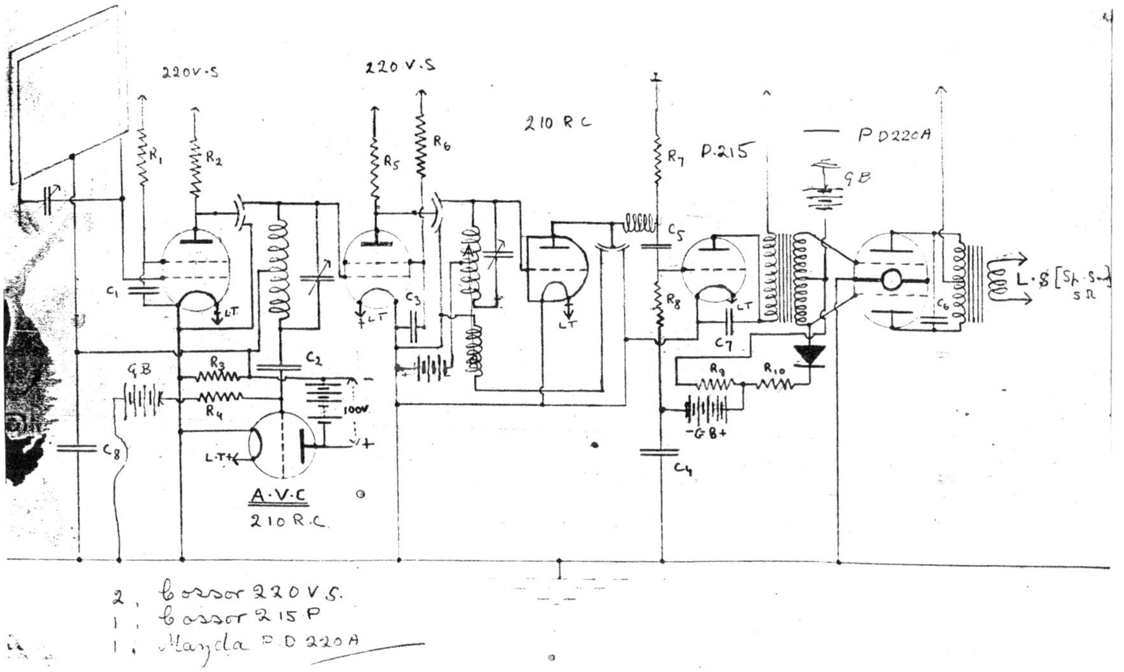

Fig. 9. A portable receiver incorporating the Lanchester-Jones "Class B" circuit. Drawn by Tony Jones about 1933. From Sketch book 7, back cover

was incorporated is in one of the sketch books. Mr. Jones recollects that the set was probably one carried in cars in direction finding "hunt the transmitter" games when, he told us, "Direction finding measurements and intersections would sometimes enable one to find the transmitter!"

The end of Lanchester's Laboratories Limited

Lanchester's Laboratories Limited made only a small profit between June 1931 and June 1932, because of the depression. Lanchester wrote that a shut down in February 1932 could have saved him, but he was too considerate of the workers and struggled on until January 1934 when illness put him, "out of action for the whole year, and nearly for good". (Lanchester, 1935).

At a meeting of the Board of Lanchester's Laboratories Limited on 3rd May 1934, Fred Lanchester was authorised to negotiate the sale of the land, buildings and contents of the Spring Road property. The company continued in existence for some years longer. (LLL).

References
Lanchester, F.W.: *Concise History of My Career*, 1933 (LC and B.M.S.I.)
Lanchester, F.W.: *Summary of Dispute with Daimler*, 1935 (B.M.S.I)
LLL: *Lanchester's Laboratories Limited Minute Book – 1925–1938* (LC)

Chapter Fourteen

LANCHESTER'S *SPAN*: AN APPRECIATION

by Joe Hastings

Considering Dr. Lanchester from a viewpoint of nearly six decades after his paper *Span* was presented as the second annual lecture to the Manchester Association of Engineers, it is difficult not to stand in awe of his scope of interests and influence.

He comes across as a true polymath having a deep understanding of engineering, ranging from structural/civil through municipal to mechanical and probably electrical disciplines of the art.

I get the feeling that Dr. Lanchester may have been in that mould of engineers that at times seek to lead the design team, somewhat pre-dating the great multidiscipline teams that evolved in the 1960s.

Dr. Lanchester's incisive grasp of domes and arches must have been outstanding in his time, for his tenets on masonry are echoed today, in for example, a paper presented to and published by the Institution of Structural Engineers: R.J.Mainstone (1990), wherein the author confirms Lanchester's observations that old structures remain structurally adequate without the need for meddling with high-tech reinforcements, and must be appraised sympathetically. Similarly W.J.Harvey and F.W.Smith (1991) backs up Lanchester's philosophy.

Dr. Lanchester's comments on the Forth Bridge are still valid today, vide the paper by D.G.McBeth (1990) which follows in his footsteps and observations.

Lanchester's amazing insight into all aspects of inflatable buildings was so far ahead of its time that it was hardly given credence by his contemporaries. It is only in the last few decades that his pioneering work has been allowed to bear fruit, and the many inflatable buildings are testimony to his foresight. In current design concepts we are using membranes, in multi layers with inflated spaces between, combined with cable suspension systems that were not available to Lanchester.

He would have revelled in the recent attitude since the early sixties whereby certain enlightened clients have been prepared to accept structural/architectural solutions on their visual impact and interest, rather than opt for the unimaginative.

One must also consider him in the light of design expertise generally prevalent at the time, when one was considered an analytical expert in the structural field during the 1930s if one had the capability to use and understand Moment Distribution. The buzz words were: "I have the Moment Distribution". This was commonly used by old boys in the early fifties when reminiscing about their achievements before the Second World War.

Dr. Lanchester had a free thinking attitude and was of such an inquisitive nature that he would have been pre-eminent in any age, no doubt equal to, say, the late lamented Peter Rice of Arup, or Professor Bolton, for example.

He was admittedly constrained by the limitations on reiterative design calculation techniques in his day, but he would have certainly enjoyed the greater insight into the actuality of structural behaviour and mechanism that computers offer.

In conclusion, Dr. Lanchester must be regarded as a latter day renaissance-type individual to whom all aspects of design were of interest, and a man to whom no branch of engineering was inaccessible: virtually a man out of his time in certain fields of engineering.

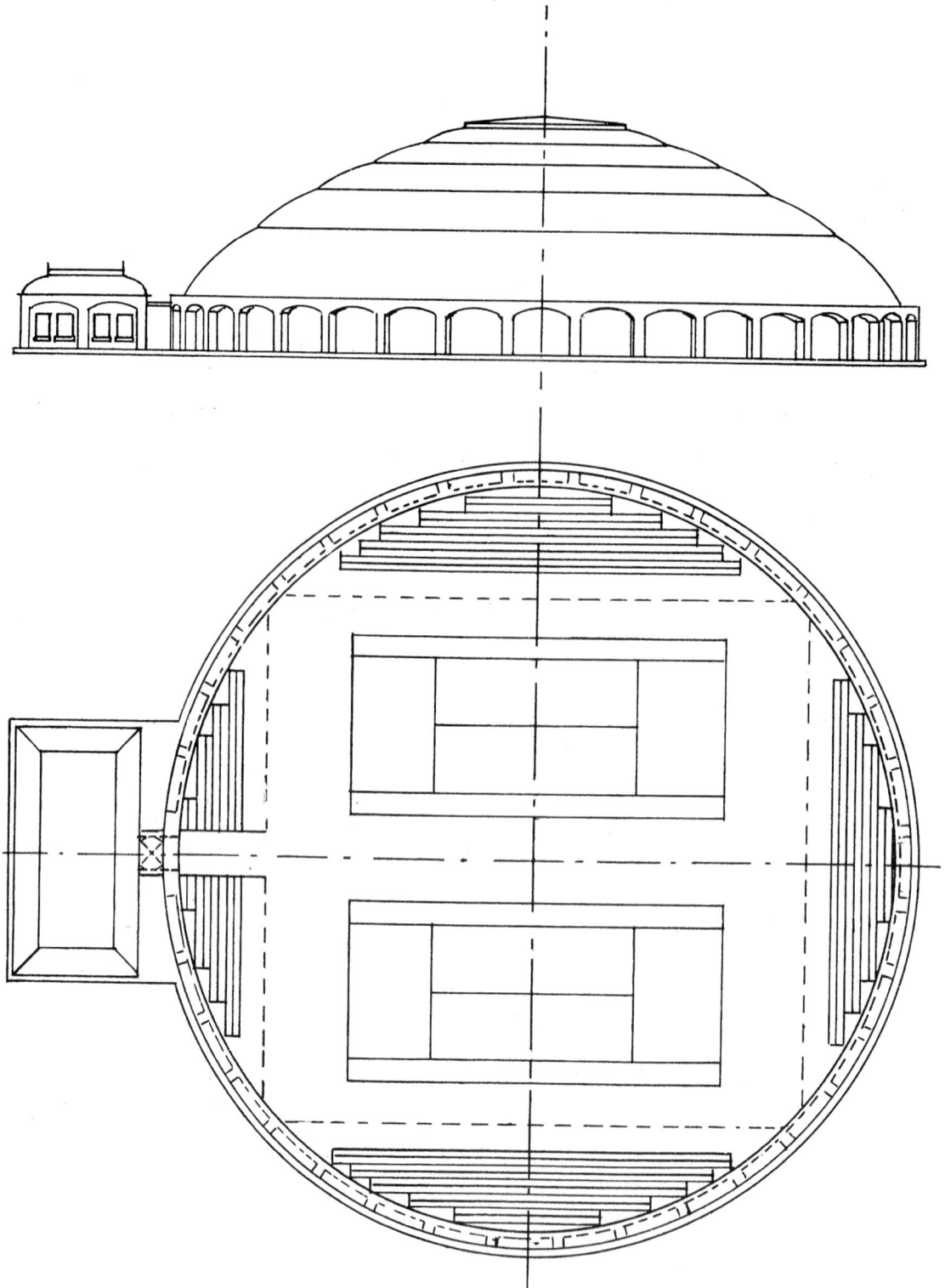

Figs 1–2. A design for an inflated canvas dome: Figs. 26 and 27 from Span: *"... represent in elevation and block plan an outline design for a covered lawn tennis court; the diameter of the circular arena, in relation to the courts as drawn, is 150 ft."*

References

Harvey, W.J. and Smith, F.W. (1991): The Behaviour and assessment of multispan arches. *Structural Engineer*, **69,** no. 24, 17 Dec. 1991, 411–17

McBeth, D.G. (1990): The Forth Bridge, *Structural Engineer*, **69,** no. 5, 6 March 1990, 93–100

Mainstone, R.J. (1990): Hagia Sophia: Justinian's Church of Divine Wisdom, later the Mosque of Ayasofya, in Istanbul, *Structural Engineer*, **68,** no. 4, 20 Feb. 1990, 65–71

Chapter Fifteen

OPTICS

by Dr. F.W. Fitzke

In the paper whose full title is "Directional fixity and the transitory visual impression or fleeting image", Lanchester describes an ingenuous experiment which allows the subjective appreciation of eye movements (Lanchester, 1936). If a source of light surrounded by darkness is viewed while the eye is moved, then a streak of light may be seen. Interestingly, he observed that with a neon lamp flickering at 50 cycles, a regular spacing within the streak becomes visible. This provided a clever means to calculate the speed of eye movement from the number of images seen.

The phenomenon described may be appreciated when a display illuminated by light emitting diodes (L.E.D.s) is viewed in otherwise dark conditions. Modern bedside clock/radios may provide the necessary conditions. When viewed during an eye movement, a series of repeated images may be seen which correspond to the flickering of the L.E.D.s. From the frequency of the alternating voltage supply and the distance the eyes move, the speed of movement can be calculated in an analogous manner to that employed by Lanchester.

In "Discontinuities in the normal field of vision", Frederick Lanchester addresses the question of why we are generally unaware of the blind spot (Lanchester, 1934). Today there is widespread understanding of the cause of the blind spot, particularly among those who have seen the demonstrations in textbooks on vision. However, understanding of the mechanism which underlies this visual phenomenon was initially mistaken and many individuals remain unaware of their own blind spots. This is partly because the structure of the eyes means that the blind spot is not generally detectable with binocular vision and requires the use of monocular vision to demonstrate. In this paper, Lanchester provides several diagrams which clearly demonstrate the disappearance of some elements when positioned in the appropriate manner with respect to the blind spot. He addressed in this paper some outstanding issues concerning the development of our understanding of vision.

When Edme Mariotte first discovered the blind spot in the location of the visual field corresponding to the optic nerve head in the late seventeenth century, it was greeted as a "heretical suggestion" because it was thought that the optic nerve was the site of visual reception (Duke-Elder and Weale, 1968). Instead of giving rise to the blind spot, it had been thought that this would be the location of highest visibility. The discovery of the blind spot was subsequently communicated to The Royal Society which had been founded by King Charles II. This led to the popular story that King Charles amused himself by "beheading" his courtiers when viewing them at a distance with one eye so that their heads disappeared. Rather than optic nerve fibres as visual receptors, Mariotte proposed the incorrect interpretation that visual perception was mediated by the layer of blood vessels at the back of the eye. It was not until the mid nineteenth century that H.Mueller correctly showed that the rods and cones were the photoreceptors.

Lanchester addressed the remaining problem that in some conditions, when light was directed onto the blind spot, a visual sensation could be elicited despite the absence of photoreceptors. The conditions under which this could be demonstrated were when a bright light source which was surrounded by darkness was focused on the blind spot or when the pattern of light extended across the blind spot onto surrounding photoreceptors.

The discussion in Lanchester's papers centred around whether there could remain a residual direct response of the blind spot to light which could be demonstrated under some conditions. This would be contrary to all our understanding of the role of photoreceptors in vision and he sought an alternative explanation for the phenomenon in higher brain mechanisms with a process of "filling-in". Lanchester clearly demonstrated that light falling on the blind spot is only detectable by the scattered light onto the surrounding sensory retina and that this was particularly evident in conditions of twilight vision. In a remarkably concise paragraph he described observations to rule out possible alternative interpretations to demonstrate clearly that the blind spot does not have a direct sensory response to light.

This led to the conclusion that there must be some "filling-in" process when certain patterns of light were viewed so that within the cortical representation in the higher centres of the brain the "... deficiency (is) made good ..." before the information is "passed to the final sensory centre whatever that may be." Here the understanding of how we see reflects conceptual limitations of the time. The idea of an *homunculus* inspecting the input from the nerve cells has an understandable appeal but is incompatible with modern concepts of distributed processing within the brain. Modern anatomical and physiological understanding frees us from this concept. It makes meaningless the question of why we see objects right side up instead of inverted as would be expected from the optical inversion of the image on the retina. Similarly, the concept of an *homunculus* confused the development of our understanding of the role of the blind spot in visual function.

The way in which we "fill-in" the blind spot extends to other regions of blindness within our visual fields of which we may be unaware. Individuals may have quite extensive areas of visual loss in the periphery before this is detected. Partly this inability to detect peripheral regions of blindness may be due to "filling-in" which occurs in these non-physiological blind spots. There is currently great interest in how the brain fills in different aspects of our visual experience such as colour or texture (Ramachandran and Gregory, 1991). The demonstration of "filling-in" is nicely made in Fig. 4 of Lanchester's paper for the left eye although the phenomenon is more easily seen with a greater spacing between the cross and the broken bar.

Although we are generally unaware of the blind spots in our visual fields, their effects on visual function are easily measured and they are routinely used in clinical tests of perimetry. Recently it has been shown that in some conditions, locations of blindness in the visual field of which the individual was unaware, became visible when a field of small, rapidly changing elements was viewed (Gruesser and Landis, 1991). Of concern is the finding that in some individuals, extensive areas of blindness can be measured of which the individual was unaware. This is particularly important in eye conditions where the spread of the blind spot or new locations of blindness in the visual field may presage later, more severe losses of vision. Of concern too is the question of the effects of unrecognised regions of visual loss on performance in such tasks as driving. It has been reported that drivers with binocular visual field loss have accident rates twice as high as those with normal visual fields (Johnson and Keltner, 1983).

The converse has also been found. It is sometimes the case that detection is possible without awareness, where objects can be visually detected and followed without awareness on the part of the individual. This is a phenomenon which has been referred to as "blindsight". This is currently thought to involve pathways in the visual system of the brain other than the primary visual pathways and is the subject of much research activity (Cowey and Stoerig, 1995).

Lanchester's contributions in this area show how progress in scientific understanding builds on earlier work to lead to conceptual advances. Questions which were investigated purely for their basic research interest provided the basis for later work which has practical

Test card (full size) as used by Lanchester in his experiments on "the blind spot"

applications. This rich heritage of learning within a framework of logical thinking provides a continuing source of inspiration for future generations.

References

Cowey, A. and Stoerig, P. (1995): Blindsight in monkeys, *Nature*, **333**, 247–9

Duke-Elder, S. and Weale, R.A. (1968): *The Physiology of the eye and vision*, Vol. 4, System of Ophthalmology, ed. S.Duke-Elder, London: Henry Kimpton, 1968

Gruesser, O.J. and Landis T. (1991): *Visual agnosias and other disturbances of visual perception and cognition*, Vol. 12, Vision and visual dysfunction, ed. J.R.Cronly-Dillon, London, Macmillan Press

Johnson, C.A. and Keltner,J.L. (1983): Incidence of visual field loss in 20,000 eyes and its relationship to driving performance, *Arch. Ophthalmol.*, **101**, 371–5

Lanchester, F.W. (1934): Discontinuities in the normal field of vision, *J. Anatomy*, **68**, 224–38

Lanchester, F.W. (1936): *Directional fixity and the fleeting image*, Royal Society of Medicine, Paper

Ramachandran, V.S. and Gregory, R.L. (1991): Perceptual filling in of artificially induced scotomas in human vision, *Nature*, **350**, 699–702

Chapter Sixteen

THE MUSICAL SCALE

by Dr. Norman A. Dyson

F.W.Lanchester's *The Musical scale* illustrates that his interest and expertise extended far beyond the topics for which he is most remembered and applauded. In his monograph we find a detailed treatment of the musical scale, mainly as used in Western civilisation, and it is clear that no survey of Lanchester's life would be complete without a detailed examination of this monograph. In the present article Lanchester's monograph on this subject will be reviewed, and an attempt will be made to appraise its value today. We must ask straight away whether the material contained within it is relevant to today's analysts of musical theory, and the answer, so far as much of it is concerned, must be "yes". In this review, Lanchester's order of presentation is generally preserved, and it is hoped that the reader will not find any difficulty in separating the reviewer's comments from Lanchester's original material. Towards the end will be found some general comments which may illuminate his work, and point to its usefulness in today's context.

The present writer is, in effect, reviewing a work which was written fifty years ago. This unusual situation seems to put the original author at a considerable disadvantage. When a review is written, the text under review is normally available to all who wish to see it. One cannot but express the hope that Lanchester may at some time in the future be allowed to speak for himself, through a reissuing of his monograph, but in the meantime the present reviewer will endeavour to look at his work (a) in the context of the time at which it was written and (b) with present-day eyes, seeking its relevance to today's student and analyst of musical theory.

Lanchester's monograph was produced and published by him in 1941. The original edition consisted of ten chapters and two appendices cyclostyled on foolscap paper (32 cm ×20 cm) upright and single-sided. Thirty copies only were issued, but the work was reissued in 1942, with the addition of appendices III, IIIa, IV and V, and with the inclusion of some additional pages at various places in the script. Forty more copies were issued, and it is a copy of this re-issue, consisting of some 105 pages in all, which the author has before him, in the format just described.

The new issue contains a note by the author to the effect that the first issue was well received by friends and acquaintances. It also contains an extract of a letter received from Sir Granville Bantock dated 21st January 1942, commending the work as a "most valuable part of the curriculum for the first year's course of the musical degree" i.e. at Birmingham University, where Bantock occupied the chair from 1908 to 1934, relinquishing it in that year to become Chairman of Trinity College, London.

In Chapter One Lanchester sets out some elementary acoustics regarding the relation between frequency and pitch, and then proceeds, in Chapter Two to discuss the major scale as we know it today. A historical starting-point may well be thought to date from the time of Pythagoras, but the author leaves this early formulation, based on the musical interval of the perfect fifth, with its frequency ratio of $\frac{3}{2}$, until a little later in the monograph (we shall come to it) and takes up the thread at the point at which Western music had developed to include the free use of major thirds and sixths in harmonic writing as an advance on the 'organum' of parallel fifths. We are now approaching the fifteenth century (although

THE LANCHESTER LEGACY

substantially earlier examples of the use of thirds exist) and the system of tuning which had been adopted by that time was known as "just intonation". It is based on the premise that the frequencies of the intervals most commonly encountered in music, whether in the form of melodic sequences or in the simultaneous sounding of notes constituting rudimentary harmony, bear the ratios of small whole numbers, and we may pick out (regarding the note C as the "tonic") the notes C, E, G, with frequency ratio $4:5:6$, as a starting point.

A further triad may be constructed on G, viz. G B D, and, by setting F as an exactly tuned fifth below C (i.e. $F:C = 2:3$), a third triad F A C. We now have all the notes of the diatonic scale and if we multiply or divide by two to bring all the notes to within the compass of one octave (a process known as "clearing octaves") we obtain frequency relationships as follows:

C	D	E	F	G	A	B	c*
1	$\frac{9}{8}$	$\frac{5}{4}$	$\frac{4}{3}$	$\frac{3}{2}$	$\frac{5}{3}$	$\frac{15}{8}$	2

Table 1

The fractions in each case represent the frequency of the note in relation to the lower C in Table 1.

We may regard this scale, duly modified as described subsequently as the basis of all music in Western civilisation from the fifteenth century onwards.

Lanchester then looks at the earlier Greek scale (the Pythagorean scale). This scale is built up from the principle that the interval of a perfect fifth is constructed from two notes having a ratio of $1:\frac{3}{2}$ exactly, and then developed by the superposition of exactly tuned fifths until the twelve notes necessary for a complete chromatic scale have been generated, i.e. $C=1$, $G=\frac{3}{2}$, $D=\frac{9}{4}$, etc., which, when cleared of octaves, generates the following sequence of frequencies for the naturals ("white notes") of the piano keyboard.

C	D	E	F	G	A	B	c
1	$\frac{9}{8}$	$\frac{81}{64}$	$\frac{4}{3}$	$\frac{3}{2}$	$\frac{27}{16}$	$\frac{243}{128}$	2

Table 2

Although Lanchester pays due homage to the Pythagorean scale, he acknowledges that "a scale so constructed will not have the attribution of tonality [because] the relations between the reference note C and the mediant E, the sixth A and the seventh B are obscure", i.e. not sufficiently close to more modern tunings to be satisfactory (even recognisable) to modern ears. We may note, however, that Pythagorean tuning would probably be deemed satisfactory for, e.g. Gregorian chant (unison singing) where the very wide major third $C:E$ would not be a disadvantage. The system did not, however, become quite extinct, and an interesting account of the later history of Pythagorean tuning has been given by Barbour (1951).

At the beginning of our attempt to outline the development of tuning from the fifteenth century onwards, we note straight away that the two methods of construction so far outlined, the Pythagorean scale built entirely on perfect fifths, and Just Intonation built on fifths and major thirds suffer from two inconsistencies. First, the superposition of twelve perfect fifths will indeed provide (when octaves are cleared) the twelve notes required for Western music, but some adjustment would inevitably be needed. The system does not "close" because of the impossibility of fitting twelve perfectly-tuned fifths into an integral

---

* The lower case c represents Helmholz's notation for what we often refer to as tenor C. C is of course an octave below this – the C to be found two octaves below middle C, and produced by an open-ended pipe approximately eight feet in length. Lanchester makes much use of Helmholz's classic work "On the Sensations of tone", in the English translation of Ellis, including Ellis's additions (1885).

number of octaves (7). That is, $\left(\dfrac{3}{2}\right)^{12}$ is not equal to 2^7. It exceeds it by a factor of $\dfrac{129.746}{128}$ or 1.0136432. This small interval is known as the *Pythagorean comma*. Secondly, four perfect fifths (whether in Pythagorean tuning or Just Intonation), corresponding to a frequency ratio of $\left(\dfrac{3}{2}\right)^4$ or $\dfrac{81}{16}$, do not exactly equal the two octaves and a major third as needed for the satisfactory tuning of thirds in just intonation outlined above. This error is expressed by the fraction $\dfrac{81}{80}$ or 1.0125. this is known as the *syntonic comma*. These unavoidable discrepancies have to be examined if a scale which is satisfactory for more modern music is to be developed.

Clearly we have become involved with intervals which are very small compared with those tabulated above. We must immediately take advantage of a unit which is widely used for measuring small intervals – the "cent", which is defined as follows:

Two frequencies f_1 and f_2 are said to differ by n cents if $\dfrac{f_2}{f_1} = 2^{n/1200}$.

$$\text{i.e. } \log_{10} \frac{f_2}{f_1} = \frac{n}{1200} \log_{10} 2$$

$$\text{or } n = \frac{1200}{0.3010} \log_{10} \frac{f_2}{f_1}$$

We find that
    the Pythagorean comma = 1.0136432 = 23.460 cents
    the Syntonic comma = 1.0125 = 21.506 cents.
Our understanding of this very convenient unit will develop with a little use.

We note that if the octave is divided into twelve *equal* semitones, 100 cents = 1 semitone. The definition of this unit therefore relates most directly to the *equally-tempered* scale; we shall defer any discussion of equal temperament until later in this review.

The system of just intonation may be developed to include all twelve notes of the chromatic scale by means of an extension of the above principles, and this is done by Lanchester, in Chapter Four of the monograph. A similar procedure is applied to the Pythagorean scale.

It will have become clear that the development of the additional notes required for the twelve-note chromatic scale can be carried out in more than one way. After twelve notes have been generated, further notes can be obtained by onward extension of the procedure, i.e. by adding further fifths. If this is carried out *upwards* in frequency the new notes will be one Pythagorean comma *sharp* with respect to the original twelve notes, and if carried out downwards the new notes will be *flat* by this amount. Lanchester developed a chart expressing this "open circle" of fifths, and also a set of templates by which notes forming a diatonic scale (major or minor) may be selected. His theoretical work here is extremely comprehensive.

It will be clear that we have arrived at a situation where notes found to have the same pitch on a twelve-note keyboard (e.g. G# and A♭, D# and E♭) now have different frequencies. These *enharmonic equivalents* are discussed in Chapter Five of the monograph. We may note immediately, by a simple method, that this situation must arise in any system in which the major thirds are tuned exactly. Table 3 will make this clear:

$$C \quad \tfrac{5}{4} \quad E \quad \tfrac{5}{4} \quad G\# \quad \tfrac{32}{25} \quad c$$

$$A♭ \quad \tfrac{5}{4} \quad c$$

Table 3

Clearly an additional note (A♭) is needed in order to provide an exactly tuned major third with the upper c. We see that the A♭ is sharp with respect to the G# by a factor $\frac{128}{125}$, or 41 cents. This is more than two fifths of a semitone, and is known historically as the *diesis*. We shall develop this matter later. It is implicit in Lanchester's tables (Fig. T.21, Chapter Five of his monograph).

At this point Lanchester makes a jump from the two scales so far described, which are of largely theoretical interest, to the tuning system very widely used today. His Chapter Six is devoted to the "equal temperament" system of tuning. This has been in wide use since c. 1800, though its introduction in England was delayed until somewhat after that time. Before referring to equal temperament in any detail, we should, however, examine the evolution of musical scales during the sixteenth, seventeenth and eighteenth centuries for which the subsequent establishment of equal temperament might be seen as a logical end-point.

The principal shortcoming in the scale of just intonation is that difficulties arise when tonalities not close to C major are used. Even the interval D—A is unsatisfactory as a fifth. The accidentals (the "black notes" on the piano keyboard) will not do service in tonalities other than those for which they were tuned. G# as tuned in just intonation may be satisfactory in the key of E major, but cannot be used as an A♭ in the key of F minor or A♭ major. Only a very limited number of tonalities may be used. Modulation from one key to another would give rise to intervals or chords the musical quality of which is unacceptable. This was certainly realised by the beginning of the sixteenth century, when the "mean-tone" system of tuning was described. Lanchester relegates a discussion of this system to an appendix, in the second edition of the monograph. We shall look at it in more detail now. The mean-tone system recognises the importance of the major thirds in virtually all harmonised music of the sixteenth, seventeenth and eighteenth centuries, and this supplants the priority given to the perfect fifths before that time. We have seen that four perfectly tuned fifths together make up approximately two octaves and a major third, exceeding it by a small interval (21.506 c) known as the syntonic comma. *The essence of mean-tone tuning is that these fifths are slightly compressed so as to ensure the exactness of the major thirds.* Each fifth must be compressed by one quarter of a comma, or approximately five cents. This will ensure the exactness of the C—E major third if the compressed fifths are C—G, G—D, D—A, and A—E.

By an extension of this procedure, accurately tuned thirds may be superimposed on D, A, and E, and "suspended" from D and G. We therefore have no fewer than eight accurately tuned major thirds, but we find that the remaining intervals C#—F, G#—C, B—E♭, and F#—B♭ are unserviceable. We have already seen that G#—C is too wide by 41 cents; the remaining three are wide by a similar amount. The dissonance inherent in the G#—E♭ fifth is unusable, and is known as the "wolf".

A scale tuned in this way provides for the performance of music in six major keys – those keys with no more than two flats or three sharps in the key-signature. To make available a further range of tonalities, a logical step was to provide additional physical keys at the keyboard, i.e. to provide A♭ as an alternative G#, and D# as an alternative to E♭. That this was actually done bears testimony to the importance of this matter. To pursue the subject, we must enter the realm of the organist and the organ-builder. The sustained sound of the organ tends to expose errors in tuning and the organ was, of course, in wide use for the performance of a rapidly expanding repertoire. The renowned English organ builder Bernard ("Father") Smith (c. 1630–1708) provided an organ in 1684 for the Temple Church, London, and supplied split keys for these notes. The organ at Durham Cathedral was furnished with a similar arrangement. The organ at the Foundling Hospital, London, which was played by Handel, was fitted with a mechanism by which D♭ and A♭ could be substituted for C# and G#, or D# and A# for E♭ and B♭, thereby increasing the number of notes available to sixteen per octave without increasing the number of physical keys at the keyboard. That a musician of the stature of Handel should require such arrangements points to the importance

Fig. 1. Split keys on the organ of the Temple Church, London (Norman, 1986)

of the matter as seen by a leading musician during the first half of the eighteenth century. The expense of the additional pipes and mechanism was, apparently, considered justified because of the wider range of tonalities being used by composers and performers.

The term "mean-tone" is derived as follows: by referring back to Table 1, (for just intonation), we see that the intervals C—D and D—E are not the same, being $\frac{9}{8}$ and $\frac{10}{9}$ respectively. A similar relationship applies to the intervals F—G and G—A. These intervals were known as the major tone (C—D) and minor tone (D—E). In mean-tone tuning the narrowing of the fifths lowers the frequencies of G, D and A by factors of

$$5^{1/4} \times \tfrac{2}{3}, \ 5^{1/2} \times (\tfrac{2}{3})^2 \text{ and } 5^{3/4} \times (\tfrac{2}{3})^3$$

respectively, E being lowered in frequency by a factor of $\frac{80}{81}$. The frequency of D (when octaves are cleared) is thus

$$\tfrac{9}{8} \times 5^{1/2} \, (\tfrac{2}{3})^2 \text{ or } \tfrac{5^{1/2}}{2}.$$

The ratio of D : C is now the same as that of E : D ($\frac{5}{4} \div \frac{5^{1/2}}{2}$), being equal to $\frac{5^{1/2}}{2}$ in each case. This is the *geometrical mean* of the justly tuned ratios $\frac{9}{8}$ and $\frac{10}{9}$.

We have seen something of the difficulties attendant upon the use of mean-tone tuning in the remoter keys. An important development during the eighteenth century may now be turned to. It consisted essentially of reducing the compression of the fifths somewhat, so that the discrepancies in the remoter keys became less obvious. The thirds became slightly wider than perfect, but not so much as to lose the advantages of quarter-comma mean-tone tuning, as the above system is called. The tuning in the remoter keys became more acceptable. Fifth-comma and sixth-comma mean-tone therefore came into use. Sixth-comma mean-tone is of particular interest, as it was used by the illustrious Silbermann family of organ builders, and their work is of particular interest, because of their activity during the life-time of J.S.Bach. It is on record, however, that J.S.Bach found himself to be dissatisfied with this tuning method, and it is not difficult to build up a picture of musicians pressing for the remoter keys for composition and improvisation, with the organ builders and tuners endeavouring to satisfy these needs as best they could by progressive changes to the system of tuning currently in use.

Further developments therefore took place. These consisted essentially of a *redistribution of the errors in the tuning of fifths and thirds*, so as to make accessible the remoter keys, whilst at the same time not destroying the advantages of mean-tone tuning, at least so far as the "nearer" tonalities are concerned. Space does not permit of a detailed consideration of these systems. Those devised by Werckmeister and by Kirnberger are the best-known. They are extensively used today on newly-built organs designed on classical lines. The purity and solidity of the major triads is manifest, and the change of character of sustained chords as one moves from keys close to C major to, say, F# minor, is also discernible, but not objectionable.

We have seen the usefulness of the cent in expressing quantitatively the departures from exact tuning. This can be represented in tabular form, and in Table 4 tables for two intervals

ERROR FROM TRUE INTERVAL
Measured in Cents (% of a semitone)

Fifth from Tonic

Key	Db	Ab	Eb	Bb	F	C	G	D	A	E	B	F#
Mean-Tone (¼ comma)	5	35	5	5	5	5	5	5	5	5	5	5
Silbermann	4	20	4	4	4	4	4	4	4	4	4	4
Werckmeister III	0	0	0	0	0	6	6	6	0	0	6	0
Kirnberger III	0	0	0	0	0	5	5	5	5	0	0	2
Vallotti	0	0	0	0	4	4	4	4	4	4	0	0
Young	0	0	0	0	0	4	4	4	4	4	4	0
Niedhardt	0	2	2	0	0	4	4	4	4	2	2	0
Equal Temperament	2	2	2	2	2	2	2	2	2	2	2	2

Major Third from Tonic

Key	Db	Ab	Eb	Bb	F	C	G	D	A	E	B	F#
Mean-Tone (¼ comma)	41	41	0	0	0	0	0	0	0	0	41	41
Silbermann	29	29	6	6	6	6	6	6	6	6	29	29
Werckmeister III	22	22	16	10	4	4	10	10	16	16	16	22
Kirnberger III	22	22	16	11	5	0	5	11	14	20	20	20
Vallotti	22	18	14	10	6	6	6	10	14	18	22	22
Young	22	22	18	14	10	6	6	6	10	14	18	22
Niedhardt	18	18	16	14	10	6	8	10	14	18	18	18
Equal Temperament	14	14	14	14	14	14	14	14	14	14	14	14

Table 4 (Norman, 1986)

(the perfect fifth and major third) are shown for each of the twelve notes of the keyboard. The mean-tone systems show the purity of the tonalities close to C major, and strong divergences as we move to remoter keys. We can see from Table 4 that later systems redistribute these errors in a somewhat different manner. So far as the fifths are concerned, the Werckmeister system favours the flat tonalities. We now appreciate that a temperament can be devised to favour any group of tonalities at the expense of the remainder. The table displays well the different "properties" of several different systems of temperament; and also illustrates the almost inevitable trend towards the system of equal temperament, in which the equal distribution of errors ensures the availability of all twelve keys, major and minor. That the pressure for developments in temperament came from the composers (rather than theoreticians or instrument-builders) is well established by reference to historical sources.

We have digressed in order to provide a coherent account of systems of temperament in use from earliest times to the present day, and to extend Lanchester's treatment where necessary. We must now return to *The Musical scale*. Chapters Seven and Eight contain very detailed discussions of two other scales found in Western music: the whole-tone scale and the pentatonic scale. The whole-tone scale (e.g. C, D, E, F#, G#, A#, C) is familiar to us from the music of Debussy onwards. It affords a ready way of breaking loose from the chains of conventional tonalities, and is available for both melodic and harmonic writing. (Earlier examples are also given). The pentatonic scale is of much wider use and more ancient origin. In the monograph we find several examples of its use. Lanchester's treatment of the highland bagpipes is worthy of mention because accurate measurements of pitch, determined by Helmholz/Ellis exist. Those measurements support strongly Lanchester's contention that the notes of the "chaunter" are extremely close to Pythagorean intervals in their tuning.

In Chapter Nine we find an interesting digression into the matter of the "absolute pitch" to which musical instruments are tuned. This has varied significantly over four centuries or more, and the comprehensive tabulations of Ellis (in Helmholz/Ellis, 1885) may be consulted. We may refer to "pitch" in terms of the frequency to which the note A (in the treble stave) is set. At the present time (1995) A = 440 Hz, but this was not always the case. We may look for a moment at some relatively recent matters: in 1859, the assignment of a frequency of 435 Hz to A, known as *Diapason Normal*, was adopted as a legal standard in France. In Great Britain the standards of pitch were not so clear, and during the latter half of the nineteenth century the pitch was allowed to rise by almost a semitone above this value. There were many extremely soundly-based objections to this situation. Sir Walter Parrott's strictures might, with advantage, be quoted here:

"Such a change was attended with many evils; it altered the character of the best compositions, it tended to spoil the performance, and spoil the voices of the best singers; and it threw the musical world into confusion from the uncertainty as to the practical meaning of the symbols used, and all for no object whatsoever."

This somewhat uncontrolled rise in pitch was therefore strongly resisted, and "Diapason Normal" gradually prevailed in Great Britain. It has, of course, been supplanted by somewhat higher standards A = 439 Hz and more recently A = 440 Hz, which prevails today. It has to be said that the pitch in certain other European countries is slightly higher than this, and a small rise in the standard of A, above 440 Hz, to conform with other countries in the European Community, has been predicted with some displeasure.

Perhaps pitch standards will never be absolutely universal. Today, baroque music is not infrequently played at a standard of A = 415 Hz, approximately 1 semitone below A = 440 Hz. Many stringed instruments and harpsichords are manufactured to this standard. Owners and players of these instruments are very wary of *upward* retuning of these instruments, which involves an additional 12% of tension on soundboards. In the case of harpsichords and chamber organs, a mechanism to shift the keyboards sideways by one note seems more practical, and is occasionally encountered.

The breadth of Lanchester's knowledge of musical scales becomes clear when we move to the tenth and final chapter of his monograph. This is devoted to the ancient Greek modes and also to the Ecclesiastical modes of the Middle Ages. A careful comparison between the terminology of the two systems is given, along with a discussion of tonality in this context. The author points out the evolution of our present major and minor scales from the ancient modes, and gives much detail relating to the scales associated with these modes. The author adds some appendices which cover other relevant topics, in addition to the mean-tone temperament already discussed. A detailed discussion of the methodology of tuning in equal temperament (on the piano) is included with much practical wisdom on piano tuning. There is also a discussion of the tuning of stringed instruments and the orchestral harp. A wide-ranging discussion, ranging from Arabic scales to Tonic Sol-fa notation is to be found here.

How can we summarise the qualities of this monograph? It is a massively detailed work, showing the application of a first-class intellect to matters which always threaten to get out of hand unless firmly checked. Granville Bantock's "testimonial" illustrates that the music students of the 1930s and 1940s were expected to know something about these matters. Today's students of music (and their teachers) are inclined to look for "relevance" when considering the values or otherwise of physical and mathematical aspects of music. Indeed, it is germane to take a look at the contents of modern textbooks on the subject to ascertain what topics might be considered useful adjuncts to a modern course in music to degree level. Take, for example *The Acoustical foundations of music* by Backus (2nd ed., New York, 1977). This admirable work starts with the physical and acoustical background of musical sound, and then proceeds to the "reception" of musical sound, including much useful information on the functioning of the human ear, and on tone-quality. There is indeed a chapter devoted to intervals, scales, tuning and temperament in which some of the material discussed at length above is treated. But the main emphasis of Backus' book is on quite different subjects, particularly auditorium acoustics and the technology of musical instruments, woodwind, brass and the piano in particular. The production of sound by electronic means is included. We see that there is less emphasis on scales and temperament than Lanchester (and perhaps Bantock!) would have liked, but Backus' work provides a good treatment of topics which have obvious career potential for the musically trained graduate. The textbook by Campbell and Greated (*The Musician's guide to acoustics*, Dent, 1987) follows somewhat similar lines.

It is worth reiterating that the tempered tunings of the eighteenth century are of great significance to the modern organist. Many modern recordings are made on instruments tuned to Werckmeister or Kirnberger systems. Different tonalities do indeed sound different: the concept of "key colour" is a reality again, being somewhat dubious when equal temperament is universally used. It is now almost unnecessary to point out that Bach's "Das Wohltemperirte Klavier" should properly be translated as "The well-tempered clavier" not "The equally-tempered clavier". The existence of these two sets of Preludes and Fugues (twenty-four in each, dating from 1722 and 1742 respectively) *under this title* clearly shows us the importance of the eighteenth century tunings which displaced quarter-comma mean-tone tuning. Research during the half-century since the publication of Lanchester's monograph adds much to what was known at that time. His work leans heavily on Helmholz/Ellis (what text on musical scales written at this depth of detail could not?) and the student who wishes to study the subject from the point of view of scholars of the late nineteenth century may consult this monumental work of 1885. Barbour in his *Tuning and temperament* (1951) has provided a modern specialist text on the subject. Blackwood (1985) in his *The structure of recognizable diatonic tunings* goes even further in his detailed mathematical approach.

Those who wish to find examples earlier than the eighteenth century of music in remote keys may examine the oft-quoted example of the "Fantasia Ut, re, mi" by John Bull in the *Fitzwilliam Virginal Book* (c. 1600). This composition includes enharmonic modulations, and the editors of the 1899 edition (Fuller Maitland and Barclay Squire) noted the difficulties which unequal temperaments would cause in performance, in fact mentioning equal temperament in this connection. The history of equal temperament goes back a good deal further in time than its gradual introduction at the end of the eighteenth century would imply, because it was already recognised as a theoretical possibility much earlier than this. Its introduction on keyboard instruments was no doubt held back for reasons which an earlier section of this review makes clear, but it may have been used in special circumstances. It would not be difficult to alter the tuning of a harpsichord to a different temperament, if particular tonalities, or groups of tonalities, were being investigated experimentally by a composer-performer. So far as non-keyboard instruments are concerned, instrumental temperaments were not always the same as keyboard instruments. The tuning of "fretted" instruments (lutes and viols), in which the "proportional ratios" are necessarily the same for all strings, has to be equal temperament, if any consistency between the same note sounded on different strings is to be maintained.

Another development which inevitably creates problems of dissonance if an unequal temperament is used is the developing use of the chord of the diminished seventh, particularly in music involving keyboard instruments. This chord may be regarded for our purposes as four superposed minor thirds, which occupy the span of an octave. It is clear that these intervals will necessarily be unequal in any tuning system other than equal temperament, and that the function (may one say the "utility") of the chord in any modulating sequence will be modified. This chord is commonly found in the instrumental music of J.S.Bach, particularly in passages where the completion of the chord is left to the continuo (figured bass) part. It is also found in the solo keyboard repertoire of that period, but its prominence does not reach its zenith until much later (e.g. in the music of Liszt) by which time the call for equal temperament was irrefutable. The "well-tempered keyboard" has indeed become the "equally-tempered keyboard", and the composition of the great corpus of romantic and modern keyboard music became possible.

To summarise, then: Lanchester's monograph is wide-ranging as well as comprehensive.

It was an excellent work at the time of its appearance, and Granville Bantock's approbation was not misplaced. It is an interesting historical document, but much has happened since it was written, and there are now some publications of more recent date which treat the various fields which Lanchester described so thoroughly. Also, the needs of music students have changed in fifty years, and this renders Lanchester's work a fascinating period piece, rather than a text for today's students. The days of the medieval Quadrivium, in which music was ranked equally with arithmetic, geometry and astronomy, for a course of natural philosophy, are long past, but the practicalities of musical study, in at least some of its branches, still require some understanding of the subject-matter reviewed here, and Lanchester's volume will always occupy an important position in the literature contributing to the gradual development of such studies.

References

Backus, J. (1977): *The Acoustical foundations of music* (Norton, New York and London)

Barbour, J.M. (1951): *Tuning and temperament* (Michigan State College Press)

Blackwood, E. (1985): *The structure of recognizable diatonic tunings* (Princeton U.P.)

Campbell, M. and Greated, C. (1987): *The Musician's guide to acoustics* (Dent, London)

Fitzwilliam Virginal Book, The (ed. Fuller Maitland and Barclay Squire, 1899, Dover reprint)

Helmholtz, H.L.F. (trans. Ellis, A.J., 1885): *On the Sensations of tone* (Dover Publications, New York)

Norman, J. (1986): in *Organists' Review*, Nov. 1986, p.251ff.

Chapter Seventeen

LANCHESTER'S POETRY

by C.B. West

Fifteen years after Lanchester's death, in 1961, the English literary world was agog with the ferocity of the debate over the Two Cultures. The originator of this literary war was the critic F.R.Leavis, who attacked C.P.Snow for daring to have any pretensions of bridging the arts–science divide. Snow was a scientist who wrote a series of popular novels which have been all but forgotten, so perhaps Leavis was right to pour scorn on his literary achievements. Nevertheless, the controversy did highlight once more the deep distrust that exists in the British psyche of anyone who tries to be good at everything.

Lanchester refers to this in his postscript written in 1943 to *A King's prayer*, first published in 1936, explaining why he wrote his verse under a nom-de-plume. After describing very briefly his engineering achievements, he continues:

"'Versatility' is not regarded with favour by the British public ('Jack of all trades, etc'). So, when writing my Lakeland Story, I attributed the authorship to a 'ghost', Paul Netherton Herries, a fitting nom-de-plume, and one that has proved useful on other occasions."

How much Lanchester considered his poetry to have literary merit we can only speculate. That he took his efforts seriously, however, cannot be doubted. He had written the occasional poem as early as 1913, and continued to write well into the 1930s; by far his most ambitious work, the narrative poem *The Centenarian*, appears to have been written in 1934–5. This runs to over a hundred stanzas, most of which are above ten lines in length, so in terms of tenaciousness alone, his efforts deserve respect. In addition, he wrote two unpublished essays on poetry, the manuscripts of which reside in the Lanchester Library.

It has to be said that the essays themselves are a disappointment. They concentrate on a sterile argument about what constitutes acceptable rhythm in English poetry, compared with Greek, together with frankly unsupportable theories about the nature of rhyme. What is worse is that Lanchester shows the same conviction of the absolute rightness of his theories on poetry that he had (with far more validity) for his engineering theories. There is evidence of the egoism of the literary critic without the depth of knowledge, and the effect on the reader is somewhat similar to that created by a precocious sixth former, full of opinions but mostly facile and lacking true awareness of what constitutes the essence of great poetry, that is the imaginative expression of feeling, thought and ideas within a particular rhythmic framework. Perhaps we should be grateful that his theories on poetry do not unduly influence his own writing, and as a result there are a sufficient number of worthwhile examples in his comparatively small output to refute the charge that his poetry is of no merit. If nothing else, they serve as an example of the wide range of Lanchester's interests and his facility to bring unusual concepts together in a meaningful way. His linkage of bird flight and mechanical flight, noted by John Fletcher in his biography, was an early example. Here he brings together music and poetry. The concept may not seem particularly original, but the manner in which Lanchester brings them together certainly is. He suggests that musical notation, specifically the symbol for the rest, should be introduced into poetry in preference to arcane expressions such as the caesura. He then proceeds to use the device himself in *The Centenarian*. It looks odd, and indeed can only be regarded as a harmless eccentricity, but it shows how Lanchester continually searches to integrate disparate ideas into a meaningful framework.

Lanchester's poetry, as we have indicated, is something of a curate's egg. It is at its best when at its slightly nostalgic. This is best illustrated towards the end of *The Centenarian*. The story is a sentimental one, with echoes of Peer Gynt, Byron's Don Juan and the Picaresque novel. It is told in retrospect by the hero, David Voegt, returning in extreme old age to the romantic haunts of his youth and meeting an old sexton in the churchyard. The sexton unwittingly recalls his story, which Voegt takes up with relish, recalling how he won the heart of a girl much above his station, Lady Bess, and the progress of their romantic involvement, before being tracked down at their favourite tryst by her angry father. In the ensuing struggle, Lady Bess is accidentally struck down and Voegt falsely accused and found guilty of being the perpetrator. He is then sprung from gaol by friends and runs off to sea, having first taken refuge, like the future Charles II, in a hollow oak tree. Despite numerous adventures on sea and land, his heart remains steadfastly loyal to his lost love, and the sexton has to break the sad news to him that she only survived for a year after the accident, dying with his name on her lips. At this point, David Voegt drifts into a reverie, recalling the simple pleasures of the rustic way of life, (in a passage rather reminiscent of Goldsmith's *The Deserted Village*) before expiring.

It says much about the character of Lanchester that he should take on a poetic genre which nearly all twentieth century poets have eschewed. The epic poem is the most difficult to sustain, and only the greatest poets have succeeded with it. Lanchester would not have claimed to be a Byron, a Wordsworth or a Browning, and it is therefore hardly surprising that the quality of the poem is very variable. It is frequently banal:

> "Nothing pleased him! he must insist
> That I should part with my own chestnut mare
> What a wrench! but how could I resist?
> He paid the price I asked without demur
> It was only 'lucre', I declare
> That tempted me my bestest friend to sell
> Naught could I refuse that lady fair
> It almost seemed, the horse I loved so well
> Caressed and loved by her, my tale of love might tell."

Only the last two lines can remotely claim to rise above the trivial, but this is the danger of a long narrative poem. At other times, irritating poetic archaisms intrude, such as "for the nonce", "full two days", "I kenned it well" and "strepent" (meaning "noisy" – even the Shorter Oxford lists this as "rare"). And despite the two essays Lanchester wrote on the importance of rhythm and accent, his own poetic ear sometimes lets him down, and lines fail to scan properly:

> "'Twas nigh on midnight when two loyal friends,
> Confederates, procured my release.
> No one knows how such adventure ends,
> Until the end has come and troubles cease.
> I conjecture someone 'lost' the keys"

The second and fifth lines jar badly, and ruin the effect of the other three. The poem becomes racier and altogether more vigorous during the sea scenes, and is faintly reminiscent of *Childe Harold's Pilgrimage*. The anapæst metre ensures the pace of the verse, and if the images lack originality, at least the sensation as a whole is invigorating. This is his description of a stampede in the Argentinian Pampas:

"The cloud comes upon us, encroaching, surrounding;
A dust-screen o'er everything hangs like a pall,
And naught remains visible, – nothing at all!
The strength and the stench of those animals, – Pah!
The crashing and hoof-beat like thunder is sounding
The ground like an earthquake is trembling afar."

One might argue that the "Pah!" is a mistake and has a bathetic effect, but the overall impact is stirringly dramatic.

The nostalgic tone, referred to above, is seen at its best in the final stanzas, as the hero, sinking fast, muses and dreams of his lost youth, conjuring up an idyllic vision of the countryside and its people and the unchanging tenor of their lives. There is certainly an attractive Hardyesque quality about the following:

"He dreams that in the night he hears a call,
And from his window sees by lantern-glow
Two men, and then his father, lithe and tall,
Heavy-booted, kicking off the snow;
Then creeping down he hears the bellows blow,
And sees his mother 'whipping up' the fire;
And then the menfolk talking there below, –
(There'd been some heavy business in the byre:)
A grand bull calf, he hears them say, got by a famous sire.

He dreams he scents the pollen-fragrant hay,
With bees in tens of thousands making song;
Th'insatiate scythes that swing and swing all day,
The sighing swathes that fall the whole day long,
He dreams of days when, strenuous and strong,
He of that band of stalwarts took the lead;
He dreams of women who, with rake and prong
And merry song, are working in the mead,
And of the shade, 'neath giant elms, where work-tired women feed."

It is clear that Lanchester also appreciated that the above examples represent the best and most inspirational parts of the poem, since he reprinted them in *The King's prayer*. The remainder of the volume contains a few short pieces, written over a period of years, and again very variable in quality. The title poem, written in January 1936, is a blast against disarmament, which was a most unfashionable view at the time but subsequently proved a correct one. Its poetic merit, however, is minimal. Much better is *Steadfast*, a short but rather affecting poem written in 1917 and taking the part of a young girl awaiting in vain the return of her future husband from the war. The last two lines seem to suggest that she is pregnant, a brave stance at the time, which contrasts oddly with the real-life drama in which Lanchester took a much less honourable part in having his sister certified because of her intention of "living in sin" with a man:

"O God, though he's gone, and the gossips are 'humming'
The pledge that I'm bearing I treasure the more"

But Lanchester is on much less sure territory when he tries to write war poetry as from the trenches. The poem *Missing*, also written in 1917, begins well enough:

"Somewhere distant, where the shells are droning,
'Neath heaven's arch where circling planes fly high;
Where locked fast in sleep, beyond Death's groaning,
My stricken comrades lie."

However, after this the poem rapidly loses its way, partly due to Lanchester's maddening device of separating his phrases with asterisks. This not only looks odd, it also breaks up the meaning and impact of the poem.

Among the other short poems in the collection, *Justa Fenebria*, just six lines long, merits quotation in full. The one false note is regrettably the opening line, which both in its words and punctuation reeks of Victorian sentimentality. Subsequently, there is an exact match of words, rhythm and mood, showing that Lanchester had gained an admirable mastery of technique. The piece is precise and moving, effective in its simplicity and absence of poetic devices. It is tempting to speculate, in view of his known (and understandable) belief by this time that he had not been given sufficient recognition for his outstanding engineering achievements, that the true subject of the poem is Lanchester himself:

> "He is – no more!
> Whose countless friends, belov'd of him, stand silent now;
> Born to high achievement, yet frugally he lived,
> Bearing the yoke of service here with dignity.
> His soul was in his work, which, living on, bears fruit:
> So doth a Soul attain to Immortality."

Two poems inspired by Omar Khayyam, *Omar's Grave* and *Meditations of an aged Philosopher after a reading of the Rubaiyat*, appear to show Shelley's influence but are not marked out by any particular originality of thought or expression. *Memory* is an attractive short love poem, with the poet reflecting wistfully on a walk home with his loved one. The volume ends with the delightful *Psychologia puellae*, whimsical and amusing, and showing a side to Lanchester's character which is not evident elsewhere:

> "Look at this, Mother dear, what a suggestion!
> Marry a man that I first met last night, –
> Who suddenly kissed me: I had such a fright!
> Some men go too fast, and others too slow,
> 'To be or not to be, that is the question'
> O may I say 'yes'? Or must I say 'no'?"

Lanchester here is in firm control of his art, and the mood of the poem is splendidly enhanced by its natural internal rhythms.

What then should one conclude from Lanchester's forays into poetry? He clearly took his efforts seriously, and there are occasions when he shows both inspiration and no little ability. His poems certainly repay study, and I hope that enough examples have been given in this short essay to indicate the range of his poetic talents. Not surprisingly, there is little evidence of the kind of originality which inspired his engineering genius, but a number of the shorter pieces in particular are both enjoyable and affecting. The passages quoted from *The Centenarian* also show that at his best he was capable of producing writing of considerable lyrical quality. It is doubtful whether Lanchester would ever have developed into a writer of great merit, and we are fortunate that his genius found its true expression in the science and engineering fields. But it was in the nature of the man that he felt obliged to make the attempt to span the Two Cultures, and without doubt his name will be remembered long after other lesser mortals who have tried to follow the same path have been forgotten.

Chapter Eighteen

THEORY OF DIMENSIONS

by David R. Thomas

F.W.Lanchester published his book *Theory of dimensions and its application for engineers* in 1936. The purpose of the book, as stated in the author's preface, was to help the young engineer to acquire a sound knowledge of dimensional theory. In achieving this aim, Lanchester wrote not only on the use of the Theory, but on the philosophy which underlies it. He also added a number of chapters containing tables of physical and engineering constants in both British and Metric (c.g.s.) units; these, while undoubtedly of value, are logically distinct from the remainder of the book. Much of his argument in the book was also contained in a paper published in the Journal of the Institution of Automobile Engineers in February 1937 (Lanchester, 1937).

Lanchester himself first learned of the Theory of Dimensions while studying under Professor Rücker at the Royal College of Science in the 1880s. In 1906 he read a paper before the Institution of Automobile Engineers based on the use of the theory, and received a very mixed reception; he later remarked that at that time the theory, while well known to physicists, was little understood by engineers. He himself, however, used it throughout his career, particularly in his work on aeronautics; he appears to have discovered Reynolds' Number, although not in the context of flow in pipes, independently of Reynolds himself.

Dimensional analysis

The basic principle of the theory of dimensions is that any physical quantity (or as Lanchester himself would have said, entity), has a certain dimensionality, and any valid physical equation must connect quantities of like dimension. Certain entities are taken as fundamental, and the dimensions of all other quantities are derived from them. In mechanics, the fundamental quantities are universally taken to be mass, length, and time, represented by the letters M, L, and T. Velocity, defined by the equation $v = s/t$ (or equivalently $v = ds/dt$) then has dimension L/T, and acceleration similarly has dimension L/T^2. As an example (not taken from Lanchester), if one were to write the equation of motion for a body falling under the influence of gravity in the form $s = 16.1\ t^2$ (which is correct if s is in feet and t in seconds) this would be dimensionally incorrect, since the left-hand side has the dimensions L, and the right-hand side T^2. The dimensionless number 16.1 should be replaced by the dimensional factor $\frac{1}{2}g$, leading to the correct form of the equation $s = \frac{1}{2}gt^2$, both sides now having the dimensions of length. The significance of the difference is apparent when one considers the form of the equation in metric units; the latter is correct, given the value of g in metric units, while the former is incorrect, the factor 16.1 requiring to be replaced with 490 in the c.g.s. system, or 4.90 in S.I. units.

Errors as obvious as the example above scarcely require the use of the theory of dimensions, but in other cases the error is more subtle and the theory is of more importance. This is particularly the case where gravitational units of force are used, and confusion may arise between pounds mass and pounds force, or equally between kilograms mass and kilograms force. Lanchester quotes as an example the expression $F = mv^2/gr$, which he states is given in some nineteenth century textbooks for centrifugal force. Force is defined by the expression F = ma, and therefore has dimensions ML/T^2; the dimensions of the given equation are therefore $ML/T^2 = ML^2/T^2 \times T^2/L^2$, which does not balance, and is therefore inadmissible. The error is in using the dimensional value g, in place of a numerical constant (32.2 in imperial units) to convert the force from absolute units (poundals) to gravitational units (pounds force). The dimensional error could be corrected by rewriting the equation as $F = Wv^2/gr$, where W is the weight of the object, but Lanchester objected to this form of the equation on the grounds that centrifugal force is independent of gravity, and would be unchanged if the experiment were to be carried out in interstellar space where g is zero.

As a more advanced application of the Theory of Dimensions Lanchester uses the study he originally published in a paper presented to the Institution of Automobile Engineers in 1906 (Lanchester, 1906). The horsepower of an engine is assumed to be dependent on three quantities: the linear size of the engine l (of dimensions L), the density of the materials p (of dimensions M/L), and a pressure or stress σ (of dimensions ML/T^2). The horsepower h is of dimensions ML^2/T^3, and we require a relation of the form

$$h = l^i \times p^j \times \sigma^k \times \text{const.}$$

Dimensionally this leads to

$$ML^2/T^3 \equiv L^i \times (M/L^3)^j \times (M/LT^2)^k$$

whence
$$j + k = 1$$
$$2k = 3$$
$$i - 3j - k = 2$$

to which the solution is k = 1.5, j = − 0.5, i = 2. The required relation is thus

$$h = l^2 \times \sqrt{(\sigma^3/p)} \times \text{const.}$$

If, therefore, a series of engines are constructed having the same proportions and differing only in the actual size, the power output will vary as the square of the linear size. Lanchester pointed out that this agrees with the R.A.C. or Treasury rating rule, which rated an engine by the cylinder area, rather than the modern practice of rating by cylinder volume; the latter would be correct only if the engines were run at the same revolution speed, whereas maintaining constant maximum stress would in practice require the larger engine to run more slowly. Alternatively, assume that the permitted maximum stress were to be multiplied by 4; then the above equation indicates that the horsepower will increase by a factor of $4^{11/2}$ or 8. Clearly quadrupling the stress limit will raise the cylinder pressure, and hence power output, by the same factor; the output will also be doubled by the possibility of running the engine at twice the speed. This would require the air pressure to be raised four times by use of a supercharger, and would increase the pressure on every oil film in the engine in the same ratio. It is also of interest to note that the horsepower varies as the square of the size, while mass varies as the cube; the power to weight ratio is therefore inversely proportional to the size. Given two comparable engines the smaller will therefore have the better power to weight ratio.

The greatest value of an analysis such as the above is when it is used, as frequently in aeronautics and naval architecture, to predict the performance of a machine from that of a model tested in a wind tunnel or tank. In many cases the analysis is carried out using the concept of dimensionless groups. The classic example of this is Reynolds Number. Lanchester introduces this by the following quotation from his own work on aeronautics.

"Let us examine generally the relations of *geometrically similar systems* possessed of *homomorphous motion* – that is, under circumstances when the theory of dimensions is strictly applicable, then the quantities upon which the motion depends are comprised by – velocity = *V*, kinematic viscosity v, and a linear (scale) dimension *l*.

Let us write

$$l = c\ V^p\ v^q$$

or, in terms of dimensions

$$L = L^p/T^p \times L^{2q}/T^q$$

and we have the equations

$$p + q = 0$$
$$p + 2q = 1$$

therefore $q = 1$ and $p = -1$ and

$$l = cv/V$$

which may be taken as the general equation connecting all similar systems of flow in viscous fluids."

(Lanchester, 1936a, p.55)

The above equation can be expressed in the form $c = lV/v$, where c is the reciprocal of the Reynolds number, and is a dimensionless quantity. Dimensionless groups of similar nature frequently arise in fluid mechanics and aerodynamics.

Dimensions and units

It is necessary to draw a distinction between the dimensions of an entity, and the units in which its values are measured. The dimensions of volume are L^3, whether it be measured in cubic feet, gallons, cubic metres, or litres. The mere definition of a unit for an entity does not make it fundamental. A related problem is to decide which entities should be taken as fundamental. In mechanics the universal practice is to regard mass, length, and time as fundamental and all others as derived. In the opening paragraph of his book Lanchester quotes Professor Everett as stating that this choice is a matter of convenience, and that, for example, mass, energy, and density might equally well be chosen. Lanchester's comment on this is that it may be true, but is of no more than academic interest as established practice must be followed. Many later passages in the book are, however, intelligible only on the premise that the author believes that the choice of fundamentals is not arbitrary, but that there exists a correct answer (which it is the task of the scientist to find) rather than a mere convention. In pure mechanics, however, there is no disagreement; the following table of dimensions of entities in this field is extracted from Table IV of *Theory of dimensions*.

Surface	L^2	Volume	L^3
Velocity	L/T	Acceleration	L/T^2
Momentum	ML/T	Force	ML/T^2
Energy	ML^2/T^2	Pressure	M/LT^2
Angular Velocity	1/T	Angular Momentum	ML^2/T
Moment of Inertia	ML^2	Torque	ML^2T^2
Power	ML^2/T^3	Density	M/L^3
Viscosity	M/LT	Kinematic Viscosity	L^2/T
Elasticity	M/LT^2	Surface Tension	M/T^2

Another use for dimensional expressions which Lanchester introduces in his book is for the conversion of units, and the reduction of the number of factors which need to be tabulated to make this possible. As an example (Lanchester, 1936a, p.155) consider the conversion of poundals to dynes. The dimensional expression for force is, as given above, ML/T^2. The basic conversion factors are 1 ft = 30.48 cm, and 1 lb = 453.56 g; the second is the unit of time in both f.p.s. and c.g.s. systems. The conversion factor for force is therefore $30.48 \times 453.56/1.0^2$; i.e. 1 pdl = 13,825 dynes.

Lanchester correctly insisted that the imperial foot-pound-second (f.p.s.) system of units was no less scientific than the metric centimetre-gram-second (c.g.s.) system; he ignores the metre-kilogram-second (m.k.s.) system which is now so familiar until he deals with electrical units. However, he deals at great length with the question of technical, or gravitational, units as distinct from absolute units.

The definition of force arises from Newton's first law: unit force is that force which will produce unit acceleration when applied to unit mass; in the f.p.s. system unit mass is 1 pound, unit acceleration is 1 ft/sec^2, and unit force is therefore the poundal. It was, however, almost universal practice when using f.p.s. units, to measure force in pounds, one pound (force) being the force exerted by gravity on a mass of one pound. This resulted in occasional confusion as to whether a quantity of, say, 6 pounds was a mass or a force; Lanchester made the excellent suggestion of using the abbreviation "lb" for the mass unit, and writing "pound" in full when the force unit was intended, and thereby anticipated the modern usage of "lbf" as the abbreviation for pound force. A more serious problem arose from an attempt, which Lanchester attributes to Professor Perry, to reconcile the pound force with Newton's law in the form f = ma. This was to define a so-called technical unit of mass, otherwise called a slug, which was the mass upon which one pound force would produce an acceleration of 1 ft/sec^2. Lanchester devoted Chapter five and an appendix of his book to exposing the confused thinking of the proponents of this system, or rather of these systems, as he identifies three different versions of the slug. The first of these treats force as a fundamental entity, defining the pound force as the force of local gravity on the Imperial Standard Pound, and the slug as the mass upon which this force will produce unit acceleration. Lanchester correctly points out that the unit of force so defined is not a constant, and is quite useless for accurate work. The second definition differs from this in that the pound force is defined as the force of gravity *at London* on the Imperial Standard Pound; this at least gives the unit a constant definition, but creates a problem in reproducing the standard force at any other location. In practice the system is unworkable because it requires a knowledge of local gravity before any calculation can be made. The third version retains mass as the fundamental entity, and defines the slug as 32.2 times the mass of the Imperial Standard Pound; the pound force then becomes the force which will produce an acceleration of 1 ft/sec^2 on this mass. Lanchester concedes that this version of the system is logically acceptable in itself, but argues that it should not be used partly because of the possible confusion with the first two versions, and partly for the simple reason that it is unconventional.

His tirade against the use of the slug as a unit of mass led Lanchester into correspondence in the columns of *The Engineer* in 1937 with a Mr. Porter, whose views he had quoted anonymously as an example of the contemporary use of the slug. The exchange apparently ended with both participants satisfied they had won the argument. Porter starts, however, by asserting that a chemical balance or a steelyard is a weight (or force) comparator. Lanchester had described this as one of the prime fallacies of those who believe in the slug, insisting that such a balance is a mass comparator, admittedly one which uses force to make the comparison. In the course of the correspondence he suggests, probably correctly, that this confusion arises because the Imperial Standard Pound is referred to legally as a standard weight, although in scientific terms it is a standard mass. It is difficult for anyone brought up in the age of space travel, when the idea of solving engineering problems for any environment where gravity (if any) is wildly different from earth standard is quite commonplace, to understand how an engineer as competent as Porter evidently was could make such an error. Lanchester, as so often, saw more clearly than many of his contemporaries.

Dimensions of thermodynamic quantities

Lanchester's treatment of the dimensions of thermodynamic quantities is more problematical. He insists that temperature has the dimensions L^2/T^2, and that it is unnecessary, and indeed incorrect, to introduce a fourth basic entity. In this he disagrees with modern practice, which would take temperature as a fourth fundamental usually symbolised by θ. His principal argument in favour of this position starts from the premises that the dimensions of heat (i.e. thermal energy) should be ML^2/T^2, the same as mechanical energy, and that in calorimetry this is the product of the mass of the calorimeter and the rise in temperature. Professors Rücker and Everett take temperature as fundamental and arrive at the dimensional expression $M\theta$ for heat; Lanchester quite reasonably objects to this, as it contravenes his axiom that any given entity has a unique dimensional expression. He nowhere suggests that he has thought of the objection to his position which would be raised by any present day scientist, that heat is the product of the thermal capacity of the calorimeter and the rise in temperature, the latter having the dimensions θ and the former $ML^2/T^2\theta$. His second argument is that if temperature has the dimensions L^2/T^2 then entropy is dimensionless, and since, in view of Boltzmann's relation $S = k \log \Omega$, it is a logarithmic quantity, this is clearly correct. There is of course an ambiguity here: entropy is defined by $S = dq/T$, and according to Lanchester's argument actually has the dimensions of mass, while the entity which is dimensionless is the specific entropy (entropy per unit mass); his usage of the term is, however, quite consistent.

After reading Lanchester's dogmatic comments on the dimensions of temperature, it is interesting to turn to an example of the use of dimensional analysis he quotes (Lanchester, 1935a, p.84), taken from a Birmingham University Mechanical Engineering Finals paper. His model answer uses what he describes as the Rücker system, in which the dimensions of temperature are taken as θ; this is a necessary decision since if they are taken as L^2/T^2 the problem simply fails to come out. There is, however, another aspect of the matter which, while he never clearly mentions it in the book, may have been at the back of his mind. He had done a considerable amount of work over the years involving fluid flow, and using dimensional theory as a means of scaling down a problem. If, for example, the linear dimensions of an aircraft wing are doubled then to maintain geometric similarity the velocity of air flow, of dimensions L/T, must also be doubled. If, however, the temperature is held constant the molecular velocities of the air due to thermal motion will be unchanged; to maintain the same ratio between velocities of internal and bulk motion the temperature must indeed be increased four fold exactly as suggested by his dimensional expression. There would appear, therefore, to be some logic in his suggestion.

Dimensions of electrical quantities

When dealing with mechanics it is possible to treat the dimensions of entities as being quite independent of the units in which those entities are measured, primarily because all systems of units are based on mass, length, and time as fundamental. In thermodynamics, provided one accepts Lanchester's insistence that thermal energy has the same dimensions as mechanical energy, the choice between mechanical and calorimetric heat units also has no influence on dimensions. In electrical theory, however, three systems of units were used in Lanchester's day, each in their own subdivision of the field: c.g.s. electro-static, c.g.s. electro-magnetic, and practical. It is impossible to reconcile the two competing sets of c.g.s. units with a single table of dimensions, although the practical system, essentially the modern S.I. units, can be regarded as being in accordance with the c.g.s. electro-magnetic units in this respect. Lanchester devotes three chapters of his book, and no less than seven appendices, to consideration of the dimensions of electrical entities.

A modern reader must take care to avoid being confused by Lanchester's nomenclature when dealing with electrical entities. The entity which would nowadays be referred to as *capacitance*, measured in Farads, he describes as *capacity*. The entity which Lanchester refers to as *capacitance* is now known as *permittivity*. This is purely a matter of an unfortunate change in definition, there is no error or ambiguity involved other than the distinction between the contemporary and modern usage of the same terms. In the following description I have used the terms with their modern meanings in all cases.

Chapter eleven of the book deals with Electrostatic Entities and Units. These are based on the definition of unit charge in terms of the repulsion between two equal electrostatic charges separated by a vacuum. The fundamental equation is Coulombs Law, in the form

$$f = q_1 . q_2/r^2$$

in which the permittivity of free space is implicitly taken to be unity. Since this equation is to apply in any system of electrostatic units, this is equivalent to regarding permittivity as a dimensionless quantity. This results in the dimensions of charge being $M^{\frac{1}{2}}L^{\frac{1}{2}}T^{-1}$. The dimensional expressions for the remaining electrostatic entities can be derived from this by applying dimensional arguments to the customary definitions. The following table, which is somewhat adapted from Lanchester's Table V, shows the results (J in the definition of potential stands for energy).

Symbol	Entity	Definition	Dimensions
Q	Charge		$M^{\frac{1}{2}}L^{1\frac{1}{2}}T^{-1}$
φ	Flux	$\varphi = 4\pi . Q$	$M^{\frac{1}{2}}L^{1\frac{1}{2}}T^{-1}$
C	Capacitance	$C = Q/E$	L
E	Potential	$E = J/Q$	$M^{\frac{1}{2}}L^{\frac{1}{2}}T^{-1}$
I	Current	$I = Q/T$	$M^{\frac{1}{2}}L^{1\frac{1}{2}}T^{-2}$
R	Resistance	$R = E/I$	$L^{-1}T$
L	Inductance	$L = R . T$	$L^{-1}T^2$

Lanchester observes that he, like many others, is unhappy with the concept of a point charge. He therefore, as an alternative, derives the above dimensional relationships by the following ingenious argument which uses only physically realisable charges.

"Let two equal spheres carrying equal electrical charges of *opposite* sign be separated by a distance L. Now let a small change take place in L. In order to preserve the condition of geometrical similarity, let a change proportional to the change in L take place in the diameter of the spheres. Then the fields will be geometrically similar and the areas of similarly situated tubes of flux will vary as L^2 and their lengths will vary as L, therefore the capacity will vary as $L^2/L \equiv L$. Thus we derive the dimensions of electrostatic capacity, $\equiv L$."

(Lanchester, 1936a, p.106)

If charge is then defined in terms of capacitance, by the expression $Q = V . C$, the table of dimensions again comes out as above.

The second traditional system of electrical units is the electromagnetic system; Lanchester deals with this in Chapter twelve of his book. The chapter opens with the statement "Nothing has ever been postulated by the physicist more unreal than the 'unit pole'" (Lanchester, 1936a, p.117); the electromagnetic system of units is, nevertheless, based on the equation for the force between two such poles. Since this equation, i.e.

$$f = P_1 . P_2/r^2$$

is functionally identical to the expression for the force between two point charges in electrostatic units, the dimensional expression for pole strength in e.m.u. is necessarily identical to that for charge in e.m.u., although Lanchester only mentions this parallel at a much later stage. The dimensions of the derived entities can again be worked out in terms of the standard definitions; the table below is adapted from Lanchester's Table VI. Note that Induction (B) is now more usually referred to as Flux Density, and Magnetic Force (H) often as the Gradient of the Magnetomotive Force; also that the expression H = B used to define the latter embodies the fundamental assumption that permeability is a dimensionless quantity. W in the expression for potential represents power.

Symbol	Entity	Definition	Dimensions
P	Pole Strength		$M^{\frac{1}{2}}L^{1\frac{1}{2}}T^{-1}$
B	Induction	$B = P/L^2$	$M^{\frac{1}{2}}L^{-\frac{1}{2}}T^{-1}$
H	Magnetic Force	$H = B$	$M^{\frac{1}{2}}L^{-\frac{1}{2}}T^{-1}$
I	Current	$I = H . L$	$M^{\frac{1}{2}}L^{\frac{1}{2}}T^{-1}$
E	Potential	$E = W/I$	$M^{\frac{1}{2}}L^{1\frac{1}{2}}T^{-2}$
R	Resistance	$R = E I$	LT^{-1}
L	Inductance	$L = R . T$	L
C	Capacitance	$L = T/R$	$L^{-1}T^2$

Lanchester was clearly dissatisfied with the need to treat electrical quantities as having two different sets of dimensions according as to whether e.m.u. or e.s.u. were chosen. This contravenes his axiom which he states to be "That no one Physical Entity can rightly have assigned to it more than one dimensional value or expression" (Lanchester, 1936a, passim). He clearly put considerable effort into an attempt to reconcile the customary systems of units with this axiom; unfortunately this problem has no solution. He quotes Professor Gisbert Kapp as stating "In the present state of physical science there is no conceivable experiment which could enable us to determine the dimensions of k and μ separately; all we know is that their product as dimensions … $k\mu = 1/v^2$" (Kapp). In this quotation k and μ represent the permittivity and permeability of free space. Electrostatic units are derived by treating permittivity as dimensionless and assigning the dimensions $L^{-2}T^2$ to permeability; electromagnetic units treat permeability as dimensionless and assign the dimensions $L^{-2}T^2$ to permittivity. The third alternative which Lanchester considered when he wrote *Theory of dimensions*, which he ascribes to Professor G.F.Fitzgerald, assigns the same dimensions (i.e. $L^{-1}T$) to both permittivity and permeability. This has the interesting consequence that current and potential have the same dimensions, and resistance becomes dimensionless. The drawback of such a system is that the unit charge, defined by the requirement that the force between two unit charges separated by unit distance (i.e. 1 cm) shall be one unit (i.e. 1 dyne), corresponds to neither the electromagnetic nor electrostatic c.g.s. unit. The difference between the various units can be exemplified by the differing values of the unit of current when expressed in amps:

electrostatic	1 unit (statvolt)	= 300 Amps
electromagnetic	1 unit (abvolt)	$= 10^{-8}$ Amps
Fitzgerald	1 unit	$= \sqrt{3} \times 10^{-5}$ Amps.

While the Fitzgerald system is of theoretical interest its adoption would be impractical, a situation which Lanchester seems to accept. An alternative means of treating the dimensions of electrical quantities is to take electric charge as a fourth fundamental quantity. He quotes from a letter by Professor W.Cramp to *Nature* in 1932 describing such a system.

As Professor Cramp observes, this results in a table of dimensions in which, unlike the previous three systems, no fractional indices occur. Lanchester dismisses this system by observing that the table of dimensions can be derived from that for the electromagnetic system by substituting Q for $M^{\frac{1}{2}}L^{\frac{1}{2}}$; and states that the dimensions of charge could equally as well be written MLQ^{-1} or $Q^3M^{-1}L^{-1}$ as Q, and that "it would not be possible to play tricks of this kind if Q were a true fundamental." This of course is nonsense; Lanchester is multiplying or dividing by a factor of MLQ^{-2}, which has the dimensions (in Cramp's system) of permeability, and adopting charge as a fourth fundamental is inconsistent with the assumption that permeability is dimensionless. Unfortunately Professor Cramp's own letter encourages Lanchester's objection by indulging in speculation about the dimensional identity of charge and mass.

It is apparent that at the time he was writing *Theory of dimensions* Lanchester's views were changing. In the body of the text he declines to make a definite choice in favour of any of the competing systems of electrical dimensions. He reproduces in an appendix a paper by Sir James Henderson (Henderson) expounding the view that the electromagnetic system should be adopted universally and the electrostatic system discarded. Henderson based this view on an analysis of the consequence of accepting Ampere's theory that all magnetic fields are due to the motion of electric charges, possibly on a microscopic scale. Modern physics accepts this conclusion almost without comment. Lanchester himself wrote that "Ampere's Theory may be *literally* true, and if this were established it would be difficult to resist Sir James Henderson's conclusions" (Lanchester, 1936a, p.135). In another appendix to the book, however, he himself propounded an alternative in which he used an argument based on relativity to demonstrate that the dual e.m.u./e.s.u. system is actually correct. The problem with his argument is that he is prepared to assign different dimensions to the magnetic field due to a stationary bar magnet and that due to a moving electron; the essence of Einstein's theory is that the two fields are observationally equivalent, which seems inconsistent with a difference in their dimensions. In a paper entitled *Electrical dimensions and units* read to the British Association in 1936 (Lanchester, 1936b) he makes it clear that since completing the book he has decided, on the basis of the relativistic argument, that the dual system is in fact correct.

In 1935, at the time Lanchester was writing *Theory of dimensions*, the International Electrotechnical Commission formally adopted a proposal, originally put forward by Professor Giorgi in 1901, that the metre, kilogram, and second should be taken as the fundamental units of mechanics, along with the value of 10^{-7} for the permeability of free space (μ_0). When the full system is worked out from these assumptions the electrical units derived are the traditional practical units, the volt, amp, and coulomb etc. The I.E.C. proposal was subsequently adopted by other international bodies concerned with definitions of units and is clearly the origin of the present *System International* units; Lanchester's objections have some theoretical force and are therefore of great interest. He had discussed the proposal in the final appendix of his book, and in particular the report by the Symbols, Units and Nomenclature Committee (S.U.N.) of the International Union of Pure and Applied Physics; his paper to the British Association mentioned above, (Lanchester, 1936b), deals with the report at greater length.

The original proposal by the I.E.C. was that one of the seven electrical units, i.e. Coulomb, Ampere, Volt, Ohm, Henry, Farad, or Weber, should be adopted as a defined standard, the magnitudes of the remainder are then determined by experiment from the standard. The S.U.N. Commission, however, declined to decide between these alternatives, and instead reverted to the original suggestion by Giorgi that the value of μ_0 be fixed at 10^{-7}. Their official report on the matter, as quoted by Lanchester, reads:

"That the 'fourth unit' on the M.K.S. system be 10^{-7} henry per metre, the value assigned on that system to the permeability of space."

(S.U.N.)

There can be no doubt that the Commission had made an unfortunate choice, and had expressed its decision unclearly. Permeability is one of the more difficult of the electrical properties to measure accurately, and is surely unsuitable as a standard. The intention was better expressed in the form adopted by the I.E.C. in 1938, which reads:

"the connecting link between the electrical and mechanical units [be] the permeability of free space with the value of $\mu_0 = 10^{-7}$ in the unrationalized system ..."

(I.E.C.)

In passing it should be remarked that the distinction between rationalised and unrationalised units concerns the point at which a factor of 4π is included; this, being a numeric quantity, has no influence on dimensions.

Lanchester objected to the S.U.N. report on three grounds. Firstly, he complained that encouraging a new metric system alongside the established c.g.s. system would lead to confusion; he cites the problems which already arose through failure to distinguish adequately between gram-calories and kilogram-calories. Secondly, he complained that the factor of 10^{-7} mentioned in the report was an error, and should be 10^{+7}. He interpreted the statement in the report as a declaration that the m.k.s. unit of permeability was equal to 10^{-7} c.g.s. units of permeability. This is the opposite of the Commission's intended meaning, which was that the c.g.s. unit of permeability, equal to the permeability of space, should be 10^{-7} henry/metre, which is the m.k.s. unit of permeability. Unfortunately Lanchester's final complaint obscures this latter point, since he insists that permeability is dimensionless and that ideally units should be chosen such that the value of μ_0 is unity. In order to reconcile the requirement that μ_0 be unity with the practical electrical units Lanchester resurrects a proposal which originated with Maxwell, that the unit of length be taken as 10^7 metres and the unit of mass as 10^{-11} grams. This he calls the *quadrant* system, as the length unit is approximately equal to the quadrant of the earth's circumference; the original founders of the metric system intended the correspondence to be exact.

The publication of his British Association paper in *Engineering* led Lanchester into an extensive and at times acrimonious dispute in the correspondence columns of that journal with, amongst others, Professor Marchant of Liverpool University (*Engineering*, November 1936 to February 1937). In the manner of such correspondence the editor eventually had to declare the matter closed without either party being persuaded of the error of their ways. The modern formal definition of the electrical units contains no direct reference to permeability, but defines the Coulomb in terms of the force between two straight parallel conductors; Lanchester would doubtless object to the apparently arbitrary factor of 2×10^{-7} which appears in the definition. The modern formal definitions of the metre and second refer respectively to the wavelength and frequency of certain distinct atomic radiations; this leads to the apparent anomaly of the speed of light being theoretically measurable with a spectrometer. It would be interesting to know what F.W.Lanchester would have made of this proposal had he lived to see it!

References

Lanchester, F.W. (1906): The Horse-power of the petrol motor in its relation to bore, stroke, and weight, *Proc. Inst. Auto Engineers*, 155–220

Lanchester, F.W. (1936a): *Theory of dimensions and its application for engineers*, London, Crosby Lockwood

Lanchester, F.W. (1936b): Electrical dimensions and units, paper read before Section G, British Association, Blackpool, 14th September 1936; reprinted in *Engineering*, **cxli**, 347–50 and 376–7

Lanchester, F.W. (1937): Treatment of problems in engineering by dimensional theory, *J. Inst. Auto. Engineers*, **5**, part 5, 17–43, February 1937

Henderson, J. (1935): Fundamental dimensions in electrical science, paper read before Section G, British Association, Norwich, 10th September 1935; reprinted in *Engineering*, **cxl**, 348

Kapp, G.: *Principles of electrical engineering*, **i,** 232, (Arnold); quoted in Lanchester, 1936a, 120

S.U.N.: Report of the Symbols, Units & Nomenclature Committee of the International Union of Pure & Applied Physics, 30th November 1935; quoted in Lanchester, 1936a, 294

I.E.C.: Resolution of the Advisory Committee on Electric and Magnetic Magnitudes and Units of the International Electrotechnical Commission, June 1938; quoted in E.G.Cullwick, *Fundamentals of electromagnetism*, C.U.P., 1938, 314

Chapter Nineteen

LANCHESTER'S AVIATION INVENTIONS

by Norman S. Ricketts

Introduction

A sound knowledge of aerodynamic theory plus an inventive analytic mind would be the assets for the work in aerial machine propulsion carried out by Frederick Lanchester in those early pioneer days of man's efforts to fly.

It was recognised by the inventors of that time concerned with powered flying machines that the unbalanced rotational torque of an engine-driven propeller had to be counteracted and this was accomplished by wash in and wash out on the aircraft's wings.

Early propulsion units also suffered from excess noise and vibration and the ground clearance required for propellers limited their diameter as well as the aerodynamic effect of one rotating blade upon another, restricting the number of blades at that time to four. A further complication on faster rotating propellers of large diameter is that toward the blade tip a loss of efficiency occurs if the tip speed approaches sonic speed. Some means of slowing that rotational speed without a loss of propulsive efficiency was required. It was to the solution of some of these problems that Frederick Lanchester applied his special talent.

His first known invention centred initially on, to quote "Improvements in and relating to aerial machines" (1897) which followed Otto Lilienthal's fatal flight in a non-powered flying machine in 1896. Lanchester's description of an aircraft wing cross-sectional and planform shape was "preferable to that of a soaring bird with convex upper and concave under surfaces, the intensity of curvature diminishing and the plan contour of the wing tapering towards its extremity". This echoed Daniel Bernoulli's total energy concept of fluid dynamic venturi flow. To counteract the unbalanced torque encountered with a single two-bladed propeller, Lanchester designed a screw propeller "in the manner of a cycle wheel with a rim to act as a flywheel with blades of fabric or thin plate stretched or fitted between pairs of spokes". (Fig. 1a).

The construction of fuselage, wings and tailplane of the aerial machine would be a framework covered with silk or other fabric or maybe thin sheet aluminium or other metal. He envisaged its propulsion unit as an engine powered pair of contra-rotating propellers

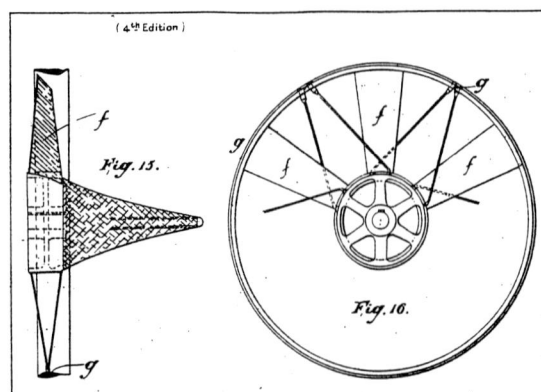

Fig. 1a. Lanchester's Screw Propeller Specification No 3608 (1897)

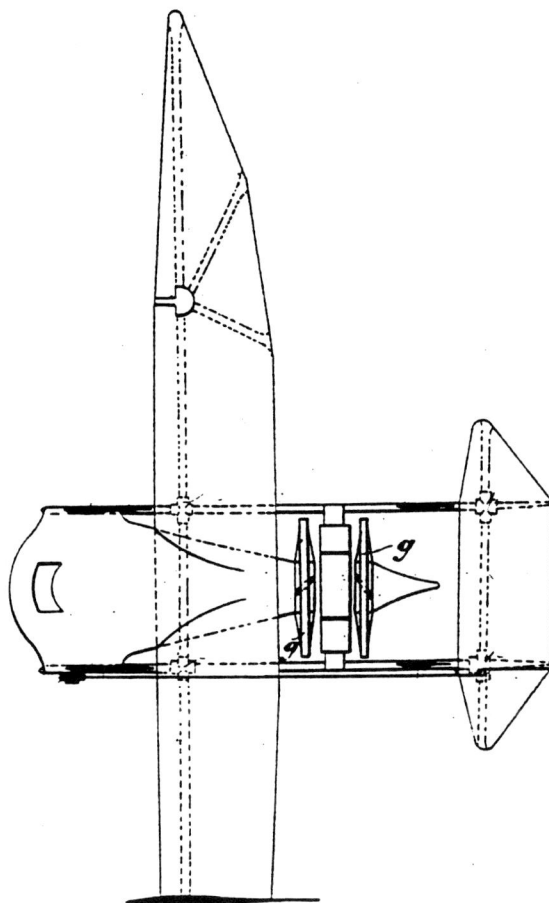

Fig. 1b. Dr. Lanchester's envisaged Aircraft with propulsion unit (Contra-Rotating Propellers)

rotating on a hollow shaft system which according to Lanchester would balance out the unbalanced rotating torque associated with single rotating propellers (Fig. 1b).

After the successful powered flight by Wilbur and Orville Wright in 1903 a greater activity in the search for a design for a more efficient aerial machine was pursued. Lanchester turned his attention to the driving mechanism for both marine and aeronautical propellers, (1907). He set out to "relieve Aerial Machines of all unbalanced torque by ensuring a greater efficiency than hitherto had been obtainable" by the increase of the effective diameter of the propeller compared with his earlier 1897 propeller design.

The newly designed propeller incorporated a blade twisted along its length to keep the effective angle of attack constant from root to tip and he would apply this propeller design to "reverse rotating propeller layout on a hollow shaft system". (Fig. 2). He determined that the propeller diameter would be equal to one quarter to one third the distance from wing root to wing tip when applied to aerial machine propulsion, this diameter being limited by the ground clearance of the propeller during take off etc. Following this design he carried out an investigation into a smoother, slower running, more efficient propulsion unit. Aerodynamically two forces exist on a rotating propeller of thrust in the direction of flight and rotational torque in the plane of rotation resisting the turning effect of the engine to whose output shafts the propellers are attached.

A further aid was to use balanced gearing where a set of epicyclic gears with sun, planet and ring gearing would be fitted for each propeller, (Fig. 3). Using this gearing with a petrol-driven engine in aerial machines the propeller speed was reduced when compared with engine crank speed, which contributed to a smoother, quieter propulsion system.

In low powered engines a two-bladed propeller would be used but with the development of power plants with higher power output, the aircraft speed increased and so it could fly

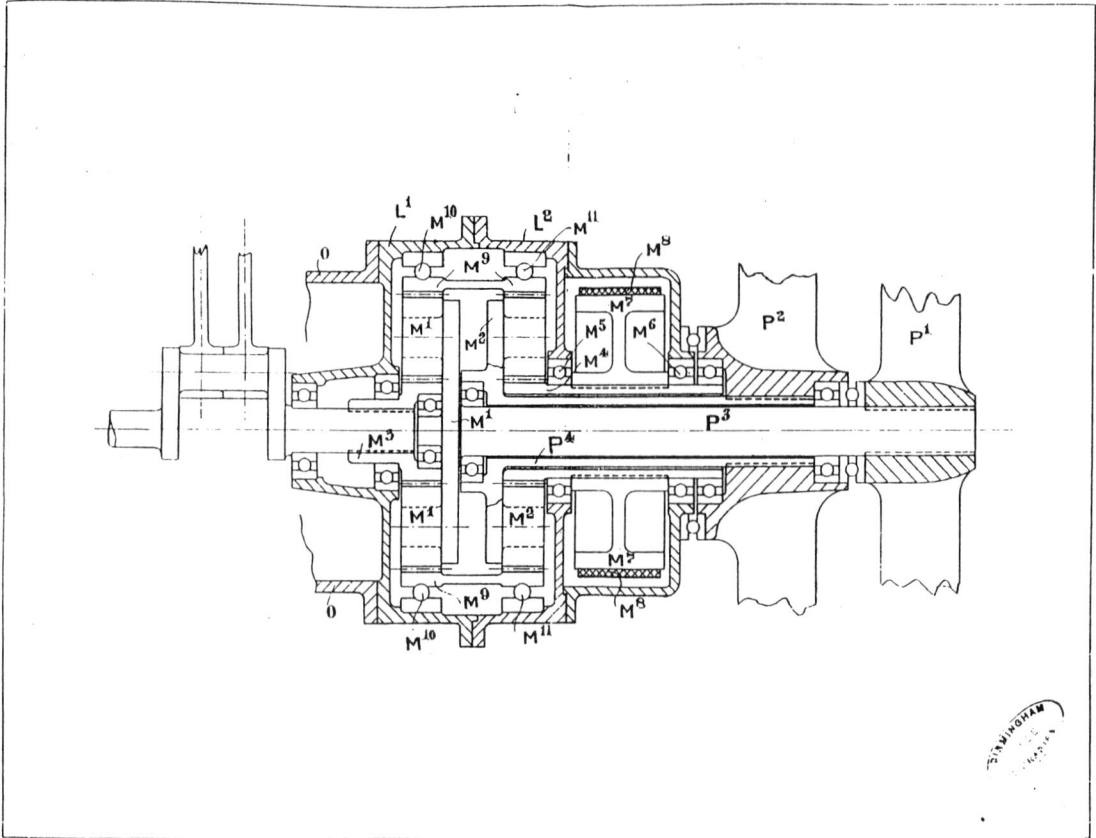

Fig. 2. New design of Propeller with Blade Twist applied to a Reverse Contra-Rotating Propeller system. Dr. Lanchester's Specification No. 9413 (1907)

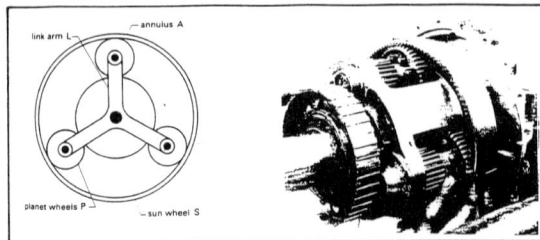

Fig. 3. Typical Epicyclic gear layout

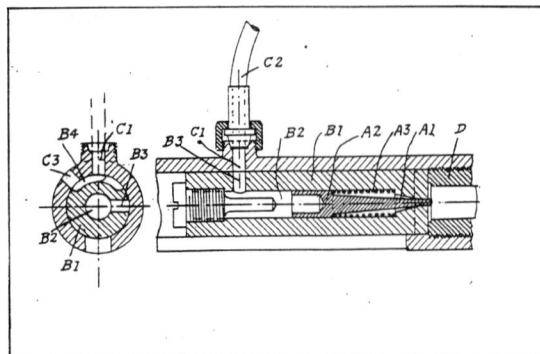

Fig. 4. Machine Gun Interrupter Mechanism. Dr. Lanchester's Specification No. 129374 (1917)

at higher altitudes where the air is less dense, the number of propeller blades would increase to three or four to increase the developed thrust and would run more smoothly.

The First World War period saw the development of military aerial machines and mounted machine guns to be fired by the pilot and located just behind the propeller with the consequence that the propeller was shot away, and so Lanchester turned his attention to a solution (1919). His invention was an interrupter mechanism synchronised to fire the machine gun through the propeller disc after each blade had rotated out of the line of fire. This was a gas operated locking mechanism whose operation was controlled by the engine rotation acting upon the firing mechanism of the machine gun, (Fig. 4). The pipe C2 supplied the pulse of gas passing through a valve synchronisd to operate a pneumatic piston firing pin A2 and a spring A3 to return the firing pin to its neutral position.

Fred Lanchester, well known in aeronautical circles by this time, submitted a paper to the Advisory Committee for Aeronautics (1918) relating to his study into the efficiency of tandem reverse rotating propellers (contra-rotating) and suggested that a full technical investigation of the subject should take place by practical experimentation along clearly defined lines. The investigation would deal with propellers of coarser or steeper pitch than those in current use which would run at slower rotational speeds with the conclusion that a tandem reverse rotational propeller system would bring about a higher propulsive efficiency and a more silent propulsion system.

He further reasoned that the slipstream generated energy loss was a combination of rotational and sternwards components and the added propeller used in tandem reverse rotating propeller systems would conserve the rotational energy loss increasing the propulsive efficiency. (This in practice did not conserve all the rotational energy loss but increases in efficiency were achieved).

The test criteria initially included:

a) That the following propeller would be of lesser diameter than the leading propeller to eliminate the turbulent effect at the tip of the leading propeller.

b) An assumption that the angular momentum of the leading propeller would be absorbed by the following propeller and both the propellers would be driven by equal and opposite torque.

c) The total rotational energy given to the forward propeller slipstream be recovered by the following propeller.

d) The blade design as far as loading and angular velocity were concerned would be based on the "least gliding angle".

The conclusion reached after the tests were carried out by Lanchester was that considerable advantage was gained from using large "gliding angles" due to heavy loading of the propeller system. The best optimum pitch/diameter ratio for tandem reverse rotating propellers was 2 to 1 or even 3 to 1 if full conservation of rotational energy loss were recoverable. Lastly, more research into the use of coarse pitch and slow running propellers would be required in order to create a more silent propulsion system.

During his experiments he concluded that the attachment of propeller bosses to the engine output shaft boss by a ring of bolts inserted parallel to the rotating engine shaft axis and keyed loosely to that shaft relied purely on frictional grip and slippages would occur during operation. He turned his attention to the solution to this problem and invented in 1917 (Lanchester, 1919) a method of locking the propeller bosses to the engine output shaft boss by placing half the bolts at an angle parallel to a right hand spiral and the other half of the bolts parallel to a left hand spiral. This locked the bolts one against the other giving a rigid drive which required no splined shaft with a more even load sharing between both bosses, (Fig. 5).

Flight (Lanchester, 1941) published in their magazine a review of Lanchester's earlier inventions in which the opening statement paid tribute to his work in aviation development

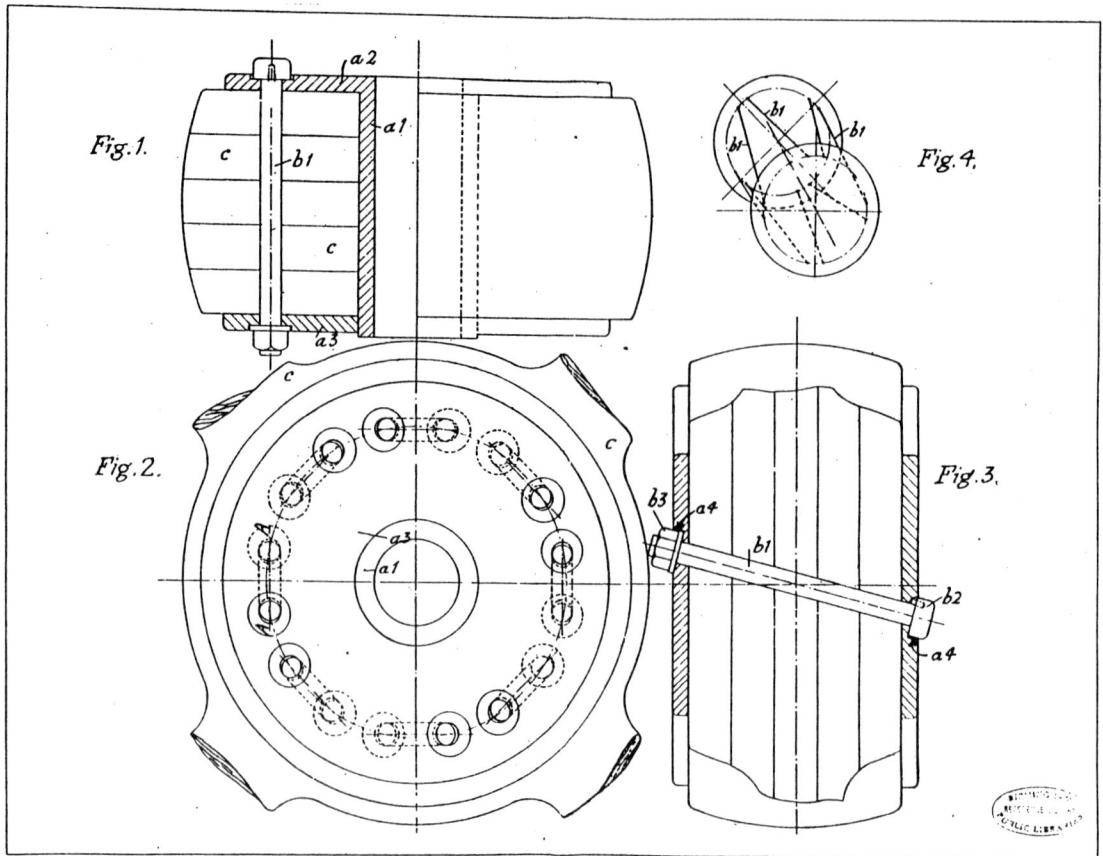

Fig. 5. Propeller Boss Locking Method. Dr. Lanchester's Specification No. 130025 (1917)

and I quote "Once again it is our privilege to publish an article by Dr. Lanchester, grand old man of aerodynamics, one is amazed at the insight into detail of the mechanisms of flight that he possessed in those very early days. As long ago as 1907 he not only visualized the advantages of contra-rotating airscrews or as he preferred to call them co-axial propellers with reverse rotation, but he patented a form of epicyclic drive whose details appear in this article".

So much interest was shown in contra-rotating airscrews that in reply to a series of letters from Mr. C.C.Walker, a director of the De Havilland Aircraft Company, Hatfield, Hertfordshire on this subject, Fred Lanchester was able to give information on the optimum diameter and pitch limits of propellers to this company who later designed and manufactured contra-rotating propeller systems for many aircraft types. Although the effect of sonic airflow was realised at that time due to the tip rotational speeds of large diameter propellers, there was no discussion of the loss in efficiency due to this problem between Lanchester and Walker at that time. He recommended that Walker obtained copies of his patents on contra-rotating propellers with reference to the elimination of unbalanced torque and greater efficiency due to the absorption of rotational energy loss using lower propeller rotational speeds.

Lanchester also remarked to Walker on the work carried out by the Italian company Macchi in relation to contra-rotating propeller systems in producing a Schneider Trophy winner in the seaplane class with the MC72 Seaplane using contra-rotating propellers, which raised the world speed record for seaplanes to 709.2 km/hr. They used two Fiat lightweight AS5 engines in tandem with a common crankcase, renaming it an AS6 engine, each engine driving its own contra-rotating propeller through a hollow drive shaft system with gearing to slow down the rotational speed of each propeller.

The De Havilland Aircraft Company continued to design and manufacture both single, variable pitch, feathering and contra-rotating propeller systems which were used on the following aircraft types to name but a few: Saunders-Roe Princess flying boat, Westland Wyvern, Avro Lincoln and Shackleton, Supermarine Spitfire Mk XIV to Mk 21 using the Griffon engine with contra-rotating propellers, Bristol Brabazon experimental aircraft using a Centaurus engine with contra-rotating feathering propeller systems and lastly the Armstrong Siddeley Double Mamba engine with contra-rotating propellers.

The development of the gas turbine engine effectively closed the door on further propeller technology development although variable pitch propellers are used on turbo-prop driven aircraft and light aircraft production to this day. Heavy lift helicopters use the contra-rotational propeller application which allows the designer to leave out the torque balancing tail rotor used on single rotor helicopters.

There can be no doubt that the early pioneering work of Dr. Lanchester with his forward thinking and his grasp of aerodynamic flow together with his inventive genius contributed to the solution of many of the problems besetting aircraft propulsion designers and that *he was a man ahead of his time*.

References

Bazzocchi, Ermanno (1972): Technical aspects of the Schneider Trophy and the world speed record for seaplanes, *Aeronautical Journal*, **76 (2),** 65–81

Beaumont, R.A. (ed.) (1942): *Aeronautical engineering*, Odhams Press

Lanchester, F.W. (1897): *Improvements in and relating to aerial machines*, G.B. patent specification 3608:1897

Lanchester, F.W. (1907): *Improvements in driving mechanisms for aeronautical or marine propellers*, G.B. patent specification 9413a:1907

Lanchester, F.W. (1918): *Investigation on the efficiency of reverse rotating propellers in tandem*, Reports and Memoranda, No. 540, Advisory Committee for Aeronautics

Lanchester, F.W. (1919): *An improved device for controlling the firing of machine guns*, G.B. patent specification, 129374:1919

Lanchester, F.W. (1919): *Improvements in the attachment of aeronautical propellers and in bosses therefore*, G.B. patent specification 130025:1919

Lanchester, F.W. (1941): Contra-props: recollections of early considerations by the Advisory Committee for Aeronautics; a pioneers 1907 patent; suggestions for further research, *Flight*, **40,** 11th December, 418–19

Chapter Twenty

SOME OF LANCHESTER'S OTHER INTERESTS

by Roy L. Thomas

F.W.Lanchester's interests were wide ranging, as can be seen from the diversity of inventions which he patented. His considerable work on motor car design and his contributions to aviation are described elsewhere in these volumes. His work on sound reproduction equipment has a separate chapter. In this chapter we shall try to show, roughly in chronological order, many other aspects of work in which Lanchester was involved.

Early inventions

Fred Lanchester's first invention, which was Patent No. 16432 of 1888, arose from making drawings of inventions in his first job, working in a patent agent's office. Becoming weary of the difficulty of drawing many parallel lines, often close together, Lanchester invented the Isometrograph, a device for hatching and shading drawings and aiding in geometrical detail. It consisted of a straight edge and a guide connected by levers so that they could be moved across a sheet of paper rather in the manner of a caterpillar: when the straight edge was held down for a line to be drawn, the guide moved away from it, and then when the guide bar was held down the straight edge moved back, the distance moved being constant but adjustable, so that parallel lines could be ruled at fixed intervals. A scale on the straight edge could be used to assist in drawing broken lines.

Fred joined his brother Henry in two other early inventions which appeared as Patents Nos. 650 and 7382 of 1889. The first was "An Improved Apparatus for Facilitating Rapid References to Books" and consisted of a piece of tape with a clip at each end. A clip was attached to each cover of the book so that the tape was stretched between the clips. On closing the book the tape fell between the leaves, so marking the place. The other was "An Improved Mechanical Contrivance for Advertising Purposes", which was intended to make advertisements appear and disappear alternately, or change colour, or two advertisements or portions of an advertisement to appear. The device worked by placing a sheet of metal with a number of parallel slots in front of a screen carrying the display material in alternate parallel lines of the same width as the slots. The screen and the grating were then moved in relation to one another so that one or other of the displays appeared.

In 1892 Lanchester turned his attention to the bicycle; Patent No. 9703 of that year, is entitled "Improvements in and Appertaining to Velocipede Pedals", and is concerned with enclosing the pedal spindle and ball bearings so that dust was effectively excluded. This drawing is remarkably like a modern bicycle pedal. In 1892 the brothers were manufacturing bicycle pedals at 41 St. Paul's Square, Birmingham.

The pendulum governor

Before Fred Lanchester went to work for Forward Gas Engine Company, for £1 a week "for experience", he met T.B.Barker, the manager, to discuss conditions. In the agreement with which he was presented any improvements were to be the property of the company. Lanchester objected. Barker said, "You don't suppose you are going to teach us how to make gas engines?". Replying, "Oh, evidently you don't, and cannot therefore regard this clause of any value", Lanchester crossed out the clause in both copies of the agreement.

THE ISOMETROGRAPH.

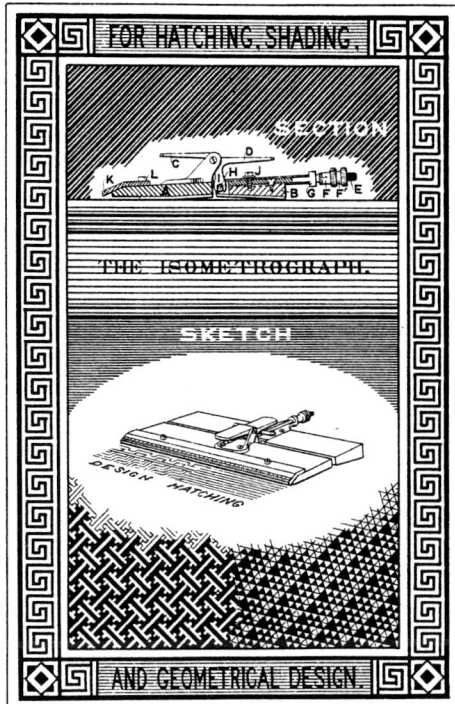

CONSTRUCTION.

The instrument consists of two main portions; a straight edge A, and guide B; connected by a steel rod at right angles to them.

This rod is fixed firmly to A, and slides in bearings formed on B, one of which is visible at G: it terminates in the screw E, on which work the graduated nut F, and lock nut F'.

The lever C. D. works on a pivot carried by A, having an arm I which gears with the pin H fixed to B.

Two screws are provided, (one of which is figured at L), with which to attach a scale or curved template, indicated by K in the figure.

ACTION.

The instrument is actuated by pressure alternately applied to the extremities of the lever C.D. A force applied to C, by bearing most heavily on A, prevents it from moving, and the forked arm I, by acting on the pin H, causes the guide B to move as far as permitted to it by the nut F.

In like manner a force applied to D keeps the guide B at rest; the pin H then becomes the fulcrum, and the motion of the lever draws A towards B.

METHOD OF USE.

HATCHING AND SHADING.—For plain hatching, the number of lines to the inch is regulated by the graduated nut F. The end C. of the lever may be depressed while the line is being drawn, and thus the rate at which the instrument can be worked is only limited by the speed with which the pen can be used.

Ornamental Hatching can be executed in any variety, by omitting lines at regular intervals.

GEOMETRICAL DESIGN.—A large variety of geometrical designs may be obtained by cross-hatching, and subsequently filling in with a fine brush or pen: an example of which is given in the bottom right-hand corner of the plate.

Other designs, such as that shown at the bottom left-hand corner, require that the lines should be dropped intermittently, in which case a scale is fixed at K, to indicate at what points the lines are to be interrupted.

All the hatching and geometrical pattern shown in the plate, is drawn by the aid of the Isometrograph, without preliminary work.

Fig. 1. The leaflet on the Isometrograph (reduced)

Soon after this he invented and patented an inertia, or pendulum, governor and Barker was paying 10s 0d royalty on each engine.

The gas engine had separate valves for the admission of air and gas, and the pendulum governor, Patent No. 12502 of 1889, was a device which controlled the speed of the engine by varying the amount of gas admitted. The air admission lever was operated directly by a cam, but the gas admission lever was pivoted to the air admission lever. A loosely pivoted weight, or pendulum, lifted by the exhaust lever, carried a blade which at low or normal

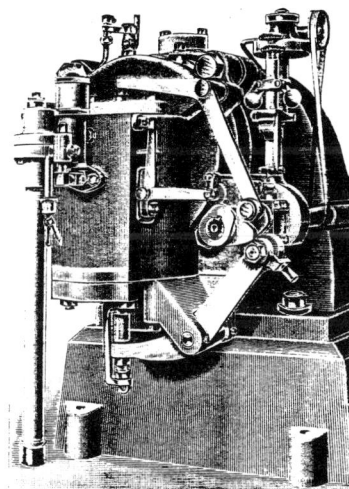

Fig. 2. A "Forward" gas engine, fitted with Lanchester's pendulum governor
(The Engineer, 6th April 1894)

speeds acted as the fulcrum for the gas admission lever, enabling it to admit gas; at high speeds the weight held the blade away from the gas admission lever so that the stem of the gas admission valve became the fulcrum for the lever. The speed was varied by varying the tension on a spring controlling the return of the weight.

The gas engine starter

Another of Lanchester's inventions while with the Forward Company is described by Archie Millership, who joined him when he began building motor cars.

"One of the early difficulties in starting a gas engine was the danger in pulling over a flywheel by hand. This was a dangerous business, especially so with a large engine when it required two men to do the work. Frequently it was reported that the men were injured or killed by being thrown over the flywheel through a backfire. Dr. Lanchester took out a patent, number 5479, in 1890 which was novel and a very simple idea. He fitted a tap leading to the valve chamber. The flywheel was pulled by hand back to the compression stroke, the gas was turned on and the valve chamber tap was lighted. As the oxygen was being consumed the flame was firstly white in colour, changing to a greenish-blue. Directly this happened the tap was turned off and the gas in the cylinder exploded, starting the engine. This was followed by a number of low pressure impulses until enough speed was reached to overcome the compression." (Millership).

In addition to being used by the Forward Company, the starter was adopted by Messrs Crossley Brothers Ltd. with whom Lanchester entered into an agreement, the basis of which was a minimum royalty of £1500 per annum. The payments were continued by Crossleys for about two years, at which time they bought a patent for a high pressure combustion type starter invented by Dugald Clerk (in which Lanchester had acquired a half-share interest). But Forward and Crossley's were not the only companies to want to use the new starter, Millership tells how,

"Lanchester asked me to go round the larger factories to see if there were any infringements of this patent. This was quite easily achieved, one had only to listen for the puff-puff of a gas engine, to contact the engine driver and ask him, to see the engine. If the patent was infringed then the owners were prosecuted. He called it his 'fighting patent', and in every case the infringers lost the day." (Millership).

Lanchester subsequently sold his patent to Harry J.Lawson, of the British Motor Syndicate and Great Horseless Carriage Company.

Gas calorimetry

In order to determine the efficiency of a gas engine it is necessary to know the amount of heat available in the gas; this is measured in a calorimeter. In 1893 Lanchester, being dissatisfied with the slow and inconvenient standard type of calorimeter of the day, thought up what he called "an extremely simple idea", which should give results more directly and rapidly. He therefore joined a course at Birmingham Technical College, to secure the use of their facilities, and attempted to construct such a calorimeter. The results were very irregular and the attempt was written off as a failure. But in 1894 Professor J.H.Poynting allowed Lanchester to experiment in the physics laboratory at Mason College (which later became Birmingham University) and, with the assistance of P.L.Gray, to make a more sophisticated instrument. While the results were more consistent, they still gave a low figure. Unfortunately, although Lanchester had intended to continue, he could not allow more time or money and, from one cause or another, all the equipment disappeared. Nevertheless he was so sure that the idea was good that forty-four years later, in an appendix to a paper to the Institution of Mechanical Engineers, Lanchester gave an outline of the design and said that he would like to see someone take up the problem for a Ph.D. thesis, being welcome to use Lanchester's work but also welcome to take all the credit. (Lanchester, 1939, p.231).

The perfect radiator

At about the same time, 1894, his attention was drawn to another type of physical measurement, that of the efficiency of different surfaces in radiating heat. This is done by comparing the heat radiated by the surface under test with that radiated by a "black-body". P.L.Gray had asked Lanchester to supply him with a cylindrical block of copper which would be covered with lamp-black and used as a standard "black-body" radiator. It occurred to Lanchester that, although this was the universally accepted black-body, it was scientifically unjustified. At that time it was known that inside an unvented enclosure the radiant energy would rise to a maximum. Lanchester reasoned that if the inside of the enclosure was a dense black colour a reasonably sized hole could be cut which would constitute an absolute standard, that is, it would behave as a truly black body. In August 1895 Lanchester was invited to a house party at Daramona, Westmeath, in Ireland, by Mr. W.E.Wilson. One day there was such a downpour that outdoor activities were impossible, so Lanchester and some of the other guests went out and bought a large biscuit tin and a half-gallon oil can and, using the facilities of their host's well-equipped workshop and laboratory and with his assistance, made up an apparatus to try out the idea. Using the apparatus they were able to measure the radiation efficiencies of a polished tinplate surface and of a lamp-blacked surface and thus show that the cylinder covered with lamp-black was not a true black-body radiator. (Lanchester, 1939, p.324).

Although by 1899 German and other physicists were experimenting with cavities in the measurement of radiation, it is probable that Lanchester was first in the world to use a cavity with a small aperture as a perfect black body.

The pendulum accelerometer

Having an interest in the performance of railway locomotives, in 1889 Lanchester invented a pendulum accelerometer in order to measure the acceleration and braking of locomotives. The pendulum accelerometer works on the basis that a pendulum suspended in a vehicle will move off vertical when the vehicle is accelerating and, whatever the size of the pendulum bob, the vehicle acceleration is given in terms of acceleration due to gravity by the tangent of the angle of deflection; hence the pendulum can be fitted with a pointer recording on a scale graduated to show the acceleration directly at every instant. In Lanchester's instrument the pendulum is suspended on knife edges carried on a frame provided with levelling screws, and a dash-pot is provided in order to prevent the setting up of oscillations by the vibration of the vehicle. The pendulum is as short as possible, in order that its oscillation period shall be rapid and therefore easily damped, and also that its motion shall differ as little as possible from that of the vehicle. In a paper to the Institution of Automobile Engineers, Lanchester says that the original instrument was home-made in 1889, but a more modern and effective machine (one of which is in the library of the Institution of Mechanical Engineers) dates from 1904. (Colour plate 4)

In the course of investigations using the instrument he noted that careful attention to releasing the brake just before stopping could obviate the jerk which otherwise always occurs. He pointed out that the jerk was due to a change in acceleration and proposed to describe it mathematically as the rate of change of acceleration. Lanchester used the improved 1904 model to measure the acceleration and braking of motor cars, for which it was more suitable than the original model. (Lanchester, 1910).

The Tapley Brake Testing Meter used in Department of Transport testing stations for the regulation testing of brakes is a pendulum-type decelerometer working exactly like Lanchester's accelerometer. It is a more compact machine and has a simple dial read-out, not a recording chart, but whereas Lanchester's chart model would record acceleration and

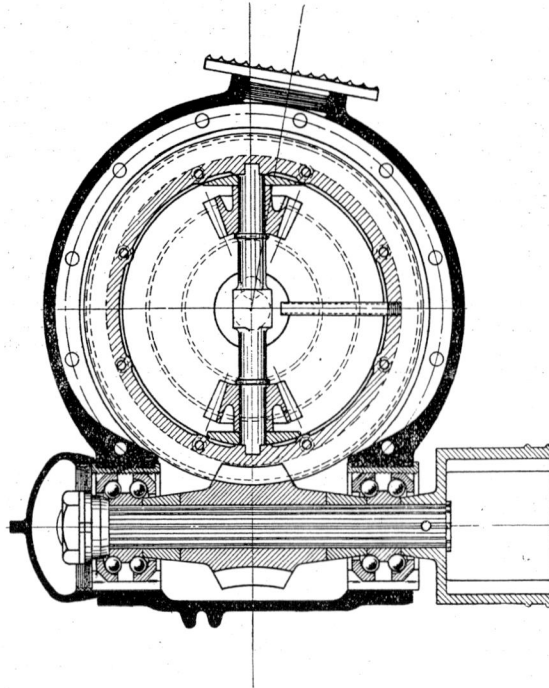

Fig. 3. First worm gear used on a Lanchester car

deceleration at the same time, the Tapley Meter has to be turned round in order to measure acceleration.

Worm gear

The chain drive to the back axle which Lanchester used on his first car, while reasonably efficient, was noisy, difficult to lubricate, and liable to need frequent adjustment. After careful examination of alternative drives, in 1896 Lanchester designed and built a special

Fig. 4. Lanchester's patent worm gear cutting machines. The one on the left is set up to cut a worm, the one on the right is set up to cut a wheel

machine for the production of a worm gear. The next year the machine was patented (Patent No. 13433, 1897) and the worm gear fitted to a motor car.

Although he had at that time neither seen nor heard of worm gear of the type in operation, Lanchester chose the hollow or Hindley type of worm, known at the time as a text-book device invented by a sixteenth century clockmaker. In this form of gear not only is the worm wheel hollowed, but the worm is curved, forming a saddle to fit with the worm wheel.

The hobbing machine was designed with a view to the avoidance of anything liable to introduce backlash. It consisted essentially of two parallel spindles connected by change gears, and a third spindle with its axis perpendicular to these and mounted in a swivelling frame which was pivoted about the axis of one of the parallel spindles. The worm was mounted on the other parallel spindle and the worm wheel on the third spindle. The action of the machine was the same whether cutting worm or wheel; with a cutter on one spindle and a blank on the other, and appropriately set up, the machine would run automatically until the cut was finished, the cutter and blank being kept in contact by a weight attached to the swivelling frame. To maintain the accuracy of gears a pair of "master cutters" was made and kept solely for making "working cutters", each of which was good for from 1,000 to 2,000 sets of gears before rejection.

In the Lanchester works, and later in the Daimler works, these machines were usually set up in threes so that one worm or wheel could be passed from one machine to another for roughing, rough finishing and finishing. Although set up for different jobs all the machines were identical, but other manufacturers used a completely different machine for cutting the worm from that used for cutting the worm wheel. American rights were held by the Warner Gear Company of Indiana and they supplied the demand for Lanchester gears in the U.S.A. and Canada.

Lanchester was keen to ensure that his gears gave the highest possible efficiency so he designed and built a worm gear testing machine. With this machine it was possible to determine the effect of small differences in tooth clearance or surface condition and of different lubricants. The machine was based on the principle of measuring the input and output torques of the gear box. The gear box under test was mounted on double gimbals

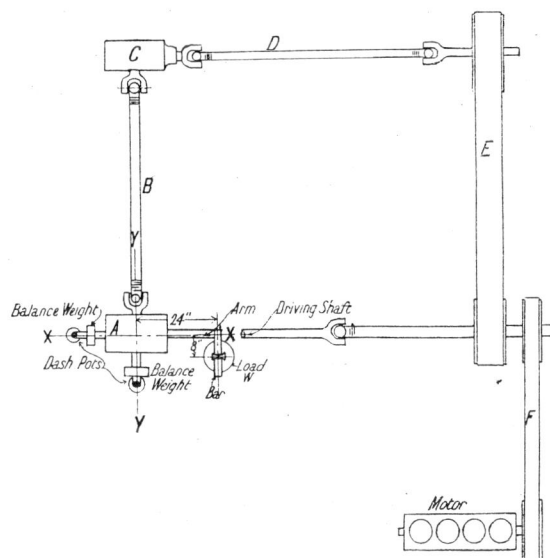

A Worm gear under test
B Drive from worm gear
C Bevel gear
E Slipping belt to balance losses in worm gear

Fig. 5. Diagrammatic sketch of Lanchester worm gear testing machine

and the power input and output were taken by universally jointed shafts, the power transmitted by the gear being restored to the driving shaft by a bevel gear and slipping belt. A weight could be moved along a bar to level the gear box and so give a measure of the torque. The National Physical Laboratory certified the accuracy of the machine to be to within one-fifth of one per cent. In the discussion on a paper on worm gear performance given in 1935 a member of the National Physical Laboratory referred to the Daimler-Lanchester testing machine as, "usually regarded as giving the most accurate measurement of the efficiency of a worm gear." (Merritt).

As well as testing gears on his machine, Lanchester cut sections of the gears to find how closely the worm meshed with the wheel. To do this he had to grout the gear teeth together with solder, cut away the unwanted parts and then slice the teeth in mesh to obtain sections showing how closely the teeth came together and hence the thickness of lubricant between them.

Other manufacturers used a parallel or cylindrical worm in their worm gear. Lanchester tested and examined these gears and came to the conclusion that his were of higher efficiency and generally superior, but when he published the results in papers to the Institution of Mechanical Engineers (Lanchester, 1913, 1916) there was much disagreement from other manufacturers. The first appendix to his paper on "Worm gear" was a copy of a "Report on the Efficiency of Daimler-Lanchester Worm Gears Tested for Messrs. The Daimler Motor Co. Ltd., of Coventry, by the National Physical Laboratory" (Lanchester, 1913, p.242), but regrettably the other manufacturers did not submit their products to the National Physical Laboratory for testing.

For the Daimler Company, Lanchester designed a range of worm gears based on standard shaft centres and wheel thicknesses. The range covered "private pleasure cars" to "the heaviest road vehicle at present in ordinary commercial use, namely, the five-ton lorry". The twelve types are tabulated and described in his paper on "Worm gears and worm gear mounting" (Lanchester, 1916, pp.119–21), and design calculations for some of the gears are in one of the sketch books (LC. SB 10, pp.45, 58, 70, 80, 81, 85, 92).

In his paper "Worm gear" (Lanchester, 1913) Lanchester says that he lays no claim to the introduction of worm gear for locomotive purposes because tram cars had been successfully propelled by worm gear for some time but he goes on to assert, "without fear of contradiction", that it was because of the success of the Lanchester worm gear that so many motors were using worm gears at that time.

Although he was convinced of the superiority of the worm gear for automobile drive, Lanchester was aware that this did not apply to all applications. In 1924 he wrote an article in *Engineering* in which he said that he was "not impressed with the prospects of the general use of worm gear as a turbine reduction gearing" in ships. Not only were the costs of worm gearing higher because of the large sizes and small numbers required, but there was a slight advantage in efficiency in favour of well-cut and well-mounted helical gearing. (Lanchester, 1924).

Colour photography

In 1895 Lanchester was granted Patent No. 16548 for an invention which was a method of producing photographs in natural colours. This consisted of a grating of parallel opaque bars placed as close to the object as possible and a prism fixed close to the lens. The print was a black-and-white lantern slide which was viewed by placing it in the camera in the place otherwise occupied by the photographic plate. On shining a white light through the slide a coloured image could be projected on to a screen.

The patent was not developed, but when J. and E.Rheinberg reinvented and improved the process in Germany they acknowledged Lanchester's earlier work, describing the differences between his apparatus and their own. (LC. Baxter 9–7).

A petrol tank level gauge

Early motor cars carried the petrol tank in a more exposed position than modern cars, so Lanchester was able to design a level gauge which was operated by hand. Although the description in the specification applied to use in the petrol tank of a motor carriage, it was not just for motor cars. Patent No. 12235 of 1902 was entitled "Improvements in Level Gauges for Tanks Containing Fluids", and the invention, while being especially for tanks which are not under pressure, could "by special arrangement be used for tanks or boilers in which moderate pressure exists". A float consisting of a saucer-shaped thin sheet-metal bowl had attached to its centre a rod, marked in gallons, passing through a tubular guide in the top of the tank. The empty bowl floated on the surface of the liquid, but when full it would sink to the bottom. To measure the level of liquid the float was lifted by a knurled knob at its upper end, spun to discharge the contents, and the bowl then lowered to float on the liquid. After taking a reading the projecting stem was pushed down into the tank and out of the way. An advantage claimed over "ordinary" floats was that the float could not become water-logged and so give false readings.

The *Lanchester 20 & 28 H.P. Owner's Handbook* says that the main petrol tank, capacity 17 gallons, was filled through a large diameter inlet under the driver's seat cushion. "Our patented level gauge is an open float attached to the lower end of a stem which projects through a guide in the top of the tank."

An improved device for the firing of machine guns

This is the title of Patent No. 129374 of 1919 which covered a device for firing a machine gun from an aeroplane. The firing of the gun was controlled by operating the firing pin by compressed air or gas admitted from a pump by way of a valve so timed as to cause the pin to be actuated in synchronisation with the propeller blades, thus avoiding hitting the propeller. (See pp. 228–229)

Aids to calculations

Fred Lanchester was always prepared to invent a device, like the Isometrograph, which would help with repetitive jobs or which would simplify complicated procedures. The usual mechanical aid to calculation was the slide rule, useful and simple to use for straightforward multiplication and division but less so for calculations concerning the expansion and compression of gases. Lanchester realised that, using a simple geometrical construction, a "radial cursor" could be fitted to a slide rule in such a way that calculations on the expansion and compression of gases could be performed with ease. If the slide is inverted (not turned over) the B and C scales of a slide rule become reciprocal scales to the A and D scales and calculations in which pv is constant can be carried out. But Lanchester proposed using a cursor which could be used at an angle other than a right angle to the

Fig. 6. Petrol tank level gauge (Lanchester 20 & 28 h.p. Owner's Handbook)

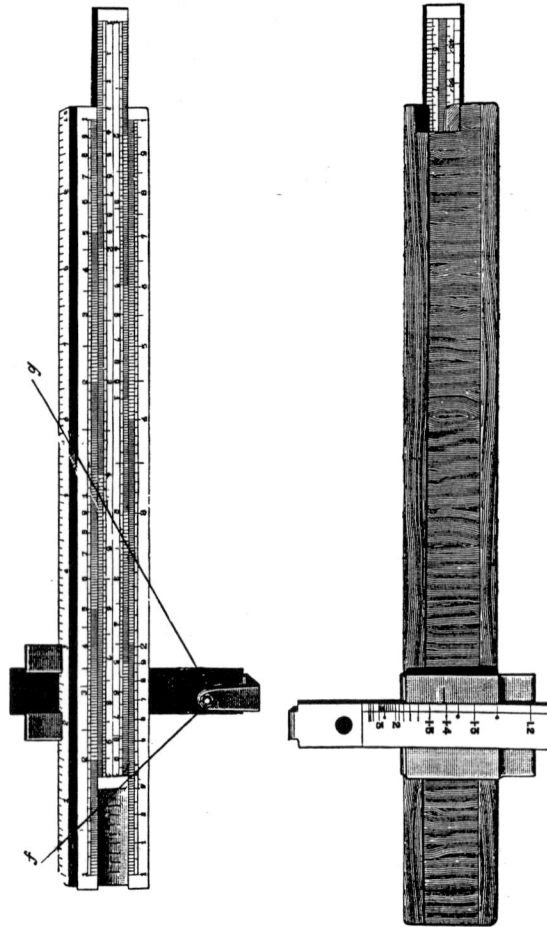

Fig. 7. Lanchester's Radial Cursor attachment to a slide rule

scales by using a pivot which could be set at a position determined by the index n so that the A and (inverted) B scales could be used directly for solving calculations in the form $p_1 v_1^n = p_2 v_2^n$. (Lanchester, 1896). The radial cursor was nevertheless a cumbersome attachment to the slide rule and of specialised use and so, although it was manufactured by Cambridge Scientific Instrument Company, it did not become a commonly-used device.

Another aid to carrying out calculations, *Lanchester's "Potted Logs"*, subtitled *A Concise Tabulation (Slide Rule Auxiliary) for Engineers*, was published in 1938 by Taylor and Francis. By tabulating equal increments of logarithms the first differences of the tabulated numbers bear a constant relation to the number (percentage differences are equal) so that a slide rule can be used to obtain differences of constant accuracy. In *Potted Logs* logarithms were listed to three places and numbers tabulated to seven places which enabled logarithms to seven places to be determined by the use of a slide rule, yet keeping the tables to only five small pages (whereas, as Lanchester comments, with some exaggeration, in his introduction, "a straight table of logarithms of equal capacity would occupy about fifty goodly tomes and require a wheelbarrow for its transport.") The publication of this booklet illustrates Fred Lanchester's way of thoroughly investigating anything which interested him. The tabulation arose from his personal need of anti-log tables of which he could find none of recent publication, so he set to work to calculate his own. At the same time he searched for earlier log tables, discovering that the Chinese invented and made use of logarithms well before their use in the seventeenth century. He suggests that because Henry Briggs (who published tables entitled *Arithmetica Logarithmetica* in 1624) decided on log10 = 1, logarithms to base ten should be called "Briggs".

Spectacle manufacture

By 1918 F.W.Lanchester's expertise as a consulting engineer was well known and so he was called upon by William Gowlland, manufacturing opticians, of Croydon, to investigate problems they were encountering in the manufacture of biconvex spectacle lenses. These were made by polishing one side of the flat blank to the required shape, then turning the blank over, cementing it to the same flat support and polishing the other side. Lanchester suggested that when polishing the second side of the lens a support shaped roughly to that of the first side should be used, thus improving the holding of the blank and also saving cement. He also recommended modifications to the tools and procedures and to the way heat was removed while grinding.

Writing to Lanchester the manufacturers said that, after implementation of only some of his suggestions, they had improved the average of good lenses per batch to 70% and they anticipated achieving better than 80% when all the suggestions were implemented. (LC. Vol. 9–9 to 9–17).

Two railway inventions

Fred Lanchester also had novel ideas about devices for use on railways. In 1917 he, together with the Daimler Company, was granted Patent No. 111170, "Improvements Relating to Railway Train and Coach Propulsion": a means of driving a train consisting of a motor carriage and a number of coaches from a single driving platform which could be at either end of the train. The invention was intended for use with an electro-mechanical drive and the controller, which was rather like a ship's wheel, had positions for starting, charging the battery, electric drive and mechanical drive. These positions were, in the words of the patent specification, "repeated in each half turn for forward and reverse, although forward and reverse are a matter of convention since the system is symmetrical". The controllers were linked by a longitudinal shaft below each car.

Later, in 1929, Lanchester was granted Patent No. 312968, "Improvements in Buffers for Locomotives, Coaches and other Vehicles". In this patent cylindrical buffers formed of moulded india-rubber were attached to the vehicle by a flange, the other end having a saddle-shaped face. The buffers were so arranged that the planes of symmetry of each buffer on one vehicle were disposed at 45° to the horizontal and the two buffer faces on any diagonal line were mutually identical in disposition but at right angles to the two faces in the other diagonal. Each vehicle had identically arranged buffer faces so that any vehicle could be reversed end for end with preservation of the correct engagement of the buffers. If this had "caught on" shunting would be much quieter.

An early moped

Although much of this time was spent on improving light weight internal combustion engines, Fred Lanchester's fertile mind turned at one time to the possibility of a mechanical assistance for a bicycle, but not in the form of a two-stroke petrol engine as in the later moped or motor-scooter. Lanchester's idea was to use a steam engine attached at the rear of the bicycle, and he made sketches and calculations for a charcoal-fired boiler supplying steam to two cylinders, one on each side of the rear wheel. He estimated that a speed of thirty-seven miles per hour could be achieved and that the bicycle would travel thirty-seven miles on one pennyworth of charcoal. He noted that, "the scheme seems *possible* but depends on stationary experiments." We have no record of any such experiments. (LC. SB 13, pp.70–72).

Pianos and player-pianos

Fred Lanchester was a competent player of the piano and player-piano and turned his inventive mind to improvements in these instruments. In dealing with these ideas he estimated the stresses in the hammer (20 lbs maximum force on the pivot) and measured the energy of the hammer blow, and from this calculated the speed of movement of the

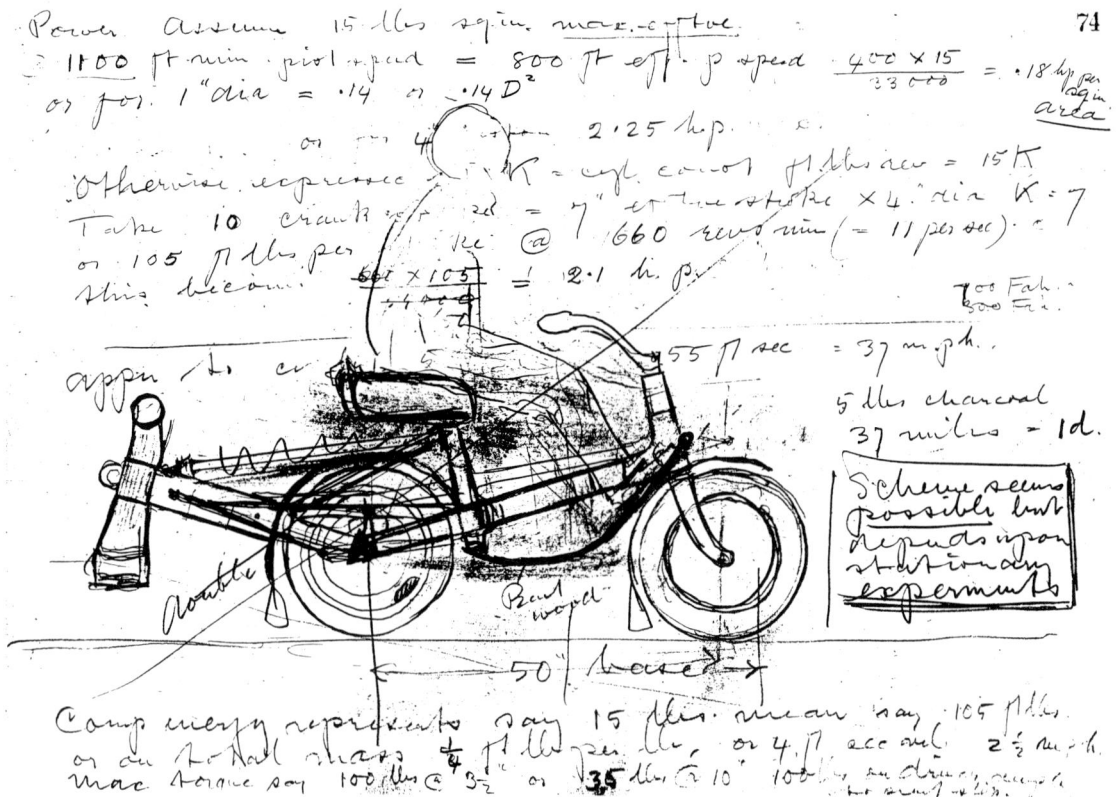

Fig. 8. Is this a moped? Page 74R from Sketch Book 13

hammer. In one of his note-books is his estimate of costs and work times for each part of the player-piano action.

Lanchester took out several patents for his piano and player-piano improvements. The first was Patent No. 9557 of 1912 (improved in Patent No. 13635 of 1913) which provided improvements in the foot operated mechanism and in the operation and control of the "loud" or "sustaining" pedal and was intended to improve the degree of control exercised by the performer especially in execution of forte and sforzando effects. The mechanism was caused to operate by the simultaneous operation of both bellows. This was followed by Patent No. 9610 of 1912, an invention to allow the melody, or any desired part of the harmony, to be suppressed so as to be suitable for accompaniments. The invention provided a means of reducing the pressure in the striking pneumatics so that the note was struck with less force so as to be inaudible, or nearly so. This may seem to be an unlikely requirement, but Mr. Richard Cole, curator of The Musical Museum at Brentford, has said that such a device would help him a lot when he wishes to accompany himself when he plays the Theremin, an electronic instrument played by moving a hand near to a rod. However, since most people who bought player-pianos were less interested in the finer points of making music, and would not want to pay the extra cost of these refinements, they were never put into commercial production.

Two other inventions of 1912 were designed to assist in piano playing. Patent No. 11016 improved the action of the soft pedal mechanism so that in pianissimo passages notes would be struck more evenly and Patent No. 21993 provided means of allowing rapid repeat strokes. We have no evidence of these being incorporated in their instruments by piano manufacturers.

Other pneumatic devices

While engaged in this work Lanchester began to consider the possibility of building a pneumatic organ. There is a sketch of a five manual instrument, but there were problems

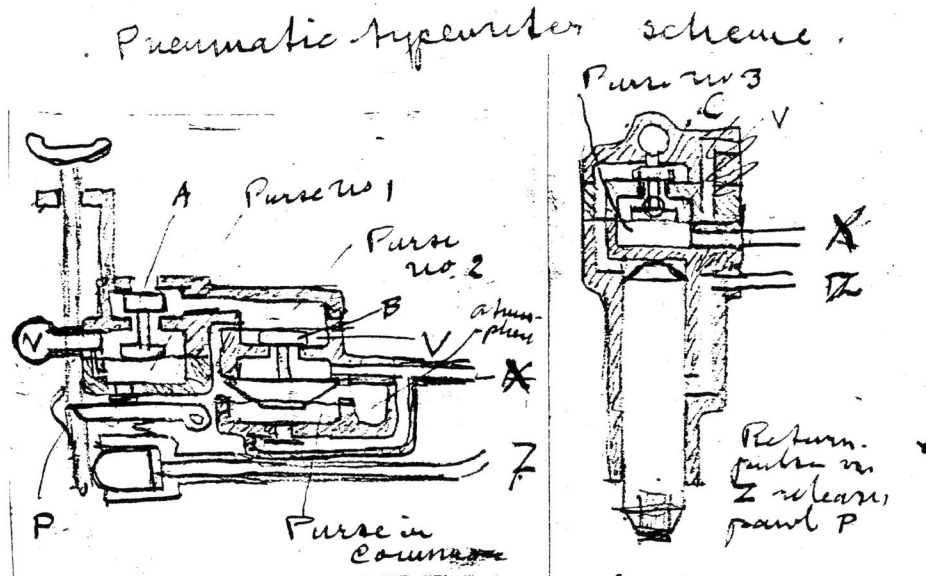

Fig. 9. "Pneumatic Typewriter Scheme". From Sketch Book 7, p.164

in enabling a note to be played on more than one manual at a time, and eventually it seems that so many possible schemes had disadvantages that the idea was dropped. (LC. SB 1, pp.45–47).

Two other machines which got no further than sketches and calculations were no doubt suggested by Lanchester's work on the player-piano.

One was the pneumatic typewriter. Pressing a key vented a "purse", which was otherwise under vacuum, to atmosphere, thereby operating other valves which caused a plunger to operate the character key. (LC. SB 7, p.164).

The other machine was a pneumatic calculator. This would have been a cumbersome machine with two keyboards, one for each of the two numbers to be multiplied. Ducts from each keyboard, which were vented to vacuum on depressing the keys, crossed on a rectangular plan giving precalculated products signalled by secondary ducts. The secondary ducts set appropriate stops limiting the movement of discs moved by pulleys operated by pneumatic bellows to 0, 1, 2, 3, 4, 5, 6, 7, 8, 9, equal arcs of motion. The various discs controlled, through epicyclic gearing, master wheels of which there were as many as there were numerals in the product. (LC. SB 3, pp.77–79).

Heat pump

Heat pumps have been used for many years for the heating and cooling of buildings, but they work by utilising the latent heat of the change from gas to liquid and vice versa as an intermediate fluid is alternately compressed and expanded, as are refrigerators. But as long ago as 1876 a Prussian patented a means of cooling or warming air using the thermodynamic principle that increasing the pressure of air heats it, and vice versa. Four other such machines were patented between 1886 and 1905. Then in 1921 Lanchester patented a heat pump working on this principle which was made up of one or more units, each consisting of two cylinders with pistons arranged to compress air on one side and expand it on the other. Patent No. 158305 describes an ingenious system in which valves, controlled by eccentrics, caused the air to be drawn into one side of a piston, expanded to lower the pressure by about three pounds per square inch and thus cool it, then passed to a heat exchanger to bring it back to normal temperature, and finally recompressed to atmospheric pressure, thereby being warmed.

From the sketch books we find that the main cylinders were 22in. diameter and 20in. stroke, raising the temperature of 8 cubic feet of air by 14°F on each stroke. The pump ran at 80 revolutions per minute and the heat exchanger required a cooling area of 250 square feet. (LC. SB 3, pp.17–19).

Measurement of work

The work a motor is capable of doing can be measured by a dynamometer. While present-day dynamometers can measure the work done under any conditions this was not always so: early dynamometers would only measure the work done at a steady speed. In 1929 Lanchester patented "An Improved Dynamometer for Testing Motive Power Engines" (Patent No. 321328), which was designed to measure the work done by the motor under test when it was accelerating under load. This was done by making the motor accelerate a flywheel to which extra weights could be added to vary the load. An electrically maintained tuning fork was arranged to make a trace on a smoked paper fitted to a drum, the tuning fork being moved axially across the drum by a lead screw driven from the dynamometer shaft. Finding the work done must have involved some very careful measurement of the trace and considerable calculation.

A "hot-and-cold" box and a wheelbarrow

As an alternative to the Dewar (or "Thermos") vacuum flask, and more like some present-day "hot-and-cold" boxes, Lanchester designed an aluminium vessel with one inch thick balsa wood insulation surrounding it. In measuring the effectiveness of this vessel, Lanchester found that two pounds of water took 160 minutes to cool from 15°C to 11°C or two pounds of ice took 26 hours to melt. The flask does not seem to have been developed: we have only the design sketches and experimental results noted in one of Lanchester's sketch books. (LC. SB 6, p.165).

When war came in 1939 Fred Lanchester decided to grow vegetables on his tennis court and concluded that the easiest way was to dig holes in the hard court and fill them with good soil in which to grow the plants. While thus "digging for victory" he found using a barrow hard work so he designed one in which, instead of one wheel, there were two wheels close together, pivoted on an inclined plane so that on being tilted the barrow steered itself. His brother George thought the idea came from roller skates. (LC. Baxter 1/48).

References

Lanchester, F.W. (1896): The Radial cursor: a new addition to the slide rule, *Philosophical Magazine*, January 1896, 52–9

Lanchester, F.W. (1905): The Pendulum accelerometer, an instrument for the direct measurement and recording of acceleration, *Philosophical Magazine*, August 1905, 260–8

Lanchester, F.W. (1910): Tractive effort and acceleration of automobile vehicles on land, air and water, *Proc. I. Auto. E.*, **4**, 123–31

Lanchester, F.W. (1913): Worm gear, *Proc. I. Auto. E.*, **7**, 215–306

Lanchester, F.W. (1916): Worm gear and worm gear mounting, *Proc. I. Auto. E.*, **11**, 79–170

Lanchester, F.W. (1924): Worm gear: its production and efficiency and its application to turbine reduction gearing, *Engineering*, 1st February 1924, 131

Lanchester, F.W. (1939): The Energy balance sheet of the internal combustion engine, *Proc. I. Mech. E.*, **141**, 289–338

Merritt, H.E. (1935): Worm gear performance, *Proc. I. Mech E.*, **129,** 163

Millership, A.J.W.: M.S. note, B.M.S.I. Millership collection.

PATENT APPLICATIONS AND SPECIFICATIONS

compiled by June Dawson

Coverage

The period covered is 1888 to 1949.

The applications and specifications listed are primarily those of Frederick William Lanchester, but for the sake of completeness, I have also included those of Henry Vaughan Lanchester (H.V.L.), George Herbert Lanchester (G.H.L.), Francis Lanchester (F.L.), and their father, Henry Jones Lanchester (H.J.L.). Unless otherwise noted, the patents listed are in the sole name of F.W.Lanchester.

Applications vs Specifications

Items in the list marked with an X are patent applications which were not proceeded with. Copies of the applications were not retained by Lanchester, or the Patent Office, so the Lanchester Library has no copies. However, in many cases, it is clear that patents granted later incorporate some of the content of the earlier applications which were not proceeded with. After the First World War, the Patent Office numbering system changed: patent applications which were not proceeded with were in a different sequence from granted patents. Applications of this type are indicated by the words " – application number".

Items marked V were void, i.e. a patent was not granted by the Patent Office, so copies are not available. Copies of all other specifications listed are held in the Lanchester Collection in the Lanchester Library of Coventry University.

Titles

The titles of most of the early patents begin with the phrase "Improvements in and relating to ...", or "Improvements relating to ...", and these words have been omitted in the list. Titles in this format have been listed under the first significant word of the full title. Titles of applications which were not proceeded with are descriptive of the subject matter of the application, rather than its title.

Number		Title
		1888
16432		Apparatus for ruling parallel equidistant lines and for like purposes
		1889
650		Automatic bookmarker (F.W.L. and H.V.L.)
651	X	Bookmarker and leaf holder (F.W.L. and H.V.L.)
652	X	Bookmarker
5840	X	Bookmarkers (F.W.L. and H.V.L.)
7382		Mechanical contrivance for advertising purposes (F.W.L. and H.V.L.)
12502		Apparatus for governing gas and other motive power engines
19868		Gas-motor engines
		1890
5479		Gas motor engines
14381	X	Governing gas and oil engines
15936	X	Gas engines
19513		Igniting and starting arrangements of gas and hydrocarbon engines
19775		Igniting device for starting gas motor engines
19846		Igniting and starting gear for gas engines

1891

1689	X	Igniters for starting gas engines
2504	X	Wheeled appliance for the feet
4222		Gas engines
4230	X	Gas engine starting arrangements
5072	X	Gas engines
5226		Valve and other arrangements of gas bags for gas engines
5956	X	Wheels of cycles
6094	X	Gas engines
11861		Gas engine starting arrangements
14945		Engine governors
16147	X	Gas engines
16384	X	Bicycles
18711	X	Power hammers, stamps, rock drills
18880	X	Holders for cutting tools
21406		Gas engines
21592	X	Changing word or figure apparatus

1892

4210		Gas engine details
4374		Gas and petroleum engines
5602	X	Starting gas engines
8099	X	Gas engine starters
9703		Velocipede pedals
13943	X	Starting gear for petroleum, etc. engines
14153	X	Friction gearing and friction clutches
18691	X	Friction gear
18696	X	Velocipede pedals axle
18795	X	Gas and oil engines
21978	X	Driving gear and bearings for bicycles
23696	X	Gas and like engines

1893

3538	X	Aerial machines
4628	X	Gas, petroleum, etc. engines (F.W.L. and Dugald Clerk)
11407	X	Gas and oil engines
14175	X	Gas and petrol engines (F.W.L. and Dugald Clerk)
19759	X	Dynamo machines
20152	X	Gas and like engines (F.W.L. and Dugald Clerk)
23168	X	Fireproof floors (H.V.L.)

1894

1579	X	Fireproof floors (H.V.L.)
2876	X	Casings and fittings for electric wires (H.V.L.)
19112	X	Gas and oil motor engines (F.W.L. and Dugald Clerk)
22500	X	Gas engines (F.W.L. and Dugald Clerk)
22946		Gas and like engines (F.W.L. and Dugald Clerk)

1895

6849	X	Gas and like engines (F.W.L. and Dugald Clerk)
15045		Gas and oil motor engines
16548		Photography in colours
18908		Gas and oil motor engines
21984		Driving mechanism of velocipedes and other road vehicles
24335	X	Brake mechanism (F.W.L. and a.n.o.)

1896

5748		Mechanism of power propelled vehicles
5814		Gas and oil motors

7603		Gas and other motive power engines
12744	X	Aerial machines
13960		Fluid pressure engines
14917		Friction clutches
16721	X	Gas and like engines
18829		Friction clutches
21697		Power propelled vehicles
21772		Power propelled road vehicles
22935		Gas and oil motor engines
22941		Keys for shafting
24805		Igniting arrangements for gas and oil motor engines
27590	X	Clutch mechanism
29543	X	Gear for transmitting power

1897

3608		Aerial machines
10043		Power propelled road vehicles
10044		Jointed shaft couplings
13433		Gear cutting machines
18197	X	Gas and oil motor engines
20758	X	Transmitting power
28418	X	Blowing and exhausting apparatus
30592	X	Gas engine starting gear

1898

5613	X	Screw propellers
6272		Gear for the transmission of power
8076		Power propelled road vehicles
10836		Fluid pressure engines
11738		Igniters for gas and oil motor engines
13915	X	Tyres
18959	X	Gas generators
24771		Gas generators
26702	X	Screw propellers and shafts

1899

281		Ignition arrangements for gas and oil motor engines
12243		Power propelled vehicles
12244		Motive power engines
12245		Starting arrangements of gas and oil motors
13467	X	Power propelled vehicles
15081		Inertia governors
15082	X	Motors
20570		Ignition arrangements for gas and oil engines
23341	X	Oil engines
23342		Packing of piston rods and plungers
24395	X	Speed gearing

1900

4806		Gear for the transmission of power
4951	X	Power propelled vehicles
7909		Power propelled road vehicles
9983		Power propelled vehicles
13965	X	Oil motor engines
15037	X	Oil motor engines
16454	X	Change speed gearing
17001	X	Gear cutting
17503		Cooling arrangements of gas and oil motor engines
18060		Gear for the transmission of power

1901

1793	X	Self propelled vehicles
2844	X	Oil motor engines
5904	X	Tyres
7739		Road motor vehicles
10500		Internal combustion engines
11130		Chimney pot or ventilator (H.J.L.)
13491		Grinding cylindrical parts for roller bearings and like purposes
17032		Thrust bearings
18354		Cooling arrangements of internal combustion engines
19070		Motor vehicles
19159		Fluid pressure engines, in part relating to power propelled vehicles
23227	X	Roofing for marquees, etc.
23412		Lifting jacks
23500		Clutch and brake mechanism of motor vehicles
23501		Activating mechanism for self-propelled vehicles
23617		Change speed mechanism for self propelled vehicles
23692		Shaft couplings
23693		Method of fixing dust-proof and oil-retaining screens for bearings and such like purposes
25121		Dashboards and wind protection for self-propelled vehicles

1902

10375	X	Engine governors
10376	X	Gas and oil engines
12235		Level gauges for tanks containing fluid
12715		Ignition arrangements for gas and oil motor engines

A.D. 1902. JULY 2. N°. 14,792.

LANCHESTER'S COMPLETE SPECIFICATION.   '1 SHEET'

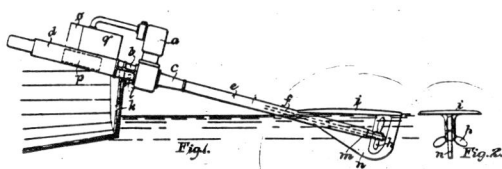

Fig.1.   Fig.2.

[This Drawing is a reproduction of the Original on a reduced scale.]

N° 14,792   A.D. 1902

Date of Application, 2nd July, 1902

Complete Specification Left, 2nd May, 1903—Accepted, 11th June, 1903

PROVISIONAL SPECIFICATION.

"Improvements in Mechanism for the Propulsion of Boats"

I, FREDERICK WILLIAM LANCHESTER, of 53, Hagley Road, Edgbaston, Birmingham, Engineer, do hereby declare the nature of this invention to be as follows :—

This invention relates to improvements in mechanism for the propulsion of boats and refers more particularly to an improved appliance for the propulsion of boats of small dimensions or as an auxiliary means of propulsion for sailing boats and larger craft.

This invention has for its object to provide a self contained propelling apparatus of a convenient and portable form which may be applied to the propulsion of any boat or vessel and may be transferred from one vessel to another with little or no preparation.

This invention consists in brief in the application of a self contained motor and screw propeller to an oar or scull of special construction, the propeller shaft being arranged longitudinally and the motor with its appurtenances being built into the handle portion of the oar and the screw propeller being protected and having its immersion regulated by the oar blade. It will be seen, of course, that the motor will be entirely self contained in the case of a petrol engine, but where the motor is of another type it may have its power supplied by means of flexible connection.

In one mode of carrying my present invention into effect I construct a high speed oil motor of small dimensions and forming part of the crank chamber I arrange tubular extensions co-axial with the crank shaft so as to form sockets into which the two halves of an oar may be fixed and the blade portion of the oar is bored its entire length to accommodate the propeller shaft which at the one end is coupled to the crank shaft of the motor and at the other end carries a screw propeller, the blade of the oar being bent to one side to allow sufficient clearance. The handle portion of the oar may conveniently be made hollow and contain the battery and coil for the purposes of ignition. The oil reservoir may also be accommodated in the handle portion or in a special saddle shaped tank suitably fitted. The angle at which the blade is set to the shank of the oar is such as will allow it to lie approximately flat on the water when the oar is in position in its rowlock and it will be understood from the foregoing description that the thrust of the screw is longitudinal to the oar and the position of the oar when in use is over the stern of the boat which is being propelled.

Small sea boats are customarily arranged with a rudimentary rowlock for the purpose of propulsion with a single oar, the said rudimentary rowlock usually taking the form of a semi-circular bite in the transom. I preferably arrange to make use of such rowlock and the oar is arranged to drop into the said bite and has a plate which rests on the top of the transom to prevent the oar from having a tendency to rotate axially which it otherwise would have owing to the reaction of the propelling torque.

A suitable projection or button is also provided on the under-side of the oar to

[*Price 8d.*]

12847	X	Gas and oil engines
12848	X	Power propelled road vehicles
13445	X	Screw propellers
13446	X	Internal combustion engines
13760	X	Gas and oil engines
14792		Mechanism for the propulsion of boats
22768	X	Gas and oil engines
23744		Turbine motors
26407		Brake mechanism of power propelled road vehicles
26408	X	Power propelled road vehicles
27286		Mud-guards and dash-boards for power propelled road vehicles
28743	X	Road vehicles

1903

477		Vaporising arrangements of oil motor engines
478	X	Gas and oil engines
566	X	Gas and oil engines
659	X	Spindle bearings
1618	X	Gas and oil engines
1619	X	Internal combustion engines
1722	X	Power propelled road vehicles
2422		Steering and controlling mechanism of power propelled vehicles
2625	X	Gas and oil engines
5105	X	Power propelled road vehicles
5106	X	Screw propellers
7064		Electrical connections
9850	X	Gas and oil engines
9937	X	Internal combustion engines
11159	X	Electrical generators
12599	X	Motor car fittings
13258	X	Pocket card filing device
17259	X	Mechanically propelled road vehicles
18189	X	Internal combustion engines
18632	X	Road vehicles
18959		Road wheels of power propelled vehicles
20968	X	Gas and oil engines
21228	X	Internal combustion engines
21229	X	Gas and oil engines
22234	X	Gas and oil engines
22936		Condensers and coolers for power-propelled vehicles (F.W.L. and John Vernon Pugh)
26514	X	Screw propellers

1904

878	X	Motors and driving gear
879	X	Reciprocating engines
1342	X	Starters for internal combustion engines
1343	X	Power propelled vehicles
10929	X	Gas and oil engines
12979	X	Road vehicles
16864		Power propelled road vehicles
16865		Motor cars
16959		Clutch, brake and change gear mechanism of power propelled vehicles and vessels
17047		Governing arrangements of internal combustion engines
17048	X	Oil valves
17049	X	Vehicle drive gear
17061	X	Engine ignition
23839	X	Transmission gear

| 24725 | | Motors and driving gear suitable for motor road vehicles and the like |
| 24726 | X | Fluid pressure engines |

1905

186	X	Lubricating engines
187	X	Brakes
7949		Mechanically propelled road vehicles
7950	X	Motor cars
17935		Aerodromes
20149	V	Sound reproducing equipment

1906

1340	X	Concrete building construction
27136		Goggles for the use of motorists and cyclists (G.H.L.)
28236	X	Vehicle speed mechanism

1907

5835		Change speed mechanism for power propelled vehicles
9413		Aerodromes
9413A		Driving mechanism for aeronautical or marine propellers
9451	X	Internal combustion engines ignition
9452		Valve gear of internal combustion engines
10140	X	Clutch gear
16427		Grinding and polishing of rollers and bushes for roller bearings and like purposes
22431	X	Projectiles

1908

3605	X	Sound recording discs
8178		Gear actuating mechanism for motor vehicles (G.H.L. and Lanchester Motor Co.)
17273	V	Motor vehicle steering mechanism
18065	X	Imparting stability to aerodrome in flight

1909

2145		Valve gear of internal combustion engines
2428	X	Internal combustion engines
2958		Manufacture of piston packings (F.W.L., James Courthope Peache and Willans and Robinson Ltd.)
8849		Steering of flying machines or aerodromes
9259		Motor vehicles (F.W.L. and Daimler Motor Co.)
10421	X	Speed reduction gearing
10422		Flying machines
18303	X	Aeronautical machines
18384		Alighting mechanism for flying machines
19108	X	Internal combustion engines
23106		Internal combustion engines
22325	X	Intergearing mechanism
23820	V	Internal combustion engine valves
23867	V	Propelling vehicles
24113	X	Lubrication of vehicles
30192		Construction of bridges (H.V.L.)

1910

5676	X	Aerial machines
13310		Resilient or spring arrangements
14700		Valve gear for internal combustion engines
15851	X	Automobile brake
18630		Detachable wheels for road vehicles
18631		Propulsion of vehicles by combustion of prime mover and electrical storage
19452	X	Internal combustion engines

21139		High speed reciprocating engines
26123	X	Piano-players
26399	X	Electric lamps

1911

3682		High speed reciprocating engines
8162	X	Power transmission change speed gear
8745	X	Power transmission change speed gear
9714	X	Screw propellers
9873		Two-stroke cycle internal combustion engines more particularly for aeronautical machines
15810	X	Worm gearing
15811		Method and apparatus for cutting worm gear
16815	X	Window cleaning apparatus
21322		Power and electrical installation of power propelled vehicles
21360		Alighting mechanism for aeronautical machines
21361	X	Automobile vehicle bonnets
22502	X	Aeronautical machines
22545	X	Change speed gear box
23271		Stabilisation of flying machines
26038		Balancing of reciprocating engines
26777		Balancing of reciprocating engines
28538	X	Change speed gear

1912

3604	X	Internal combustion engine reversing gear
7041	X	Rectifying gear wheels
9557		Mechanical pianoforte players
9610		Mechanical piano players and like instruments
11016		Mechanism of pianofortes and mechanical players
12418		Starting mechanism for internal combustion engines employed for the propulsion of automobile road vehicles
12522	X	Pianofortes
13284	X	Starting internal combustion engines
21993		Actions of pianofortes and like instruments
25074	X	Measuring efficiency of transmission
27375		Starting of internal combustion engines
27398		Balancing of reciprocating engines
28517		Detachable road wheels for automobile vehicles (F.W.L. and Daimler Motor Co.)
28817		Dash pot or retarding apparatus

1913

179	X	Musical instruments
346	X	Musical instruments, pneumatic players
7162	X	Projectiles
9427		Change speed gear for motor vehicles
13634	X	Internal combustion engines
13635		Mechanical pianoforte players
15423	X	Lighting, etc. for vehicles
15424	X	Vehicle brake
15429	X	Vehicle windows
15430	X	Shaft-coupling joint
15544		Folding seats, tables and the like (G.H.L. and Lanchester Motor Co.)
16060	X	Engine starters
16832	X	Engine starters
16974	V	Wheels
17026	X	Engine starters

17636	X	Vehicle windows
21444	X	Projectiles
22508		Means for storing and supplying liquid fuel to internal combustion engines
23349	X	Musical instruments
24210	X	Musical instruments
25275	X	Musical instruments
29407		Cylinders of internal combustion engines (F.W.L. and Daimler Motor Co.)
29461	X	Vessel propellers

1914

1563	X	Vehicle windows
2320	X	Pianofortes
2321	X	Pianofortes
3255	X	Spur gearing
5030	X	Ignition
13418	X	Chairs, etc. (G.H.L.)
17450	X	Signalling
17451	X	Bearings
17544	X	Rotating cylinder engines
17545	X	Rotating cylinder engines
17546	X	Radial engines
17775	X	Aeronautical engines
18064	X	Telephones
18146	X	Fuel supply
20638	X	Guns
20905		Explosive projectiles and in fuse mechanism therefor

[Second Edition.]

Nº 20,905   A.D. 1914

Date of Application, 13th Oct., 1914
Complete Specification Left, 13th May, 1915—Accepted, 13th Oct., 1915

PROVISIONAL SPECIFICATION.

Improvements in Explosive Projectiles and in Fuse Mechanism therefor.

I, FREDERICK WILLIAM LANCHESTER, of 53, Hagley Road, Edgbaston, Birmingham, Engineer, do hereby declare the nature of this invention to be as follows :—

The present invention relates to improvements in explosive projectiles and in fuse mechanism therefor more particularly adapted to the conditions encountered in attack on aircraft in which it is necessary that the fuse should be sensitive to influences other than direct impact.

The present invention consists in brief in the provision in an explosive projectile of a pneumatic chamber with passages communicating to the outside 10 by which the pressure within the chamber is caused to depend upon, and to vary according to, the circumstances attending the flight of the projectile, and in a pressure sensitive mechanism in communication with the said chamber by which the cap or primer is fired or the burster charge otherwise detonated or exploded.

15 The present invention also consists in the provision in an explosive projectile of a pneumatic piston device actuated by pressure trapped on the one side and relieved on the other side either by the stopping or obstruction of apertures in the nose of the projectile by impact with canvas fabric or other yielding substance, or by encounter with gas of relatively low density, or by both causes 20 in combination.

The present invention also consists in an explosive projectile comprising a pneumatically operated piston for the firing of the charge in accordance with the preceding paragraph, in so arranging the piston that it will operate alternatively by its own inertia instead of by pneumatic pressure in the event of 25 heavy impact.

The present invention also consists in the provision in an explosive projectile of a split spring ring as a guard to prevent the movement of a pneumatically operated piston or inertia piece the same being adapted to be thrown out of action in such a manner as to allow the charge to be exploded by centrifugal 30 force when the projectile is in rapid rotation.

The present invention further consists in the explosive projectile and fuse mechanism and relay detonating device hereinafter described.

In one mode of carrying the present invention into effect as applied to an exploding projectile or shell for aircraft attack the burster charge is arranged 35 and contained within the rear portion of the projectile and is held in position by a thin plate and screw ring forming the wall of the chamber facing forward ; immediately behind the thin plate is arranged a charge of fulminate or detonator ; the space forward of the thin plate or bulkhead is made cylindrical and fitted with a piston fitting approximately the piston being made hollow to 40 include a considerable volume of air ; the space forward of the piston communicates by a number of holes of small diameter with the region in front of the projectile a hollow or depression being formed with a projecting lip preferably furnished with a sharp edge this depression with its lip form a crater of

[Price 6d.]

A.D. 1914. OCT. 13, No. 20,905.
LANCHESTER'S COMPLETE SPECIFICATION.   (1 SHEET)

(2nd Edition)

[This Drawing is a full size reproduction of the Original.]

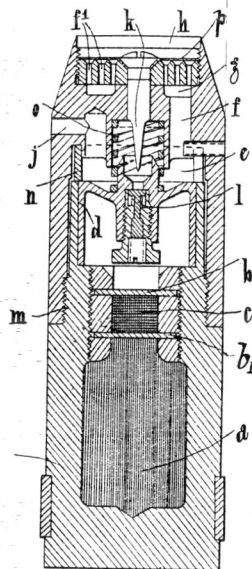

Maiby & Sons, Photo-Litho.

1915

4411	X	Signalling
4412	X	Telephones
73366		Electrical transmission for automobile vehicles (F.W.L., Daimler Motor Co. and John Lloyd Milligan)
9376	X	Fuses
10517	X	Storing liquid fuel
15066	X	Note carriers
17367	X	Engine coolers

1917

105373	Valve operating gear of petrol and other internal combustion engines
108970	Motor vehicles (F.W.L. and Daimler Motor Co.)
109529	Motor vehicles (F.W.L. and Daimler Motor Co.)
110182	Fuel feed valve and valve operating mechanism for internal combustion engines
111170	Railway coach and train propulsion (F.W.L. and Daimler Motor Co.)

1918

119140	Clutch and change gear mechanism more especially applicable to motor vehicles
119339	Construction of tent and field hospitals, depots and like purposes
121243	Lubrication of high speed reciprocating engines

1919

121765	System of springing for automobile vehicles
121767	Springing of road or rail vehicles
127620	Mechanism for the receiving and launching of aeroplanes at sea
127674	Internal combustion engines
127680	Floating mines and submarine vessels

127,859

PATENT   SPECIFICATION

Application Date, June 4, 1917. No. 7998/17.

Complete Left, Dec. 4, 1917.

Complete Accepted, June 19, 1919.

PROVISIONAL SPECIFICATION.

Improvements in Floating Mines and Submarine Defences.

I, FREDERICK WILLIAM LANCHESTER, of 41, Bedford Square, London, Engineer, do hereby declare the nature of this invention to be as follows:—

The present invention relates to improvements in floating mines and submarine defences and refers more particularly to improvements in mines of the submerged type adapted to the better regulation of the depth of submersion and in the application of such mines to the purpose of submarine defence either against submarine or surface vessels.

The present invention has for its object the employment of the energy of wave motion both for the purpose of securing a regulated depth of submersion and in order to control a group or line of mines in some desired relation as to position or formation.

The present invention consists in brief in a construction of mine and system of control in which the differential motion of deep sea waves at different degrees of depth is employed to act on horizontal or inclined fin surfaces to vary the effective buoyancy for the control of depth or give rise to forces having horizontal components for the purpose of ordering the trend of motion of a submerged body, or for both purposes in combination.

The present invention further consists in floating mines of the submerged type in which fins or blades are adapted to be acted upon by differential wave motion in a means of control by a pressure sensitive device such as to render the mine self-regulating as to its depth of immersion.

The present invention further consists in a method of sustaining in extended formation a group or line of submerged floating mines by means of mines or buoys in which mechanism arranged to operate in the manner aforesaid is adapted to give rise to forces having a horizontal component.

In one mode of constructing a submarine mine in accordance with the present invention I arrange two component bodies or vessels connected by a chain of appropriate length. The one vessel which may be the mine proper (containing the explosive and fuse mechanism) is made specifically heavier than the water; the other vessel which contains the depth regulating mechanism is made specifically lighter than the water, the whole combination including the chain coupling being preferably of slightly less density than the water in which the mine is to be employed so that there is in sum a small positive degree of buoyancy. The natural condition of flotation of such a combination which would be assumed in smooth water is that the vessel containing the depth

[Price 6d.]

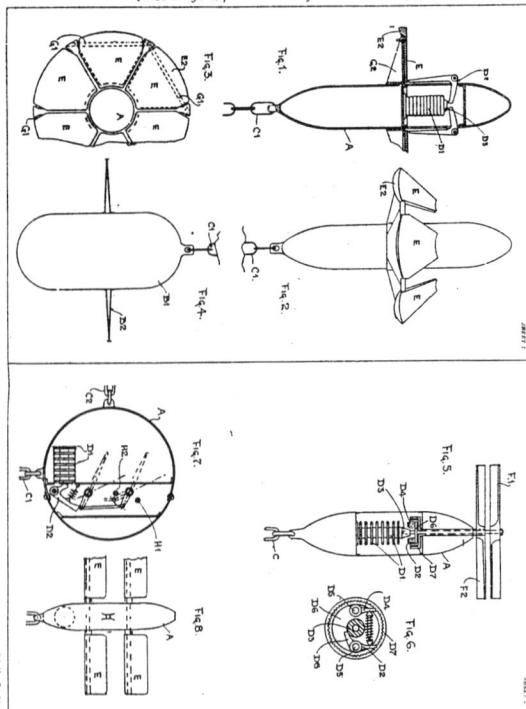

[This Drawing is a reproduction of the Original on a reduced scale.]

127686	Hot air engines
127687	Heat engines for the production of motive power
127859	Floating mines and submarine defences
128686	Internal combustion engines
129026	Fuse mechanism for explosive asphyxiating or incendiary projectiles for the use of artillery
129374	Device for controlling the firing of machine guns
129727	Lateral plumb indicator for aircraft
130025	Attachment of aeronautical propellers and in bosses therefor
130070	Aeroplanes and apparatus for launching and receiving same
131325	Internal combustion engines
131618	Fireplaces (F.L.)
132363	Suspension and power transmission mechanism of automobile vehicles
133156	Brakes and brake mechanisms of power propelled vehicles
135301	Internal combustion engines
135582	Construction of crankshaft and of bearings therefor
135756	Pneumatic tyres and wheels for road vehicles
136335	Means and mechanism for the elimination of torsional vibration in high speed engines, shafting, crankshafts and the like
136341	Vehicle bonnets (G.H.L. and Lanchester Motor Co.)

1920

137125	Spring suspension of vehicles
137679	Fuel supply systems for internal combustion engines (G.H.L. and Lanchester Motor Co.)
139630	Automatically locking lever or winch mechanism (G.H.L. and Lanchester Motor Co.)
140594	Vehicle wheels (G.H.L. and Lanchester Motor Co.)

PATENT SPECIFICATION

Application Date : Apr. 21, 1920.   No. 11,069/20.   **165,584**

Complete Left : Dec. 14, 1920.

Complete Accepted : July 7, 1921.

PROVISIONAL SPECIFICATION.

Improvements relating to Folding Seats for Motor Vehicles.

We, GEORGE HERBERT LANCHESTER, a subject of the King of Great Britain and Ireland, and THE LANCHESTER MOTOR COMPANY LIMITED, a company duly organised under the laws of Great Britain and Ireland, all of Armourer Mills, Montgomery Street, in the City of Birmingham, do hereby declare the nature of this invention to be as follows :—

This invention has for its object to provide an improved motor vehicle seat of the kind which can be folded into a recess in the floor of the vehicle.

In carrying the invention into effect, a seat portion has hinged to its rear edge a back portion, and in conjunction with each of the side edges of the seat are arranged two crossed struts which are pivotally connected together at or near their centre. The upper end of one of the struts is hinged to a bell-crank or like extension from the pivot between the seat and back portions, and the lower end of this strut is arranged to slide in a horizontal guide slot in a shallow well formed in the floor of the vehicle. The other strut has at its upper end a pivot pin arranged to slide in a guide slot extending along a side edge of the seat and arranged near the front end of that edge, whilst the lower end is pivoted in the aforesaid well. To the upper end of the second strut is also hinged one end of a tie

which at its opposite end is pivoted to a point on one edge of the back portion at a suitable distance from the hinge connection between the back and seat portions. The structure is locked in the operative position by the engagement of the pivot pin of the second strut with a notch in the inner end of its guide slot. It will be understood that a similar arrangement of parts is provided at each side of the seat.

When not required for use the seat portion is slightly raised to disengage the pivot pin at the upper end of one of the struts at each side of the seat from the recess which it engages, and the back is then folded down on to the seat. At the same time the whole of the system closes down into the recess in the vehicle floor, the movement being permitted by the various hinge joints and the sliding connections above mentioned. The back then lies flush with and forms part of the floor.

To bring the seat into use the above operation is reversed by lifting the back portion out of the floor.

By this invention we are able to provide an extra removable seat in a motor vehicle in a very simple and convenient manner.

Dated this 20th day of April, 1920.

MARKS & CLERK.

COMPLETE SPECIFICATION.

Improvements relating to Folding Seats for Motor Vehicles.

We, GEORGE HERBERT LANCHESTER, a subject of the King of Great Britain and Ireland, and THE LANCHESTER MOTOR COMPANY LIMITED, a company duly organised under the laws of Great Britain and Ireland, all of Armourer Mills,

[*Price 1/-*]

165,584 COMPLETE SPECIFICATION   1 SHEET

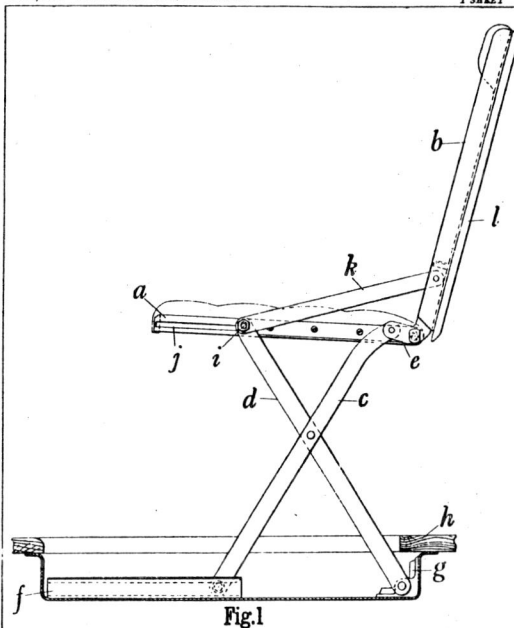

[*This Drawing is a reproduction of the Original on a reduced scale.*]

Fig.1

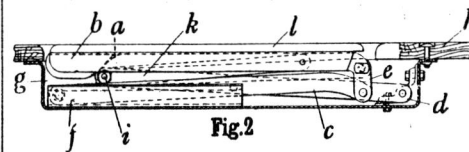

Fig.2

PATENT SPECIFICATION

Application Date : Dec. 8, 1920. No. 34,744 / 20.   **176,514**

Complete Left : Oct. 10, 1921.

Complete Accepted : Mar. 8. 1922.

PROVISIONAL SPECIFICATION.

Improvements in the Transmission and Brake Mechanism of Power Propelled Road Vehicles.

I, FREDERICK WILLIAM LANCHESTER, 41, Bedford Square, London, W.C. 1, Engineer, English, do hereby declare the nature of this invention to be as follows:—

The present invention relates to improvements in the transmission and brake mechanism of power propelled road vehicles and refers more particularly to a simplified form of frictional transmission more especially applicable to light weight vehicles to run on common roads.

The present invention consists in brief in frictional transmission gear in which the tyre or rim of the vehicle wheel itself is made the agent of transmission and in which the driving road wheels carry their load through the medium of jockey pulleys mounted on the chassis and riding on the top of, or on the upper part of the wheel, and not through an axle bearing as heretofore customary.

The present invention further consists in friction transmission gear in accordance with the preceding paragraph in which two driving road wheels are provided mounted on a single axle and in which each of the said wheels engages with two jockey pulleys mounted on the chassis, one of which is a driver and the other of which is an idler, and in a means for disengaging the road wheels from the driving jockey pulleys arranged to act on the axle so that the motor and the car shall be disconnected and run free the one of the other.

The present invention further consists in friction transmission gear in accordance with the preceding paragraph in which means are provided for applying a brake to the road wheels by a continued movement of the axle after disengagement with the driving jockey pulleys.

The present invention further consists in friction transmission gear in accordance with one of the preceding paragraphs in which the jockey and idler pulleys are furnished in duplicate or multiple sets so that the road wheels with their axle may be readily transferred from one set to another set in order to vary the gear ratio.

The present invention further consists in friction transmission gear in accordance with one or more of the preceding paragraphs in which the driving jockey pulleys are furnished with floating bands or tyres whereby any frictional slip as between the jockey pulley and the road wheel takes place between the jockey pulley and its floating band or tyre.

In one mode of carrying the present invention into effect as applied to a light two seat road automobile a body is constructed (conveniently of timber) of box form and of sufficient width to contain the driving wheels and their axle, the rear portion of the box body being closed in above and open beneath. The axle consists of a length of shafting to which are keyed at opposite ends, the hubs of the two driving wheels, or alternatively the said hubs may run free on the axle and be provided with collars by which the wheel gauge is determined. In the upper portion of the box body or chassis four jockey pulleys are arranged adapted to bear two and two on the road wheels aforesaid, in such manner that the weight of the body, or a portion of the weight of the body, is borne by the road wheels through the medium of the jockey pulleys.

[Price 1/-]

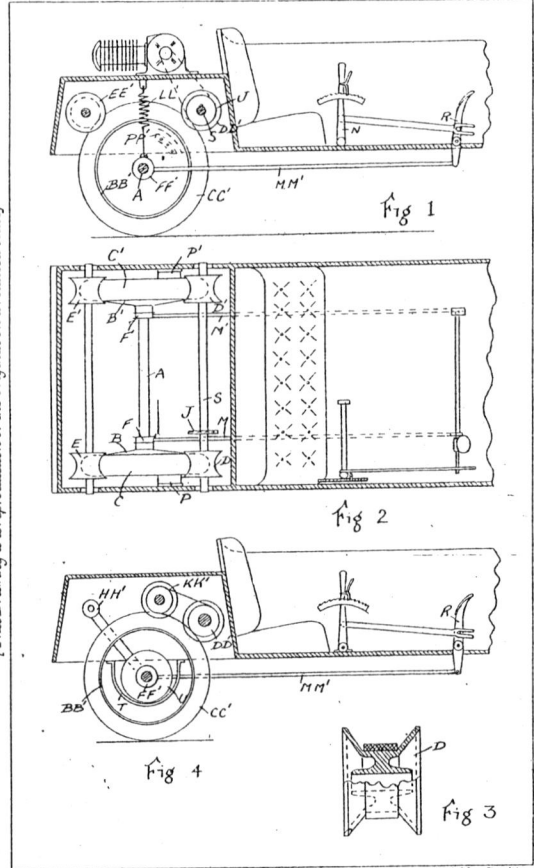

[This Drawing is a reproduction of the Original on a reduced scale.]

176,514 COMPLETE SPECIFICATION   1 SHEET

Fig 1

Fig 2

Fig 4

Fig 3

Malby & Sons, Photo-Litho.

141096	Pneumatic tyres and wheels for road vehicles
141931	Gear changing mechanisms for mechanically propelled vehicles (G.H.L. and Lanchester Motor Co.)
145193	Construction and roofing of buildings for exhibitions and like purposes
150009	Actions of pianofortes

1921

157524	Resilient wheels for road vehicles
158305	Apparatus for heating or cooling and ventilating buildings and ships
165584	Folding seats for automobiles (G.H.L. and Lanchester Motor Co.)
166017	Transmission and brake mechanism of power propelled vehicles
167616	Under carriages of power propelled vehicles

1922

174758	Magneto ignition of internal combustion engines (F.W.L. and Daimler Motor Co.)
176514	Transmission and brake mechanism of power propelled road vehicles
180383	Suspension or springing of power propelled vehicles
187364	Propulsion mechanism of vessels and of automobile vehicles of amphibious type
188541	Steering of automobile vehicles and like purposes

1923

195684	Road wheels of automobile vehicles
196643	Transmission and brake control mechanism of power propelled vehicles
198527	Road wheels for automobile vehicles
198707	Brake mechanism of power propelled vehicles (F.W.L. and Daimler Motor Co.)
200875	Motor road vehicle brakes (G.H.L. and Lanchester Motor Co.)

1924

209837	Transmission and change gear mechanism of motor vehicles

216599	Two-to-one driving mechanism of cam shafts or eccentric shafts in internal combustion engines operating on the four stroke cycle (F.W.L. and Daimler Motor Co.)
216600	Pistons of internal combustion engines (F.W.L. and Daimler Motor Co.)
224932	Brake and brake mechanism of power propelled veicles

1925

229722	Springing of motor road vehicles
229842	High speed reciprocating engines (F.W.L. and Edward Claude Shakespeare Clench (of Aster Engineering Co. (1913) Ltd.))
229889	Vehicle wheels (G.H.L. and Lanchester Motor Co.)
232661	Lubrication system of reciprocating engines
232772	Motor vehicle chassis (F.W.L., Daimler Motor Co. and Joseph Dixon)
238921	Coating internally of hollow articles with metal (F.W.L., Daimler Motor Co. and George Needle)
240540	Valve and valve gear of internal combustion engines
241965	Automobile vehicles
243304	Balancing of reciprocating engines
243786	Steering mechanism of automobile vehicles
244007	Steering mechanism of automobile vehicles

1926

253173	Operating mechanism for change gear in automobile vehicles
261536	Pedal mechanism for controlling power propelled vehicles (F.W.L. and Lanchester's Laboratories Ltd.)
261874	Steering mechanism of automobile road vehicles (F.W.L. and Daimler Motor Co.)
262174	High speed multicylinder motive power engines
262242	Construction of automobile vehicles

1927

263555	Steering mechanism of power propelled vehicles (F.W.L. and Lanchester's Laboratories Ltd.)
263904	Front wheel brake mechanism for automobile road vehicles (F.W.L. and Lanchester's Laboratories Ltd.)
266782	Construction and winding of armatures for electric generators or motors (F.W.L. and Lanchester's Laboratories Ltd.)
267190	Cylinder cooling of internal combustion engines (F.W.L. and Lanchester's Laboratories Ltd.)
268859	Armatures of electro-motors and generators (F.W.L. and Lanchester's Laboratories Ltd.)
271133	Petrol feed of carburettors for internal combustion engines (F.W.L. and Lanchester's Laboratories Ltd.)
273960	Variable speed gears for use on motor vehicles and for like purposes (G.H.L., Fred Lee Orcutt and Gear Grinding Co.)
282537	Window and window lifting mechanism more especially for motor vehicles (F.W.L. and Lanchester's Laboratories Ltd.)

1928

282880	Dynamo electric machines (F.W.L. and Lanchester's Laboratories Ltd.)
292651	Means of effecting an angular adjustment or movement between two connected parts (G.H.L.)
293148	Bonnets of automobile vehicles (F.W.L. and Lanchester's Laboratories Ltd.)
298687	Valve and valve mechanisms of internal combustion engines (F.W.L. and Daimler Motor Co.)
299552	Spindles and spindle mountings for the abrasive wheels of precision grinders (F.W.L. and Lanchester's Laboratories Ltd.)
300667	Change gear mechanism more especially applicable to automobile vehicles (F.W.L. and Lanchester's Laboratories Ltd.)

1929

304406		Steering mechanism of power propelled road vehicles (F.W.L. and Lanchester's Laboratories Ltd.)
305971		Mechanism for the rectification of gear wheels (F.W.L. and Lanchester's Laboratories Ltd.)
309931		Improved damper for eliminating torsional vibration in the shafts of high speed engines and machinery
312103		Ignition mechanism for internal combustion engines
312968		Buffers for locomotives, coaches and other vehicles
317169		Field magnets of microphones, loud speakers and the like
317339		Improved telephonic loudspeaker
318279		Moving coil microphones
320647		Moving coil loud speakers
321328		Dynamometer for testing motive power engines
321967		Gramophones and like sound reproducing apparatus

1930

325784		Loud speaker and like horns
327145		Sound reproduction apparatus
327790		Loud speakers and sound emission apparatus
337811		Mechanism for the reproduction of sound
338656		Appliance for the convenient storage and handling of pins, gramophone and other needles and the like
339307		Warning devices for motor vehicles
339949		Apparatus for the reproduction of music, speech and the like
341574		Wireless reception and apparatus therefor

1931

356246		Construction of field magnets for moving coil loud speakers (F.W.L., Ralph Howard Pearsall and Lanchester's Laboratories Ltd.)
356367		Mechanism for the reproduction of recorded music, speech and the like (F.W.L. and Lanchester's Laboratories Ltd.)
361841		Mechanism for the reproduction of sound

1932

364683		Diaphragm mountings for acoustical apparatus
368708		Loud speakers of the moving coil type
382952		Multiple ratio transformers
383376		Telephonic receivers or speakers of the moving coil type

1933

388534		Stuffing box substitutes (G.H.L. and Daimler Motor Co.)
390573		Motor vehicles (G.H.L., Birmingham Small Arms Co. and Algernon Edward Berriman)
390584		Telephonic receivers and loud speakers
11859	X	Electric amplifiers—application number

1934

409821		Clutch and change gear mechanism of power propelled vehicles (F.W.L., British-Thomson Houston Co. and Arthur Primrose Young)
410589		Electromagnetic clutches (F.W.L. and British-Thomson Houston Co.)
34038	X	Epicyclic geared transmissions—application number
35764	X	Epicyclic geared transmissions—application number

1936

34498	X	Airflow over aircraft wing surfaces—application number

1937

1796	X	Anti-icing of aircraft wing surfaces—application number
2458	X	Explosive bombs—application number

13898	X	Aircraft engine control—application number
13899	X	Cooling the charge of supercharged internal combustion engines—application number
19715	X	Valve mechanism of internal combustion engines—application number
32436	X	Construction, motive power and suspension of automobiles—application number

1938

| 175 | X | Airbrakes for aircraft—application number |

1939

509786		Suspension of motor road vehicles
6509	X	Method and means of dropping bombs—application number
6510	X	Maintaining pressure in aircraft at high altitude—application number
6511	X	Induction systems of aircraft internal combustion engines—application number

1943

| 552560 | | Trigger mechanism for automatic firearms (G.H.L., John Gordon Remington and Ronald Wallace Buckler) |

1944

561575		Electric torches (G.H.L. and Sterling Engineering Co.)
561988		Stocks for automatic firearms (G.H.L. and Sterling Engineering Co.)
562694		Trigger mechanism for automatic firearms (G.H.L. and Sterling Engineering Co.)

1945

567230		Automatic firearms (G.H.L. and Anthony Edgar Somers)
572440		Automatic firearms (G.H.L. and Anthony Edgar Somers)
572632		Firearms (G.H.L. and Anthony Edgar Somers)
572700		Manufacture of rifles, machine guns and sub-machine guns (G.H.L. and Anthony Edgar Somers)
573029		Trigger mechanisms for automatic firearms (G.H.L.)

1946

| 579219 | | Portable platforms (F.L.) |

1947

| 594056 | | Optical projectors for images carried on a film strip (G.H.L.) |
| 594057 | | Optical projectors for images carried on a film strip (G.H.L.) |

1948

| 610917 | | Rotary earth boring bits (G.H.L. and Sterling Engineering Co.) |

1949

| 628266 | | Mobile well drilling apparatus (G.H.L., John Hetherington and Sterling Engineering Co.) |
| 628267 | | Multiple-way valves (G.H.L., John Hetherington and Sterling Engineering Co.) |

BIBLIOGRAPHY

"Publications" are those publications which are known to be the words of Frederick William Lanchester. This section includes his written contributions to discussions, and reports of discussions of papers given by him.

"Reports and Discussions" list the publications which contain reports of Lanchester's words, some of which may be verbatim reports of his contributions to discussions.

PUBLICATIONS

1888

The Isometrograph, for hatching, shading and geometric design
no publisher, no date (probably 1888)
Pamphlet describing an instrument devised by F.W.L. for drawing office use (see p.233)

1895

Radial cursor: a new addition to the slide-rule
Communication III, in *Philosophical Magazine*, January 1896, pp.52–59; "A proposed addition to the slide rule to improve its calculating capabilities" read before the Physical Society, 25th October 1895 (see pp.239–240)

1905

The pendulum accelerometer, an instrument for the direct measurement and recording of acceleration
Philosophical Magazine, vol. 10, August 1905, pp.260–68.
Communicated to the Physical Society, and read 16th June 1905 (see p.235)

1907

Aerodynamics, constituting the first volume of a complete work on aerial flight
Constable & Co., 1907
Second volume, entitled *Aerodonetics . . .* was published in 1908. Second ed. published by Constable in 1909
The horse-power of the petrol motor in its relation to bore, stroke and weight
Proc. Institution of Automobile Engineers, session 1906–7, vol. 1, pp.155–220
Paper given at I.A.E., including appendices
Written communication on **The effect of size on the thermal efficiency of motors**, by H.L. Callendar
Proc. Institution of Automobile Engineers, session 1907–8, vol. 2, pp.254–70

1908

Aerodonetics, constituting the second volume of a complete work on aerial flight
Constable & Co., 1908; second ed. published by Constable in 1910. First volume was published in 1907; this volume includes appendices on the theory and application of the gyroscope, on the flight of projectiles, etc.
Note: Lanchester Collection copy ("author's copy no. 2") includes three sheets of typescript "Errata, Jan 1910, retyped Mch 1942" by F.W.L.

The rating of petrol engines: Report of the Rating Committee of the Institution [of Automobile Engineers] on the Society of Motor Manufacturers and Traders proposals for rating petrol engines

Proc. Institution of Automobile Engineers, session 1908–9, vol. 3, pp.37–40; F.W.L., although a member of the Committee had been unable to attend its meetings, so communicated his views in a separate document, published on pp.40–46; he took part in the subsequent discussion, published on pp.96–98

Some problems peculiar to the design of the automobile

Proc. Institution of Automobile Engineers, session 1907–8, vol. 2, pp.187–202; includes discussion

Paper read before I.A.E. 11th March 1908

1909

Aerial flight

Journal of the Royal Society of Arts, Proc. of the Society, no. 2971, vol. 57, pp.997–1008, 29th October 1909

Text of the first Cantor Lecture, given at Royal Society of Arts on 26th April 1909; F.W.L. did not supply copies of the full text of the second and third Cantor Lectures, given on 3rd and 10th May 1909 respectively, so they were never published; they were reported in *Engineering* (see below)

Aerodynamics, constituting the first volume of a complete work on aerial flight

Second ed., Constable & Co., 1909; first ed. published by Constable in 1907

"Except for correction of errata, revisions of a minor character, and some few references to later work, no change has been made in the work since the publication of the first edition"

Aerodynamik: ein Gesamtwerk uber das Fliegen aus dem englischen Ubersetzt, von C. und A.Runge; erster Band

Leipzig, Berlin, B.G.Teubner, 1909

German translation of first ed. of *Aerodynamics*

Mechanical flight

Times Engineering Supplement, 3rd March 1909, and 7th April 1909

Two short articles

Notes on the resistance of planes in the normal and tangential presentation and on the resistance of ichthyoid bodies

Advisory Committee for Aeronautics, 1909

Its Reports and Memoranda, No. 15

Preliminary report by Mr. F.W.Lanchester on the weights and h.p. of petrol motors for aeronautical purposes

Advisory Committee for Aeronautics, 1909

Its Reports and Memoranda, No. 10

The Wright and Voisin types of flying machines: a comparison

Aeronautical Journal, vol. 13, January 1909, pp.4–12

F.W.L.'s reply to subsequent correspondence is in *Aeronautical Journal*, vol. 13, April 1909, pp.58–9

1910

Aerodonetics, constituting the second volume of a complete work on aerial flight

Second ed., Constable & Co., 1910, first ed. published by Constable in 1908

F.W.L. note: "It differs in no important respect from the first edition, 1908"

Factors that have contributed to the advance of automobile engineering, and which control the development of the self-propelled vehicle

Proc. Institution of Automobile Engineers, session 1910–11, vol. 5, pp.33–5

Presidential address to I.A.E. given 12th October 1910

Tractive effort and acceleration of automobile vehicles on land, air and water
Proc. Institution of Automobile Engineers, session 1909–10, vol. 4, pp.123–66
Paper read to I.A.E. 12th January 1910

1911

Aerodynamics, constituting the first volume of a complete work on aerial flight
Third ed., Constable & Co., 1911; first ed. published by Constable in 1907
"Except for correction of errata, revisions of a minor character, and some few references to later work, no change has been made in the work since the publication of the first edition"

Aerodynamik: ein Gesamtwerk uber das Fliegen aus dem englischen Übersetzt, von C. und A.Runge: zweiter Band: Aerodonetik
Leipzig, Berlin, B.G.Teubner, 1911
German translation of first ed. of *Aerodonetics*

1913

Aerofoils of high aspect ratio
Advisory Committee for Aeronautics, 1913
Its Reports and Memoranda, 1913–14, No. 109

Catastrophic instability
Advisory Committee for Aeronautics, 1913
Its Reports and Memoranda, No. 114

Internal-combustion motors for railways: the internal combustion engine applied to railway locomotion, including an account of the Daimler Company's equipment
Engineering, 21st November 1913, pp.701–3

Surface cooling and skin friction
Advisory Committee for Aeronautics, 1913
Its Reports and Memoranda, No. 71

Worm gear
Proc. Institution of Automobile Engineers, session 1912–13, vol. 7, pp.215–41; pp.242–306 include appendices, discussion and correspondence
Paper read to I.A.E. 12th February 1913

1914

Aerodynamique, le vol aerien
Paris, Gauthier-Villars, 1914
French translation, by C.Benoit, of first volume of second ed. of *Aerodynamics*

Aircraft in warfare
Engineering, various issues from 4th September 1914 to 25th December 1914
Sixteen articles corresponding to the chapters of *Aircraft in warfare: the dawn of the fourth arm* published in 1916

Engine balancing
Proc. Institution of Automobile Engineers, session 1913–14, vol. 8, pp.193–259; pp.260–78 includes discussion
Paper given to I.A.E. 11th March 1914

The flying machine from an engineering standpoint
Proc. Institution of Civil Engineers, vol. 98, session 1913–14, part 4, pp.3–80; pp.80–96 include appendices and discussion; reprinted by Constable in 1915
James Forrest Lecture given to I.C.E. 5th May 1914

A note by Mr. Lanchester relating to the effect of wash on the tail of an aeroplane, as touching on Report No. T. 368
Advisory Committee for Aeronautics, 1914
Cyclostyled report to A.C.A., Technical report T. 389; probably early edition of A.C.A. Reports and Memoranda No. 161

Note on the stability of the flying machine as affected by considerations relating to propulsion
Advisory Committee for Aeronautics, 1914
Its Reports and Memoranda, No. 115

Note relating to the effect of wash on the tail of an aeroplane as touching Report No. 111
Advisory Committee for Aeronautics, 1914
Its Reports and Memoranda, No. 161

1915

A contribution to the theory of propulsion and the screw propeller
Trans. Institution of Naval Architects, vol. 57, 1915, pp.98–116
Paper given at Spring Meeting of I.N.A. 25th March 1915

The cylinder cooling of internal combustion engines, more especially as concerning automobile practice
Proc. Institution of Automobile Engineers, session 1915–16. vol. 10, pp.57–145; pp.146–58 includes discussion
Paper given to I.A.E. 15th December 1915

The flying machine: the aerofoil in the light of theory and experiment
Proc. Institution of Automobile Engineers, session 1914–15, vol. 9, pp.171–227; pp.228–59 include discussion; also reprinted by the I.A.E. in 1915, with *The Screw propeller*, under the title *The Flying machine: two papers*
Paper given to I.A.E. 10th March 1915

Gyroscopic action and propeller vibration
Advisory Committee for Aeronautics, 1915
Its Reports and Memoranda, No. 185
Earlier cyclostyled edition produced as A.C.A. Technical report T. 466

A note on the subject of skin friction
Advisory Committee for Aeronautics, 1915
Its Reports and Memoranda, No. 149

Report on high altitude flying and the development and improvement of the aeronautical motor
Advisory Committee for Aeronautics, 1915
Its Reports and Memoranda, No. 220
Earlier cyclostyled edition produced as A.C.A. Technical report T. 559

The screw propeller
Proc. Institution of Automobile Engineers, session 1914–15, vol. 9, pp.263–324; pp.325–54 include discussion and written communications; also reprinted by the I.A.E. in 1915 with *The Flying machine: the aerofoil*, under the title *The Flying machine: two papers ...*
Paper given at I.A.E. 14th April 1914

1916

A discussion on reports T. 580 and T. 585
Advisory Committee for Aeronautics, 1916
Cyclostyled paper to A.C.A. on its reports, T. 580: The Effects of v1 variation on unstable flow in the case of certain aerofoils suitable for airscrew design; and T. 585: Interference between the propeller and the body of an aeroplane

Aerodonetique
Paris, Gauthier-Villars, 1916
French translation, by C.Benoit, of second ed. of *Aerodonetics*

The Air Board
Land and Water, 15th June 1916, pp.17–18

Air defence problems and fallacies: Air Ministry or Board of Aeronautics
Land and Water, 27th April 1916, pp.13–14
Land and Water, 4th May 1916, pp.13–14

Air defence problems and fallacies: the failure of the Derby Committee
Land and Water, 20th April 1916, pp.12–14
Aircraft in warfare: the dawn of the fourth arm
Constable & Co., 1916
Aircraft policy, and the Zeppelin menace from the national standpoint
Land and Water, 23rd March 1916, pp.17–18, and 30th March 1916, pp.13–14
Two articles
Britain's aeroplane policy
Land and Water, 10th February 1916, pp.13–14, and 17th February 1916, pp.15–16
Two articles
Development of the military aeroplane: the question of size
Engineering, 3rd March 1916, pp.212–14
F.W.L.'s replies to subsequent correspondence in *Engineering*, 28th April 1916, pp.409, and 12th May 1916, pp.460
The flying machine from an engineering standpoint ... including A Discussion concerning the theory of sustenation and the expenditure of power in flight, a paper presented at a meeting of the International Engineering Congress, San Francisco, 1915
Constable & Co., 1916
Reprint of James Forrest Lecture given to Institution of Civil Engineers, 5th May 1914
The interference of superposed planes
Advisory Committee for Aeronautics, [report] T. 708a, 1916
Cyclostyled report to A.C.A.; continuation of Report T. 708
Note
Advisory Committee for Aeronautics, [memorandum] T. 673/1, 1916
Cyclostyled memo. to A.C.A. Subcommittee on the Capabilities of an Aeroplane in relation to its size.
Note on the conditions which determine the cyclical component round an aerofoil
A.C.A.?, 1916
Rise and fall of the French Air Ministry
Land and Water, 11th May 1916, pp.10–12
The sea and the air: the future of the Air Board: a lesson from naval history
Land and Water, 29th June 1916, pp.13–14
The so-called "air muddle", and some of its exponents
Land and Water, 18th May 1916, pp.17–18
A study concerning the best proportion for a stream-line body
Engineering, 19th May 1916, pp.465
Then and now: the difficulties of the pioneer designers
Autocar, 4th November 1916, pp.478–80
Torsional vibrations of the tail of an aeroplane
Advisory Committee for Aeronautics, 1916
Its Reports and Memoranda, No. 276
Worm gear and worm gear mounting
Proc. of Institution of Automobile Engineers, session 1916–17, vol. 11, pp.79–139; pp.140–72 includes discussion
Paper given to I.A.E. 13th December 1916

1917

Aerodonetics, constituting the second volume of a complete work on aerial flight
Constable & Co., 1908; reprint of second ed. published by Constable in 1910. First volume was published in 1907; this volume includes appendices on the theory and application of the gyroscope, on the flight of projectiles, etc.

The Air Ministry
Flying, 21st November 1917, pp.275–6, and 5th December 1917, p.307–8
Two articles
The air raid of July 7th
Flying, 18th, July 1917, pp.479–80
A campaign of slander: a few facts
Flying, 15th August 1917, pp.51–4
The defence of London: II: formation flying
Flying, 1st August 1917, pp.19–20
Flying as affected by the wind
Flying, 26th September 1917, pp.147–8
Flying as affected by the wind: the economic aspect
Flying, 12th September 1917, pp.115–16
***Flying* reprisals competition**
Flying, 5th September 1917, pp.99–100
Formation flying
Flying, 25th July 1917, pp.3–4
Formation flying as taught by nature
Flying, 8th August 1917, pp.35–7
The foundation stones
Flying, 19th December 1917, pp.354–56
The handicap of the wind
Flying, 29th August 1917, pp.83–4
Industrial engineering: present position and post-war outlook
Constable & Co., 1917
Presidential address to the Junior Institution of Engineers, 11th December 1916
Memorandum on the use of aluminium alloy sheet in place of fabric for aeroplane wings, etc.
Advisory Committee for Aeronautics, 1917
Its Reports and Memoranda (new series), No. 359
The military aspects of reprisals
Flying, 17th October, 1917, pp.209
Moonbeams and the larger lunacy
Flying, 10th October 1917, pp.186–187
Vertical climb or direct lift: the chaser of the future
Flying, 14th November 1917, pp.259–60

1918

Aerodynamics, constituting the first volume of a complete work on aerial flight
Fourth ed., Constable & Co., 1918; first ed. published by Constable in 1907
"Except for correction of errata, revisions of a minor character, and some few references to later work, no change has been made in the work since the publication of the first edition"
High altitude flying
Advisory Committee for Aeronautics, 1918
Its Reports and Memoranda, No. 534
Industrial economics during and after the war
Journal of the Junior Institution of Engineers, vol. 29, part 2, 1918, pp.2–32
Presidential address to J.I.E. 22nd January 1918

1919

The aftermath of the war: economic and industrial problems
Birmingham, Buckler & Webb, 1919
Presidential Address to the Vesey Club, 20th January 1919

1921

Relativity
Engineering, 2nd April 1921, pp.477–9, and 29th April 1921, pp.514–16
Spur gear erosion
Engineering, 17th June 1921, pp.733–4

1922

An investigation of certain aspects of the two-stroke engine for automobile vehicles
Proc. Institution of Automobile Engineers, session 1921–2, vol. 16, pp.3–36; pp.37–51 include discussion
Paper given to I.A.E. 11th January 1922

1924

Epicyclic gears, by F.W.Lanchester and G.H.Lanchester
Midlands Branch of Institution of Automobile Engineers, 1924
Paper given by F.W.L. to Midlands branch of I.A.E. 26th April 1923
Worm gear: its production and efficiency and its application to turbine reduction gearing
Engineering, 1st February 1924, pp.131–2

1926

Sustentation in flight
Journal of the Royal Aeronautical Society, 1926, pp.589–606
Wilbur Wright Memorial Lecture given to R. Aero Soc., 27th May 1926

1928

Automobile steering gear: problems and mechanism
Proc. Institution of Automobile Engineers, session 1927–8, vol. 22, pp.727–51; pp.752–71 include discussion and communications
Paper given to I.A.E. 6th March 1928
Automobile steering gear: problems and mechanism
Automobile Engineer, vol. 18, no. 239, March 1928, pp.102–6; with appendix by R.H.Pearsall, pp.106–8
India-rubber as an auxiliary to suspension
Proc. Institution of Automobile Engineers, session 1927–8, vol. 22, pp.575–94; pp.595–9 include discussion
Paper given at joint meeting of I.A.E. and Institute of Rubber Industry January 1928
India-rubber as an auxiliary to suspension
Automobile Engineer, vol. 18, no. 242, June 1928, pp.226–30

1929

Communication on **Automatic spark advance**, by H.S.Rowell and C.G.Williams
Proc. Institution of Automobile Engineers, session 1928–9, vol. 23, pp.623–4
Written communication on paper given to I.A.E. 9th April 1929
Coil ignition
Proc. Institution of Automobile Engineers, session 1928–9, vol. 23, pp.214–44; pp.245–66 include discussion
Paper given to I.A.E. 7th January 1929

1933

Air-gap transformer and choke
Journal of the Institution of Electrical Engineers, vol. 73, no. 442, October 1933, pp.413–18

The anti-dazzle problem: a solution to the difficulty?: details of a system which has been in use for five years
Autocar, 5th May 1933, pp.740–1

Pioneer work: a reply from Dr. Lanchester
Autocar, 17th March 1933, pp.431
F.W.L.'s reply to articles on detachable wheels published in *Autocar*, 17th February 1933, p.247, and 3rd March 1933, p.324

1934

Directional fixity and the transitory visual impression or fleeting image
Birmingham, the author, 1934
Typescript edition of 12 copies

Discontinuities in the normal field of vision
Journal of Anatomy, vol. 68, 1934, part 2, pp.224–38

The expansion of the universe
Engineering, 5th January 1934, pp.3–4

1935

The Centenarian: a Lakeland story told in verse, by Paul Netherton-Herries
Birmingham, Cornish Brothers Ltd, 1935
Paul Netherton-Herries was the pseudonym F.W.L. used for his verse publications

Relativity: an elementary explanation of the space-time relations as established by Minkowski, and a discussion of gravitational theory based thereon
Constable & Co., 1935

1936

Electrical dimensions and units
Engineering, 25th September 1936, pp.347–50, and 2nd October 1936, pp.376–7;
Engineering, 6th November 1936, pp.498–9, 4th December 1936, p.622, 11th December 1936, pp.647–8, and 19th February 1937, pp.201–2 include F.W.L.'s replies to subsequent correspondence.
Two-part article of paper given at British Association meeting 14th September 1936. A reprint contains all above, plus an appendix of matters omitted from the *Engineering* articles

A King's prayer, and other poems, by Paul Netherton-Herries
The author, 1936
Paul Netherton-Herries was the pseudonym F.W.L. used for his verse publications

Motor car suspension and independent springing
Proc. Institution of Automobile Engineers, session 1935–6, vol. 30, pp.668–726; pp.727–62 include discussion
Also published in *Journal of the Institution of Automobile Engineers*, vol. 4, part 7, April 1936, pp.17–77, without discussion
Paper given to I.A.E. in April 1936

Communication on **Needle roller bearings**, by C.H.Smith
Proc. Institution of Automobile Engineers, session 1935–6, vol. 30, pp.359–60
Written communication on paper given to I.A.E. 21st January 1936

The part played by skin-friction in aeronautics
Nature, vol. 138, 12th December 1936, pp.1022–3
Synopsis by F.W.L. of his Thomas Hawksley Lecture

A study of the dynamo based on dimensional theory
The Engineer, 27th November 1936, pp.578–80

The theory of dimensions, and its application for engineers
Crosby Lockwood and Son Ltd., 1936

1937

At high altitude
Journal of the Royal Aeronautical Society, vol. 41, no. 317, May 1937, pp.388–400, and no. 318, June 1937, pp.437–66

The future of the airship
Engineering, vol. 143, no. 3724, 28 May 1937, pp.613–14

The gas engine and after
London, Institution of Mechanical Engineers, 1937
24th Thomas Hawksley Lecture given to Institution of Mechanical Engineers, 5th November 1937

The Lanchesters
The Autocar, 13th August 1937

The part played by skin-friction in aeronautics
Journal of the Royal Aeronautical Society, vol. 41, no. 314, February 1937, pp.68–113; pp.113–31 include discussion; vol. 41, no. 316, April 1937, pp.322–3 contain errata
Lecture read to R. Aero. Soc. 12th November 1936

The treatment of problems in engineering by dimensional theory
Proc. Institution of Automobile Engineers, session 1936–7, vol. 31, pp.545–71; pp.572–92 include discussion; also *Journal of the I.A.E.*, vol. 5, no. 5, February 1937, pp.17–43
Paper given to I.A.E. in February 1937

Reply by F.W.Lanchester to criticism [by R.C.Porter] of his Theory of dimensions, in *Engineer*, vol. 163, no. 4245, 21st May 1937, pp.598
Engineer, vol. 163, no. 4248, 11th June 19376, pp.679–80; "further polemic" in no. 4249, 18th June 1937, pp.696–7

1938

Energy balance sheet of the gas engine
Engineering, vol. 145, no. 3756, 7th January 1938, pp.4–5; no. 3758, 21st January 1938, pp.55–6; no. 3760, 4th February 1938, pp.114–17; no. 3762, 18th February 1938, p.188; no. 3767, 25th March 1938, pp.319–21; no. 3770, 15th April 1938, pp.407–10; no. 3774, 13th May 1938, pp.522–5; and supplementary data and corrections in vol. 146, no. 3785, 29th July 1938, pp.137–8

Independent springing
Proc. Institution of Automobile Engineers, session 1937–38, vol. 32, pp.409–29; pp.438–69 include discussion and communications (pp.466–8 by F.W.L.)
Two papers, with the same title, one by F.W.L. (pp.409–29), the other (pp.430–7) by G.H.L. given to I.A.E. 4th January 1938

Lanchester's potted logs: a concise tabulation (slide rule auxiliary) for engineers
Taylor and Francis, 1938
First edition of Part 1 of **Potted logs**

Second annual lecture on "Span"
Manchester, Butterly & Wood Ltd, 1938
Lecture given to Manchester Association of Engineers, 9th November 1938
also published in *Trans. of the Manchester Association of Engineers*, session 1938–9, part 3, pp.59–99; pp.99–100 include discussion

1939

The Energy balance sheet of the internal combustion engine
Proc. Institution of Mechanical Engineers, vol. 141, 1939, pp.289–338
Paper given at Institution of Mechanical Engineers, 17th March 1939; "Amplification and revision of *The energy balance sheet of the gas engine*"

The Energy balance sheet of the internal combustion engine

Engineering, vol. 147, no. 3820, 31st March 1939, pp.388–9, and no. 3822, 14th April 1939, pp.448–50; *Engineering*, vol. 147, no. 3821, 7th April 1939, p.411 include Erratum

Abridged version of paper given at Institution of Mechanical Engineers, 17th March 1939

Exhaust efflux propulsion

Flight, 16th November 1939, pp.396a–d, 23rd November 1939, supp. p.d and p.417, 7th December 1939, p.460, and 14th December 1939, pp.495–6; *Flight*, 28th December 1939, pp.526–7 includes correspondence and F.W.L.'s replies thereto

Four articles; fifth was published in 1941, see below

Flying boat entrance lines

Engineering, vol. 148, no. 3944, 15th September 1939, pp.298–9

Lanchester's potted logs: a concise tabulation (slide rule auxiliary) for engineers, part 2: the natural logarithms and other tables

Taylor and Francis, 1939

First edition of Part 2 of **Potted logs**

Lanchester's potted logs, parts 1 and 2

[revised second ed.], Taylor and Francis Ltd., 1939

The maximum range of flight

Engineering, vol. 148, no. 3840, 18th August 1939, pp.203–5, and no. 3841, 25th August 1939, p.217

Rocket efficiency

Flight, 28th December 1939, pp.526–7

Replies to correspondence re. his articles on **Exhaust efflux propulsion**

Tests of steel air-raid shelters

Engineering, vol. 147, no. 3815, 24th February 1939, pp.216–17

Letter from F.W.L.

The unaccounted drag of aircraft

Engineering, vol. 148, no. 3848, 13th October 1939, pp.403–6

Engineering, no. 3855, 24th November 1939, pp.590 includes F.W.L.'s reply to subsequent correspondence

1940

"Boost": long distance flying and take-off problem

Flight, vol. 38, no. 1650, 8th August 1940, pp.109–10

Effect of windage on projectiles, with special reference to aeroplane gunnery

Flight, vol. 38, no. 1634, 5th September 1940, pp.186f–h

Formation flying

Flight, vol. 37, no. 1631, 28th March 1940, pp.286e–f, and pp.287–9

The rear engine: its pros and cons

Autocar, vol. 84, 22nd March 1940, pp.300–2

1941

Contra-props: recollections of early considerations by Advisory Committee for Aeronautics; a pioneer's 1907 patent; suggestions for further research

Flight, 11 December 1941, vol. 40, 1720, pp.418–9

Exhaust efflux propulsion – V

Flight, vol. 39, no. 1686, 17th April 1941, pp.285–6

Last of five articles in *Flight*, earlier ones were published in 1939, see above

Exhaust efflux propulsion

Mechanical Engineer, vol. 63, no. 9, September 1941, pp.672–3

The musical scale
> Birmingham, the author, 1941
> Cyclostyled edition limited to 30 copies

The propeller. How many blades?
> *Journal of the Royal Aeronautical Society*, vol. 45, no. 368, August 1941, pp.267–74

Rocket propulsion
> *Flight*, vol. 40, no. 1702, 7th August 1941, pp.78–80, and no. 1706, 4th September 1941, pp.135–6

1942

Contra-props
> *Flight*, vol. 40, no. 1720, 11th December 1941, pp.418–19

Lanchester's law
> *Engineering*, vol. 154, no. 3998, 28th August 1942, p.174
> Comment by F.W.L. on article with same title by "Sirius" in *Engineering*, vol. 154, no. 3996, 14th August 1942, pp.123–4

Relativity and radiation
> Birmingham, the author, 1942
> Typescript copy, prepared for printers in Lanchester Collection; never published commercially

1944

Preface, and **Appendix** [to **The aftermath of the war ...**, the presidential address to the Vesey Club, 1919]
> Typescript, so although bound with a reprint of the address, may not have been published.
> Appendix is **Currency and inflation**, comments on the Cunliffe Report (first interim report of the Committee on Currency and Foreign Exchanges after the War)

1945

The musical scale
> Second ed., Birmingham, the author, 1945
> Edition limited to 60 copies, numbered 81–100 (?); includes amendments from first ed., 1941

REPORTS AND DISCUSSION

1894

Report of **The soaring of birds and the possibilities of mechanical flight**
> *Reports of Meetings: General Meeting, of the Birmingham Natural History and Philosophical Society*, 19th June 1894
> F.W.L. gave an address with this title (see p.271)

1905

Communication on **The pendulum accelerometer, an instrument for the direct measurement and recording of acceleration**, by F.W.L., communicated by C.V.Boys
> *Proc. Physical Society*, vol. 19, 1905, pp.691–701

1907

Discussion of **Accessibility and cleanliness, and the best means of attaining them**, by F.L.Martineau
> *Proc. Institution of Automobile Engineers*, session 1906–7, vol. 1, pp.50–1
> Contribution to the discussion of Martineau's paper read at I.A.E. 16th January 1907

REPORTS OF MEETINGS.

GENERAL MEETING, JUNE 19, 1894.

The President, Professor T. W. Bridge, M.A., in the chair.

The following were elected members of the Society: Mrs. Bevan, Messrs. J. W. Clulow, F. H. Pepper, William B. Avery, H. G. Manle, F.G.S., H. H. Bloomer, F. E. Lloyd Dembski, B.A. (Lond.), and T. Swale Vincent, M.B. (Lond.), &c.

The following were proposed for membership: Miss M. E. Hutchinson, Miss B. J. Hay, and Mr. Fred Tibbetts.

Mr. F. W. Manchester delivered an Address entitled "The Soaring of Birds and the Possibilites of Mechanical Flight." The lecturer divided his subject into three parts: (1) Aëro-dynamics, dealing with the principles involved in obtaining a sufficiently unyielding support from so mobile a fluid as air; (2) aërodromics (Langley), dealing with the maintenance of equilibrium and with the problems and mechanism relating to directing and controlling such; and (3) sources of power available and the means of utilising the same. The lecturer proceeded to show that at the present time scientific investigation as to the phenomena of flying has to a great extent taken the place of over-sanguine and imaginative speculation; and although we cannot say definitely whether man is any nearer flying than he ever was, it is certain that we are beginning to have an insight into the underlying principles. The lecturer went on to describe the action of the soaring bird, and to distinguish between soaring and gliding. Reference was made to Professor Langley's well-known investigations on the behaviour of plane surfaces driven through the air by a whirling table, and his conclusions on the results of these experiments. It was shown that one horse-power was sufficient to maintain over 200 lb. weight in the air, and at the same time to propel it at a rate of forty-five miles per hour, which the lecturer graphically illustrated by reference to the fact that a modern torpedo boat has sufficient engine power on board to support it in the air and to propel it at a considerable speed if its mechanical flight were possible. The general principles of flight were discussed and illustrated by lantern slides, specimens, and models.

A vote of thanks was proposed by Professor Bridge and seconded by Mr. P. L. Gray. In the discussion which followed, Professor Allen, and Messrs. R. W. Chase, Dugald Clarke, Hookham and others took part.

October, 1894.   F

Discussion of **Future of automobilism**, by R.E.Crompton
Proc. Institution of Automobile Engineers, session 1907–8, vol. 2, pp.17–19
Contribution to the discussion of Crompton's paper read at I.A.E. 20th November 1907

Discussion of **The motor car considered as a carriage**, by William Gilchrist
Proc. Institution of Automobile Engineers, session 1906–7, vol. 1, pp.307–11
Contribution to the discussion of Gilchrist's paper read at I.A.E. 12th June 1907

Discussion of **The principles of the carburetter as determined by exhaust gas analysis**, by Dugald Clerk
Proc. Institution of Automobile Engineers, session 1907–8, vol. 2, pp.69–72
Contribution to the discussion of Clerk's paper read at I.A.E. 11th December 1907

1908

Discussion of **The effect of motors on roads**, by W.J.Taylor and Douglas Mackenzie; and **Notes on wheel diameters**, by R.E.Crompton
Proc. Institution of Automobile Engineers, session 1907–8, vol. 2, pp.399–400
Contribution to the discussion of two papers read at I.A.E. 13th May 1908

Discussion of **The effect of size and speed upon the performance of an internal combustion engine**, by B.Hopkinson
Proc. Institution of Automobile Engineers, session 1908–9, vol. 3, pp.239–42
Contribution to the discussion of paper read to I.A.E.

Discussion of **Front driving, with special reference to electric and hydraulic transmission**, by H.S.Hele-Shaw, R.W.Harvey Bailey and J.S.Critchley
Proc. Institution of Automobile Engineers, session 1907–8, vol. 2, pp.157–63

Contribution to the discussion of the papers read at I.A.E. 22nd January and 12th February 1908

Report of **The laws of flight**

Engineering, 11th September 1908, pp.331–4

Engineering, 25th September, 1908, pp.423–6

Report of paper read before Section G of the British Association at Dublin, 8th September 1908

Report of **Motor-car design: some problems peculiar to the design of the automobile**

Engineering, 13th March 1908, pp352–5, and 20th March 1908, pp.384–7

Report of paper read at Institution of Automobile Engineers, 11th March 1908

Discussion of **The transmission of power from the engine to the road wheels in motor vehicles**, by L.A.Legros

Proc. Institution of Automobile Engineers, session 1908–9, vol. 3, pp.362–4

Contribution to the discussion of paper read to I.A.E.

Discussion of **The weight of motor cars and motor car parts**, by M.O'Gorman

Proc. Institution of Automobile Engineers, session 1908–9, vol. 3, pp.135–8

Contribution to the discussion of paper read to I.A.E.

1909

Report of **Aerial flight**

Engineering, 30th April 1909, pp.599–601

Engineering, 14th May 1909, pp.671–2

Report of the three Cantor Lectures, given at the Royal Society of Arts, 26th April, 3rd May, and 10th May 1909

Discussion of **Chains for power transmission**, by A.S.Hill

Proc. Institution of Automobile Engineers, session 1909–10, vol. 4, pp.342–3

Contribution to the discussion of paper read to I.A.E.

Report of **Dynamics of flight**

Engineering, 30th April 1909, pp.599–601

Report of lecture given at Institution of Automobile Engineers, April 1908

Summary of **The flight of birds**

The Engineer, 19th February 1909, pp.198–9, and 26th February 1909, pp.225–6

Summary of paper given by F.W.L. to Birmingham Natural History and Philosophical Society, 19th January 1909

1911

Discussion of **American tendencies in motor-car engineering**, by H.E.Coffin

Proc. Institution of Automobile Engineers, session 1911–12, vol. 6, pp.49–54

Contribution to discussion of paper read to I.A.E.

Discussion of **Engine design for taking advantage of h.p. rating rules**, by L.H.Pomeroy

Proc. Institution of Automobile Engineers, session 1911–12, vol. 6, pp.101–4

Contribution to discussion of paper read to I.A.E.

Discussion of **Problems relating to aircraft**, by M.O'Gorman

Proc. Institution of Automobile Engineers, session 1910–11, vol. 5, pp.296–303

F.W.L., as President, opens the discussion of O'Gorman's paper read at I.A.E. 8th March 1911

Discussion of **The rating of petrol engines, a Committee Report and Discussion Paper of the Institution of Automobile Engineers**

Proc. Institution of Automobile Engineers, session 1910–11, vol. 5, pp.217–42

Paper written by Horsepower Formula Committee of the I.A.E.

Discussion of **The use of pressed steel in automobile construction**, by L.A.Legros

Proc. Institution of Automobile Engineers, session 1910–11, vol. 5, pp.371–83

F.W.L., as President, opens the discussion of Legros' paper read at I.A.E. 10th May 1911

Discussion of **Wheel and road**, by A.Mallock
  Proc. Institution of Automobile Engineers, session 1910–11, vol. 5, pp.326–44
  F.W.L. as President, opens the discussion of Mallock's paper read at I.A.E. 8th March
  1911

1912

Vote of thanks to T.B.Browne for his Presidential Address [on the recent development of
  the automobile industry]
  Proc. Institution of Automobile Engineers, session 1912–13, vol. 7, pp.30–1
Discussion on **The influence of low production cost on quality**, by D.Leechman
  Proc. Institution of Automobile Engineers, session 1912–13, vol. 7, pp.503–4
  Contribution to discussion of paper read to I.A.E. on 12th October 1912

1913

The internal combustion engine applied to railway locomotion
  Engineering, 19th September 1913, pp.381–2
  Summary of **Internal-combustion motors for railways …**
Report of **Catastrophic instability in aeroplanes**
  Engineering, 24th October 191e, pp.574–5
  Summary of paper given to Section G of the British Association meeting at Birmingham
  in October 1913
Discussion of **The wheel and the road**, by R.E.Crompton
  Proc. Institution of Automobile Engineers, session 1912–13, vol. 7, pp.427–9
  Contribution to discussion of paper read to I.A.E. on 9th April 1913

1915

Report of **A contribution of the theory of propulsion and the screw propeller**
  Engineering, 2nd April 1915, pp.392–5
  Report of paper given at Institution of Naval Architects, 25th March 1915

1916

Report of interview with F.W.L. on **Air raid precautions: why they are not yet available:
  Mr. Lanchester's views**
  Birmingham Daily Mail, 12th February 1916, p.6

1920

Discussion of **Roads and vehicle maintenance**, by R.E.B.Crompton
  Proc. Institution of Automobile Engineers, session 1920–21, vol. 15, pp.185–6
  Contribution to discussion of paper given to I.A.E. 8th December 1920

1921

Discussion of **Notes on motor car gear-boxes**, by H.F.L.Orcutt
  Proc. Institution of Automobile Engineers, session 1921–2, vol. 16, part 1, pp.305–7
  Contribution to discussion of paper given to I.A.E. 14th December 1921
Discussion of **Suggested lines of development in transmission**, by S.Bramley-Moore
  Proc. Institution of Automobile Engineers, session 1920–21, vol. 15, pp.309–22
  Contribution to discussion of paper given to I.A.E. 12th January 1921

1922

Discussion of **An account of some experiments on the action of cutting tools**, by E.G.Coker
  and K.C.Chakko
  Proc. Institution of Mechanical Engineers, vol. 1, 1922, pp.603–4
  Contribution to discussion of paper given to I.A.E. April 1922

1924

Discussion of **Balancing of automobile engines**, by H.S.Rowell
  Proc. Institution of Automobile Engineers, session 1923–4, vol. 18, part 2, pp.525–6
  Contribution to discussion of paper given to I.A.E. 25th March 1924

1925

Discussion of **A review of the rating question**, by A.E.Berriman
Proc. Institution of Automobile Engineers, session 1924–5, vol. 19, pp.374–80
Contribution to discussion of paper given to I.A.E. January 1925

1926

Discussion of **Experiments in laminated springs**, by H.S.Rowell
Proc. Institution of Automobile Engineers, session 1925–6, vol. 20, pp.628–32
Contribution to discussion of paper given to I.A.E. April 1926
Discussion of **The optical indicator as a means of examining combustion in exploding engines**, by W.Morgan and A.A.Rubbra
Proc. Institution of Automobile Engineers, session 1926–7, vol. 21, pp.200–1
Contribution to discussion of paper given to I.A.E. December 1926
Discussion of **Pneumatic tyres for heavier loads**, by C.Macbeth
Proc. Institution of Automobile Engineers, session 1925–26, vol. 20, pp.558–9
Contribution to discussion of paper given to I.A.E. March 1926

1936

Editorial review of **The part played by skin friction in aeronautics**
Engineering, vol. 142, no. 3698, 27th November 1936, pp.588–9
Review of paper given before Royal Aeronautical Society, 12th November 1936; for full text see p.000 above
Motor car suspension and independent springing
Engineering, vol. 141, no. 3665, 10th April 1936, pp.394–6, and no. 3667, 24th April 1936, pp.464–5
Correspondence re United States Patent Office refusal to accept F.W.L.'s patent application 259452 in 1918
Journal of the Institution of Automobile Engineers, vol. 4, part 9, June/July 1936, pp.4–6

1937

Contribution to discussion and written communication by F.W.Lanchester relating to paper **Profile drag** by Prof. B.Belvill Jones
Journal of the Royal Aeronautical Society, vol. 41, 1937, pp.358, and 363–5
Contribution to discussion on **Plastic materials for aircraft construction**, by N.A.de Bruyne
Journal of the Royal Aeronautical Society, vol. 41, 1937, p.580
Report of **The Gas engine and after**
The Engineer, vol. 164, no. 4270, 12th November 1937, pp.538–9, and no. 4271, 19th November 1937, pp.556–7

1938

Report of **Independent springing**, under the title **Independent rear wheel suspension for touring cars**
Engineering, vol. 145, no. 3757, 14th January 1938, pp.41–2

1939

Editorial review of *The Energy balance sheet of the internal combustion engine*
Engineer, vol. 167, no. 4341, 24th March 1939, pp.377–8
Editorial review of *Span*
Engineering, vol. 147, no. 3815, 24th February 1939, pp.223–4

COLLECTIONS OF PRIMARY SOURCES ON THE LANCHESTERS

LANCHESTER COLLECTION, LANCHESTER LIBRARY, COVENTRY
UNIVERSITY
(Much Park Street, Coventry, CV1 5HF)

In 1959 Coventry City Council was seeking a name for its new college of technology, and chose "Lanchester College of Technology", in honour of Frederick William Lanchester and the Lanchester family. Although the Lanchester early work was carried out in Birmingham, Fred had worked for the Daimler Company in Coventry, and the firm eventually owned the Lanchester marque.

The Lanchester Collection was founded in August 1961 with the donation by Dorothea Lanchester, Fred's widow, of books, documents, papers and correspondence of her late husband. This first collection was sorted, described, listed and indexed by the College Librarian, Eric Baxter, and the product, *Catalogue of the private papers of F.W.Lanchester in the Library of Lanchester College of Technology*, was successfully submitted to the Library Association as a thesis for the Fellowship of the Association in 1966.

In 1970 the Lanchester College of Technology was amalgamated with the Coventry College of Art and the Rugby College of Engineering Technology to form Lanchester Polytechnic. The choice of Lanchester's name had become even more appropriate, since, above all, Fred Lanchester was a "poly-technic" man.

In 1987 the name Lanchester disappeared from the title of the Polytechnic, to avoid confusion with Manchester Polytechnic and Lancaster University. It was now called Coventry Polytechnic, but the Librarian had already, in 1983, sought and received the approval of the Board of Governors to name the Polytechnic Library the Lanchester Library, ensuring that the Lanchester name survived in the institution. On 29th April 1984 the Daimler and Lanchester Owners' Club held their rally at the Polytechnic, to mark the official opening of the refurbished Lanchester Library, a ceremony carried out by George Lanchester's widow "Steve". (Fig. 1, and plates 5, 6 and 7) Subsequently the Polytechnic (from 1992 Coventry University) named its gallery and its restaurant "Lanchester", maintaining the tradition.

The 600 + items in the first tranch of donated Lanchesteriana comprised a wide range of primary material including typescripts of unpublished papers, originals and carbon copies of business and private correspondence, manuscripts and sketches, calculations, and some photographs. Eric Baxter organised them into seventeen, mainly subject, groups, and within these they are in chronological order. The papers were tipped into guardbooks for ease of handling, and for preservation. Baxter's *Catalogue* ... contains a very detailed index, and he noted that the material had been seen, and used by Dr. P.W.Kingsford in the preparation of his *F.W.Lanchester, the life of an engineer* (London, Edward Arnold, 1960).

Mrs. Steve Lanchester and many friends and family of the Lanchesters have given correspondence and artefacts to add to the Collection. Among these are the scale model of the first Lanchester car (as modified in 1897) made by George Lanchester at the age of eighty-three, and several of the medals won by the Lanchesters (both brothers and cars), including the gold medal of the famous "Gold Medal" Phaeton. (Fig. 5, Colour plate 8)

The second tranch of material for the Collection came in 1982, when the present author,

Fig. 1. Geoffrey Holroyde (Polytechnic Director), Mrs. "Steve" Lanchester and Neville Boothroyde, Chairman of the Board of Governors

Eric Baxter's successor, heard that some Lanchesteriana was coming up for auction. I purchased it (unseen!) for the Library and was relieved and delighted to find an outstanding collection of unique items. In addition to typescripts, manuscripts and correspondence, which linked in well with the first collection, there were thirteen of Fred's sketchbooks. These are one-hundred-page bound volumes, rather like artists' sketchbooks, in "landscape" format, which Fred used as notebooks: he seems to have had one on every desk he used. When he needed something to write on, he turned to the next blank page and used that. Few entries are dated, but the variety of content is extraordinary: new ideas for a patent jostle with the results of tests on a car component. Some pages have been amended by Fred subsequently, probably towards the end of his life.

With support from the (by then renamed) Lanchester Polytechnic, Mr. Yoichi Takeda and the British Library, the loose papers were sorted and professionally set in guardbooks. The whole Collection, including both the first and second tranches were microfiched for security, and to enable interested parties to have copies.

Following the death of Prof. R.D.Lockhart, his collection of the correspondence between himself and Fred Lanchester, stretching over a period from 1934 to 1946, was donated to the Lanchester Collection.

The final tranch (to date) was again purchased at auction in 1991: two large volumes containing almost one thousand original photographs of Lanchester cars, taken near the factory between 1908 and 1918. Again, for security, all have been rephotographed, and copies made. During the 1980s the Librarian sought publications by, and references to Fred Lanchester, and purchased or had copies made for the Collection. It now contains more than 3,000 items, including originals or copies of all the patent specifications granted to the Lanchester brothers. (Patent applications not followed up were not retained by the Patent Office).

In 1991 the present author began the task of indexing the Lanchester Collection, using the power of a modern computer software package. Work continues on this, and there are currently (January 1996) 1,850 items in the computerised index.

Access to the Lanchester Collection is by arrangement with the University Librarian, Coventry University, and assistance in locating and using material is available.

BIRMINGHAM MUSEUM OF SCIENCE AND INDUSTRY
(Chamberlain Square, Birmingham)

The Museum library has about 250 letters to and from F.W.Lanchester, mostly from the period 1914 to 1918. There are letters to and from Lord Trenchard, Neville Chamberlain, Sir David Henderson, and Lord Fisher, amongst others. Subjects range from disputes with publishers about articles he had written, to the use of blunderbusses as weapons in aircraft, and the politics of the French (and proposed British) Air Ministries, but most are on matters concerning aerial warfare, tying in with Fred's wartime work with the Advisory Committee for Aeronautics.

A further one hundred or so items in the Library include articles by and about Fred, notes relating to the Lanchester Motor Company, and catalogues of the cars in the 1920s. Especially interesting are the letters between Fred and Percy Martin, of Daimler, during their acrimonious dispute in late 1929, which resulted in the end of the twenty-year consultancy agreement.

Access to this material is normally available, by prior arrangement with the Curator of the Museum.

INSTITUTION OF MECHANICAL ENGINEERS
(1, Birdcage Walk, London SW1H 9JJ)

The Library of the Institution has a small, but unique collection of drawings and sheets of calculations by F.W.Lanchester. Their importance lies in that they are from the very earliest period of Fred's work on motor cars, and especially the machine tools to produce the parts for them. There is a volume of parts lists, and two volumes of drawings of motor car parts, all from about 1903. Even earlier is the "Manager's Office Calculation and Sketchbook", by F.W.Lanchester, with dates added later by Fred or his brother George, of 1899 and 1900.

Access to this collection is more restricted, but by arrangement with the Librarian of the Institution.

ROYAL AERONAUTICAL SOCIETY
(4, Hamilton Place, London W1V 0BQ)

The Society's Library has a collection of F.W.Lanchester's publications, donated by him in the 1930s. They include the relatively rare German edition of *Aerial flight*. There are also three folders of correspondence, mostly between Fred and the Secretary of the Society, and from the 1930s. They include Fred's proposal for an annual prize to be given to a student for model gliders, which was agreed in 1914, stopped by the War, and revived in a different form in 1920. The most poignant correspondence is from the period 1938 to 1942, when Fred's financial position was desperate, and the Society members were trying to raise funds to help him. They were successful, but Fred was unwilling to accept charity, and attempts to procure a Civil List pension had failed.

Access to the Society's Library is by arrangement with, and payment to the Librarian.

Figs. 2 and 3. Two examples of photographs from the volumes purchased in 1991

UNIVERSITY OF SOUTHAMPTON
(University Library, Highfield, Southampton SO9 5NH)

The archives department of the University Library has a few published items and some correspondence relating to F.W.Lanchester.

Access is by arrangement with the University Librarian.

PRIVATE COLLECTIONS

Chris Clark, the author of the first two volumes of *The Lanchester legacy*, has a substantial collection of material on the Lanchester brothers and their work, principally relating to the construction and production of cars. Much of it is unique, including recordings and transcripts of interviews with people who had worked with or for Fred Lanchester.

Mrs Elaine Lanchester has some more sketchbooks, and documents left by her father-in-law, George Lanchester. Some are by George, and some by Fred, from the First World War period.

Other members of the Lanchester family have private and personal family memorabilia relating to all three of the brothers.

INDEX

C. S. Clark, Hon MA

Lanchester Historian

Tel: 01531 890204

Author and Distributor of

The Lanchester Legacy

www.lanchesters.com

Dear Friend,

There's More to Come!

First of all, let me thank you for the purchase of Lanchester Legacy Volume Two. I hope you enjoy reading it as much as I enjoyed writing it and compiling the hundreds of illustrations.

In fact, I have far more photographs in my collection than it was possible to include in this Volume, and many of them include historical and contemporary images of owners and enthusiasts who have enjoyed cars within this marvellous marque.

The exclusion of photographs where the people in shot might have taken focus away from the cars has inspired me to start working on a new Volume, turning the *Lanchester Legacy Trilogy* into the *Lanchester Legacy Series*.

Volume Four, which I hope to complete in time for publication around the middle of 2017, will be a celebration of Lanchesters AND the people who loved them. Perhaps you fall into this category yourself, in which case I would be delighted to receive any photographs that you have of your Lanchester cars (past or present), especially if they are depicted in unusual settings or with interesting perspectives of the people around them.

If you do have such pictures, or can take some now if you still own a Lanchester, then please scan them and email them to me at Lanchesters@gmail.com. I cannot promise to include them in Volume Four, but if they are of high-quality and meet the criteria I've mentioned above then they stand a very good chance of selection.

Volume Four will also include the results of my recent Worldwide Census of Remaining Lanchesters. Once again, if YOU are fortunate enough to still own a vehicle, do please ensure that I have its full details by completing the Census form on my website at www.lanchesters.com. I will be listing full details of EVERY known Lanchester car in Volume Four, making it a unique source of reference as well as a joyous pictorial journey. I do hope you'll be kind enough to add the book to your collection when it becomes available.

Finally, I'd love to receive your feedback on your copy of Volume Two once you've had a chance to read it. Simply email me at Lanchesters@gmail.com with your comments or – better still – write a brief review of the book that can appear on my website.

That's more than enough from me for now; I want to let you start leafing through your new purchase right away!

With Very Best Wishes

Chris Clark

December 2016

The Lanches, Ledbury Road, Dymock, Gloucestershire, GL18 2AG. United Kingdom.